Coexistence and Confrontation of Agricultural and Food Models

Pierre Gasselin · Sylvie Lardon · Claire Cerdan ·
Salma Loudiyi · Denis Sautier
Editors

Coexistence and Confrontation of Agricultural and Food Models

A New Paradigm of Territorial Development?

Foreword by Jan Douwe van der Ploeg
With the Editorial Support of Sylvie Zasser

éditions Quæ
Éditions Cirad, Ifremer, INRAE
www.quae.com

Springer

Editors
Pierre Gasselin
UMR Innovation
INRAE
Montpellier, France

Claire Cerdan
UMR Innovation
CIRAD
Saint Pierre, Réunion, France

Denis Sautier
UMR Innovation
CIRAD
Montpellier, France

Sylvie Lardon
UMR Territoires
INRAE and AgroParisTech
Aubière, France

Salma Loudiyi
UMR Territoires
VetAgro Sup
Lempdes, France

Section Editor
Sylvie Zasser
INRAE
Castanet-Tolosan, France

Translated by
Kim Agrawal
SAICE
Pondicherry, India

ISBN 978-94-024-2177-4 ISBN 978-94-024-2178-1 (eBook)
https://doi.org/10.1007/978-94-024-2178-1

Jointly published with Éditions Quæ
Éditions Quæ, RD10, 78026 Versailles cedex, France

Translation from the French language edition: "Coexistence et confrontation des modèles agricoles et alimentaires" by Pierre Gasselin, Sylvie Lardon, Claire Cerdan, Salma Loudiyi, and Denis Sautier,

This Springer imprint is published by the registered company Springer Nature B.V.
The registered company address is: Van Godewijckstraat 30, 3311 GX Dordrecht, The Netherlands

Foreword: A Bold and Necessary Book

Agriculture and food production urgently need a transition towards a configuration that is more in line with both nature and society. Such a transition cannot but be radical, far-reaching, massive, global, multilevel and multidimensional. In particular, it will imply a reshuffling of power relations that is, at this time, hardly conceivable. The need for such a transition has been argued in convincing ways by many; there is no need to repeat their many solid arguments here.

At the same time, there is a dazzling array of questions about the possible modalities of this necessary transition. How to get it started? What will be its main mechanisms and driving forces? How will States, civil society and social movements align themselves (or not) during the transition? And, in particular, how to tie together the many small steps needed in order to achieve major moves forward? Questions like these abound, and the lack of adequate answers, even partial ones, can be discouraging.

The difficulty of arriving at answers that are adequate, as well as practically useful, resides not only in the fact that the transformation of agriculture and food production is yet to take place but also in the ineluctable awareness that the path to these answers will need to pass through unknown territories. Furthermore, there are two additional, probably interrelated, complications. First, there will not be a *single*, more or less well-delineated process of transition. There will be instead (as this book also shows) *several* partial and contrasting processes of transition that will enter into very complex, and probably unstable, interrelations. These partial processes of transition are already underway. Thus, today's agricultural sector is moving towards accelerated industrialisation and scaling up, and is undergoing a form of de-aggregation at the same time. These developments coexist with movements towards multifunctional agriculture and repeasantisation, understood as more peasant-oriented agriculture, as is happening in the field of agroecology. Such processes indeed often *coexist* alongside each other and simultaneously generate persistent confrontations. Any careful study of the local level (a basic necessity in agricultural studies) will reveal several incipient forms of transitional trajectories. This leads to the second major difficulty: the general direction of change is difficult to predict. As confusing as historical processes of transition might have been (e.g. the transition from sailing ships to steamships), the overall direction of the process was known and understood. And, whilst this change was possibly contested, both those favouring it and those opposing it knew which way the winds were blowing. Even the 'monstrosities' created during the process, such as steamships that were equipped with masts and sails,[1] showed the way forward. This is a major difference with today's situation.

The simultaneous occurrence of several partial and contrasting (if not competing and excluding) transitional processes is resulting in an often confusing and sometimes overwhelming heterogeneity. With *coexistence*, as it is aptly termed in this book, one sees a palette of diverse and parallel economies, forming a rainbow of many different colours, some of them miscible and maybe giving rise to new amazing colours, other combinations just becoming brown messes. The palette is not stable, it is changing constantly and always subject to influence—but nobody knows who is holding the brushes and palette knifes. What we do know, however, is that this chaotic medley is the starting point for the transition(s) that we need. It is from here that we have to muddle forwards. *Inter alia* this book demonstrates this to be a sound methodological principle: studying the richly chequered coexistence of different forms and expressions, some old, others new, and trying to understand how the whole is moving forwards and simultaneously changing and what can we do to influence this for the better.

This 'new' coexistence, understood as melting pot of interacting and partly conflicting transitional trajectories, clearly differs from 'old' forms of coexistence, like the *minifundia* and *latifundia* that combined in one interdependent whole throughout Latin America. Or the plantations and small-scale native farms of Southeast Asia, with the former having induced a structural decline in the latter. The same was true of the combination of large and small farms throughout Western Europe.

The oppositions between different actors can induce mutual reciprocity, overt hostility or both. There was, for instance, clear hostility in the *minifundia–latifundia* combination, but also a basic complementarity within this framework: small farms delivering labour (and sometimes food) to large farms, the owners of the latter defending, if needed, the small farmers (according to the patron–client model). There were exchanges and mutual dependencies, although their terms were constantly contested. This made for a negotiated complementarity. In Europe, there was complementarity as well (although its terms were different): small farmers cleaned the canals of the large farms, the large farmers serviced the small ones with their tractors and draught animals. Small farmers were slow to change (it was too risky), large farmers had the moral duty to innovate and pass on their findings to others, including to the small farmers. Needless to say, there was animosity and struggle as well: the weapons of the weak were never out of reach—but inside these 'old' forms of coexistence,

[1] Having both a steam engine and sails made these ships much more expensive, which is why they were called 'monstrosities'. However, until the technological support structure was fully developed (with, e.g., the possibility to acquire sufficient coal at each port of call), this somewhat awkward, if not irrational, combination was unavoidable. Nonetheless, these 'monstrosities' made the overall direction of the transition very clear. Something comparable can be encountered in some of the 'hybridities' of today.

there was complementarity, sometimes more, sometimes less, but enough to hold the system together.

This differs fundamentally from the 'new' coexistence we are facing today. Now, there are many-stranded contradictions which translate into a strong degree of competition, if not in mutual exclusion. It is not a matter of competition in the neo-classical sense of the word, at the level of prices. Instead, it finds expression, above all, as a scramble for resources, a fight to appropriate the most promising symbols ('healthy', 'sustainable', 'smart', 'strong', 'optimal', etc.), and a struggle for privileged access to markets. The associated fights and struggles are, on the one hand, very much mediated through agricultural policies—they even take place through policy or the explicit lack of it—and, on the other hand, they deeply divide the farming populations, as attested recently by the outbreak of farmers' populist protests.

In short: exploring the new coexistences is no easy task. This probably explains why most agricultural knowledge systems excel in avoiding this theme. In my country, for instance, there is a lot of easy talk about 'transitions', but any reference to the difficult coexistence as a starting point for such transitions is avoided, just as, consequently, the contested nature of the transitional process(es) is obscured. Thus, transition is suggested to be the smooth unfolding of new, comfortable realities—an unfolding that is governed by the logic of the markets and/or 'unavoidable' technological developments. Suspiciously, the State is often thought as having only a marginal role in the envisaged transitions.

Within this panorama, the French research community is a well-known and highly valuable exception. This book is a bold effort by some of this community's most respected members. It reflects the strong international orientation of the agricultural sciences in France, their diversity, the spaces offered to diverging views, and the centrality of critical debates. It is a pleasure to scan the list of authors: anyone who has participated in international debates is sure to recognise several names. The book's authors are acknowledged scholars with wide horizons and considerable interdisciplinary experience.

This book proposes four theoretical dimensions that allow for an exploration of the coexistence of and confrontation between different agricultural and food models. They are diversification (as opposed to specialisation), innovation, adaptation and transition. This exploration framework is then applied to a diverse set of empirical realities (ranging from Europe to South America, Polynesia to Asia and to Africa)[2] and culminates, through careful comparative analysis, at a set of highly relevant insights. The book is guided by a provocative hypothesis: it is the heterogeneity of agricultural and food systems (the 'confrontational coexistence' that they are part of) that strengthens their transitional capacity (and allows for meaningful theoretical elaboration). It is worth stressing that this hypothesis represents a strong contrast with the 'proposals for transition' advocated by dominant institutions. These latter proposals ignore the empirically existing heterogeneity for the most part (it is according to them, at best, irrelevant noise), whilst they focus on solutions located

[2] The book is based on 25 case studies. Some chapters provide a comparative analysis of several of them.

in (and derived from) the domain of a 'world that does not yet exist'. Doing so is easy and their proposals often look well-polished and clean. This book, however, is radically different: it shows the value of *not* being afraid of the mud, the contestations, the doubts and the setbacks of the real world. Nor does it gloss over the many disparities that exist between different countries.

The main hypothesis is buttressed by specific ones that pertain to the specified theoretical dimensions. They avoid facile dualism and introduce differences of scale (spatial, temporary, social), variations in power relations, the dynamics of actor–structure relations, the changing nature of interrelations between past, present and future, and the mediating role of agricultural policies. Their application in the empirical chapters shows that processes of transition (including the partial and as-yet-unfinished ones) are full of surprises. Transitional processes hardly ever follow the straightforward logic that modernisation theories entail. The initially marginal sometimes becomes the source of far-reaching and radical changes, whilst the strong and powerful often have to step back in order to engage in uneasy, hybrid configurations. Reciprocal cause–effect relations are more common than one-way patterns that run from clear causes to well-outlined outcomes. All this makes the empirical (and theoretically well-grounded) exploration of transitions of agricultural and food systems into an extraordinary intellectual adventure. This book demonstrates that such an adventure is far from easy—it requires courage. But by carrying out this bold enterprise and by including a thorough rethinking of the concepts of quality, agency, innovation, etc., it provides the instruments necessary to forge ongoing processes of transition into successful historical events.

Wageningen, The Netherlands Jan Douwe van der Ploeg

Jan Douwe van der Ploeg is Professor Emeritus of Rural Sociology at Wageningen University & Research in the Netherlands and Associate Professor of Agricultural Sociology at China Agricultural University in Beijing. He has conducted extensive research on processes of agricultural transition and on dynamics of rural development.

Preface to the English Edition

This English volume is the translation of the original French edition published by Quæ in 2021: *Coexistence et confrontation des modèles agricoles et alimentaires. Un nouveau paradigme du développement territorial?*

This book presents the results of the Format project 'Combining forms of agriculture and food systems at different territorial scales: coexistence, confrontation or hybridisation of models', funded between 2015 and 2017 by the interdisciplinary GloFoodS metaprogramme 'Transitions for global food security' created by INRA and CIRAD, the two leading French agricultural research institutes.

This book would not have been possible without the support and funding of INRAE's Action and Transitions Division and CIRAD's Environments and Societies Department, as well as of the Innovation Joint Research Unit (Montpellier) and the *Territories* Joint Research Unit (Clermont-Ferrand).

Thirty-six authors from France and other countries (Argentina, Belgium, Brazil, Burkina Faso, Japan, Switzerland, Vietnam) have contributed to the book. Jan Douwe van der Ploeg has done us the honour of writing the foreword.

The text underwent a demanding evaluation process, both in terms of content and form: each chapter was proofread and corrected at least four times, by an external expert in the relevant topic, by one of the scientific coordinators, by Sylvie Zasser (editor at INRAE's Action and Transitions Division) and finally by Juliette Blanchet (proof-reader). In addition, Valérie Mary from Quæ publishers was responsible for editorial coordination.

We would like, in particular, to mention the quality and rigour of Sylvie Zasser's editorial supervision. She worked tirelessly and patiently to help improve the quality of the various texts and ensured the book's timely publication.

To each of these institutions and individuals, who have contributed to make the book a success, we express our heartfelt thanks.

In addition, the coordinators would like to extend special thanks to the book's translator, Kim Agrawal, for his commitment, his uncompromising exactitude and attention to detail, and his unceasing desire to get at the true meaning of words and ideas.

A first version of the translation of each chapter was submitted to its authors for validation and modifications. Final decisions were taken in consultation with Pierre Gasselin and the translator.

Montpellier, France	Pierre Gasselin
Aubière, France	Sylvie Lardon
Saint Pierre, France	Claire Cerdan
Lempdes, France	Salma Loudiyi
Montpellier, France	Denis Sautier

Contents

Editors and Contributors

About the Editors

Pierre Gasselin geographer at INRAE (Innovation joint research unit, Montpellier), conducts research in France and Latin America on the transformation of farmers' activities, on the conditions conducive to their territorial integration and on providing support to them.

Sylvie Lardon geographer at INRAE and AgroParisTech (*Territoires* joint research unit, Clermont-Ferrand), is a specialist in participatory prospective diagnosis. She studies the transformations of rural and urban territories in the Mediterranean region, Argentina and Brazil.

Claire Cerdan geographer, researcher at CIRAD (Innovation joint research unit, Réunion), works on localised agrifood systems and the role of territorial resources in the recomposition of territories in Africa and Latin America.

Salma Loudiyi geographer, full professor at VetAgro Sup (*Territoires* joint research unit, Clermont-Ferrand), leads research on the implementation of integrated food policies in France and North America.

Denis Sautier economist at CIRAD (Innovation joint research unit, Montpellier), studies localised agrifood systems and place-based labelling in Asia and Africa.

Contributors

Christophe Albaladejo INRAE, ACT Scientific Division, AgriTerris Network, UNLP FCAyF-Conicet, Buenos Aires, Argentina

Gilles Allaire INRAE, Toulouse, France

Clémentine Antier Université Catholique de Louvain, Sytra (Transition of Food Systems), Louvain-la-Neuve, Belgium

Saydou Bancé Programme d'appui à la décentralisation et à la participation citoyenne (Depac), Ouagadougou, Burkina Faso

Philippe V. Baret Université Catholique de Louvain, Sytra (Transition of Food Systems), Louvain-la-Neuve, Belgium

Virginie Baritaux VetAgro Sup, UMR Territoires, Clermont-Ferrand, France

Pascal Bonnet Cirad, DGDRS, Environment and Societies Department, Montpellier, France

Patrick Caron University of Montpellier/Cirad, Montpellier Advanced Knowledge Institute on Transitions, Montpellier, France

Claire Cerdan Cirad, UMR Innovation, Saint-Pierre, Réunion, France

Jean-Daniel Cesaro Cirad, UMR Selmet, Montpellier, France

Emmanuelle Cheyns Cirad, UMR Moisa, Montpellier, France

Roberto Cittadini Instituto Nacional de Tecnología Agropecuaria (INTA), Universidad Nacional de Mar del Plata (UNMDP), Mar del Plata, Argentina;
UMR Innovation, Montpellier, France

Agnès Coiffard Instituto Nacional de Tecnología Agropecuaria (INTA), Universidad Nacional de Mar del Plata (UNMDP), Mar del Plata, Argentina

Nora Daoud Consorcio Andaluz de Impulso Social (CAIS), Seville, Spain

Guillaume Duteurtre Cirad, UMR Selmet, Montpellier, France

Pham Duy Khanh Rudec-Ipsard, Hanoi, Vietnam

Guy Faure Cirad, UMR Innovation, Montpellier, France

Rosalia Filippini University of Parma, Department of Economics and Management, Parma, Italy;
INRAE, UMR Territoires, Clermont-Ferrand, France

Jérémie Forney University of Neuchâtel, Institute of Anthropology, Neuchâtel, Switzerland

Pierre Gasselin INRAE, UMR Innovation, Montpellier, France

Nathalie Hostiou INRAE, UMR Territoires, Clermont-Ferrand, France

Marie Houdart INRAE, UMR Territoires, Clermont-Ferrand, France

Vanessa Iceri CNPq (Brazil), AgroParisTech, UMR Territoires, Clermont-Ferrand, France

Claire Lamine INRAE, Ecodevelopment Research Unit, Avignon, France

Sylvie Lardon INRAE and AgroParisTech, UMR Territoires, Aubière, France

Ronan Le Velly Institut Agro, Montpellier SupAgro, UMR Innovation, Montpellier, France

Salma Loudiyi VetAgro Sup, UMR Territoires, Clermont-Ferrand, France

Véronique Lucas INRAE, UMR BAGAP, Rennes, France

Nguyen Mai Huong Rudec-Ipsard, Hanoi, Vietnam

Martine Napoléone INRAE, UMR Selmet, Montpellier, France

Emmanuel Pannier UMR Paloc, IRD, Paris, France

Denis Sautier Cirad, UMR Innovation, Montpellier, France

Kae Sekine Aichi Gakuin University, Graduate School of Economics, Nagoya, Japan

Aurélie Toillier Cirad, UMR Innovation, Montpellier, France

Frédéric Wallet INRAE, UMR Sadapt, Paris-Saclay University, Paris, France

General Introduction. Questions, Issues and Analytical Framework

Pierre Gasselin, Sylvie Lardon, Claire Cerdan, Salma Loudiyi and Denis Sautier

The issue of the coexistence and confrontation of agricultural and food models is a topic that is preoccupying society and political and professional fields. In some countries, such as Brazil (Sencébé et al., 2020) and to a lesser extent Argentina (Albaladejo, 2020) or Vietnam (Duteurtre et al., 2015), agricultural development is based on models of dual social and technical forms of agriculture, with one being described as family-based or peasant-based farming (Bosc et al., 2018), and the other as industrial or corporate farming (Purseigle et al., 2017) or as an agribusiness. In France, the history and structures of agricultural production, exchange, innovation and regulation have led to less assertive or more gradual agricultural and food models (Deverre & Lamine, 2010; Hervieu & Purseigle, 2013; Gasselin et al., 2014). Agricultural and food models are sometimes expressed in terms of development issues, such as the right to food, which is embodied in the 'food sovereignty' project (Rosset, 2003; Jarosz, 2014), or climate change, which has given rise to so-called climate-smart agriculture (Caron, 2016; Karlsson et al., 2018; Oui & Touzard, 2018; Taylor, 2018). These models can also be defined in terms of technical modalities (e.g. conservation agriculture, precision agriculture or permaculture), marketing of products (short supply chains, fair trade, etc.), social forms of organisation of labour and capital (family farming, corporate farming, etc.) or socio-spatial configurations (e.g. urban farming). These agricultural and food models underpin not only forms of public action, agri-chain structures, but also configurations of territorial development, which we examine in particular in this book. The sociotechnical controversies within each of these models and between them shape alliances and confrontations between actors and ideas.

For several years now, the research community has been focusing on this issue of coexistence and confrontation of agricultural and food models, whether in France (Petit et al., 2018) or elsewhere (Argentina, Brazil, USA, Netherlands, Japan, Belgium, etc.). It provides a research agenda whose elements, preoccupations and objectives need to be refined and structured within an international research community that is itself under construction. This collaborative book is one of the first collective scientific endeavours on this topic. It is intended for researchers, teachers, students and, more broadly, all those who are involved in territorial development:

individuals (development officials, elected officials, journalists, etc.) as well as institutions (associations, local authorities, cooperatives, chambers of agriculture, government departments, etc.).

This introduction first discusses the different acceptations of the concept of the agricultural and food model. We then recall the key facts of the differentiation of agricultural and food models since the Second World War—which justifies the current interest in their coexistence and confrontation. We then proceed to characterise the major elements of the situations of coexistence of models studied in this book. Finally, we present the research and development challenges, followed by the scientific issues and the analytical framework we have used to organise the book.

The Model: Analytical Archetype, Desired Future or Standard for Action

The sociotechnical[3] and socio-ecological[4] forms observed in agriculture and food systems often differ from what are known as 'models', which are abstract, schematic and simplified representations that actors (researchers, agricultural advisors, trade unionists, elected politicians, etc.) make of a complex reality. The scientific literature uses three acceptations of the concept of agricultural or food model.[5] First, researchers and experts construct models as archetypes of a reality observed either today or in the past in order to characterise its diversity and facilitate its understanding for transformative action. These models are often described in terms of regimes (Wiskerke & van der Ploeg, 2004; McMichael, 2009), frames of reference (Muller, 1990; Gisclard & Allaire, 2012, Hall et al., 2015), styles (van der Ploeg, 2010; 2012), agricultural systems (Plumecocq et al., 2018), food systems (Sobal et al., 1998; Fournier & Touzard, 2014), sociotechnical systems (Geels, 2010; Darnhofer, 2015), etc. Second, an agricultural and food model can also be a desired future that actors demand, such as the peasant agriculture advocated by agricultural unionism or the alternative forms of consumption promoted by movements such as 'slow food'[6]. Third and finally, a model is sometimes defined as a set of standards for action in a certification and assessment process, such as organic farming or halal or kosher food. Some researchers mobilise the concept of agricultural or food model by combining these three meanings (Albaladejo, 2020). However, models always have, on the one hand, an analytical dimension and, on the other, a normative dimension that actors

[3] Sociotechnical forms combine human representations, decisions and practices with biotechnical entities (Bijker, 1997).

[4] Socio-ecological forms refer to the way in which ecological dimensions interact with sociotechnical dynamics (Holling, 2001).

[5] We do not consider here models defined as mathematical formalisms which relate variables embedded in descriptive, normative or predictive explanatory equations that deal with food and/or agriculture.

[6] International movement to raise awareness of eco-gastronomy and responsible consumption, as a reaction and opposition to fast food.

use to think and act. These two facets are in tension and must be clearly laid out, for example to show that the peasant agriculture of Mendras (1967), an analytical archetype, is not the peasant agriculture of the Confédération Paysanne (a major French farmers' union), which represents a desired future. As we will see below, an agricultural and food model refers to an overall coherence of the relationships that humans establish with activity, nature, techniques, knowledge, the State, markets and the territory (Gasselin, 2019). Furthermore, the concept of a model presupposes that a group of actors builds a minimal consensus to make it a collective reference, to be defended or criticised.

Differentiation and Diversity of Agricultural and Food Models

Why should we take an interest in the coexistence and confrontation of agricultural and food models at the territorial scale? A first reason to do so is that after several decades of public policies that encouraged their homogenisation, we have to admit that these models continue to diversify. Indeed, globalisation has not succeeded in standardising agricultural and food models, despite their spread across the planet, for example during the Green Revolution (Pingali, 2012); the industrialisation[7] of agricultural production, processing and distribution; or even the affirmation of agroecology as a shared horizon (Pimbert, 2018).

The main determinants of the differentiation of agricultural and food models have been known for some time (McMichael, 2009; Hervieu & Purseigle, 2011; IPES-Food, 2016; Allaire & Daviron, 2017; Gaitán-Cremaschi et al., 2019). To begin with, they include the massive intensification of commodity and capital flows, and major technical transformations, in particular those based on new genetic selection regimes (fixed varieties, GMOs, etc.) and the use of chemical inputs and fossil fuels (Daviron, 2019). Urbanisation, the concentration of production, processing and distribution

[7] The definition of the industrialisation of agriculture most often refers to that of conventional agriculture (Bernard de Raymond & Goulet, 2014; Galliano et al., 2017). Historians (Malassis, 1997; Mazoyer & Roudart, 1997; Daviron, 2019) situate industrial agriculture in the great movement of the industrialisation of economies with the use of non-renewable natural resources (coal for the steam engine, then later oil and phosphates). Thus, agricultural systems with technical itineraries based on fertilisers, phytosanitary products, heavy mechanisation and non-renewable energy sources are considered industrial, even if the labour is family-based. Another perspective defines an agricultural system as industrial when it serves agro-industry, irrespective of the forms of contractualisation that bind them (cooperativism, vertical integration, etc.). In this case, the industrial character refers to the nature and structure of the downstream agri-chain, and to the forms of organisation of work and the organisation of the capital of processing, distribution and catering companies. The industrial character can also refer to the idea that agricultural enterprises base their technical-economic rationales on economies of scale (division of labour, specialisation of tasks, standardised and mass production). This organisation of production aims to maximise labour productivity, which is best achieved in large-scale production units.

structures and new food demands, supported by public policies promoting international trade and lower food prices, have also favoured the industrialisation of agriculture and food systems. This consists of the production of low-priced and standardised-quality food for mass markets, using inputs that are themselves industrial and relying on economies of scale and regional agricultural specialisation. The industrialisation of agriculture and food systems is taking place across the entire planet and is dominant in terms of the quantities produced and of the balance of power between actors in food systems. Nevertheless, it has not wiped out certain peasant and artisanal forms that have transformed themselves to face new contexts, nor has it prevented the appearance of new forms of agriculture and food systems such as permaculture or the slow food movement (Hervieu & Purseigle, 2015).

This industrialisation is showing its limits in many respects (IPES-Food, 2016). Despite the growth in per-capita food production and high labour productivity, the industrial food system generates major environmental problems (soil, water and air pollution, greenhouse gases, soil erosion, loss of domestic and wild biodiversity, weeds resistant to bio-aggressors, deforestation, etc.), spatial problems (increase in cultivated areas, land grabbing, etc.), economic problems (poverty, reduction in the availability of jobs, precarious incomes, dependence on and cost of chemical inputs, volatility in prices of inputs and products, etc.), social problems (hunger and malnutrition, infringement of human and workers' rights, land conflicts, loss of knowledge and know-how, animal suffering, etc.) and health problems (producers exposed to pesticides, consumers subject to pollution, zoonoses, non-communicable food-borne diseases, etc.). Through its negative externalities, the industrial food system compromises food and nutritional security, the sustainability of ecosystems, social justice and responses to climate change. In addition, this industrial food system exhibits a high degree of vulnerability, especially due to the genetic uniformity of monocultures and livestock, and the low productivity and insufficient food autonomy of certain regions (Fraser et al., 2005; Hodbod & Eakin, 2015; Urruty et al., 2016).

In response to these shortcomings of 'modernisation' and to new development challenges, a host of proposals are emerging from research, empirical practices, social movements and public policy (Deverre & Lamine, 2010; Horlings & Marsden, 2011; van der Ploeg & Ventura, 2014; Caron et al., 2018; van der Ploeg, 2018; Gaitán-Cremaschi et al., 2019; HLPE, 2019). These 'alternatives' have long remained marginalised not only in science, public policy and the media, but also in markets and professional spheres. However, contemporary history is marked by a succession of food, environmental, climatic and health crises that keep reminding us of the urgency of finding solutions to the failures of industrial food systems. Gradually, certain 'alternative'[8] models are gaining recognition: organic farming, agroecology, local food, veganism, etc. They are now tolerated and sometimes even promoted as a response

[8] The concept of 'alternative' lacks a stable and accepted definition. It can refer to the existence of a social movement engaged in a political conflict (Pelenc et al., 2019) or in demands for justice, to actors' projects (Le Velly, 2015), to niches of innovation on the fringes of the sociotechnical system inherited from history, to a low level of institutionalisation, to marginality (social, economic, spatial, etc.), etc.

to the limitations of conventional models and to new challenges (health, environmental, food, demographic, etc.) (Beus & Dunlap, 1990). This positive reassessment of the alternative has been taking place since the 1990s in a 'quality turn' through which new qualifications in agriculture and food systems are emerging (organic, fair trade, geographical indications, vegan, etc.) (Goodman, 2003). These many and different agricultural and food models, sometimes inherited from agricultural revolutions (Mazoyer & Roudart, 1997; Regnault et al., 2012), are often examined in a dual way in a binary opposition to the model from which they diverge (conventional vs alternative, agro-industrial vs peasant, GMO vs non-GMO, modern vs traditional, *latifundium* vs *minifundium*, etc.). However, this dualism obscures the great diversity of agricultural and food models, and of their territorial interactions—which we discuss in this book.

Research and Territorial Development Challenges

Thus, researchers, statistical institutions and development entities have long been attempting to characterise the differentiation and diversity of forms of agriculture and food systems (Chayanov, 2019 [1927]; Colson, 1986; Mazoyer & Roudart, 1997). Some researchers have analysed the different currents of theoretical thought that form the basis of this substantial collection of work (Hervieu & Purseigle, 2013; van der Ploeg, 2018). In the same vein, there is no shortage of comparisons of the performance of agricultural and food models (Seufert et al., 2012; Dumont & Baret, 2017; Muller et al., 2017), making room for a wide variety of postures and methods, which could be multi-criteria, multi-scale, scenario- or modelling-based, meta-analysis based and/or participatory in nature (Binder et al., 2010; Talukder & Blay-Palmer, 2017). In contrast, studies that examine the conditions and effects of interactions between actors and systems of different agricultural models are less common.[9] Yet, in view of recent history, the heterogeneity of our agricultural and food worlds[10] and their interconnectedness compel us to analyse and govern situations of coexistence and confrontation of agricultural and food models. The main research in this area has so far focused on the coexistence of GMO and non-GMO farming (Jank et al., 2006; Hubbard & Hassanein, 2013; Kalaitzandonakes et al., 2016), primarily around biotechnical and ecological issues. However, many researchers in the human and social sciences have recently become interested in the issues of coexistence of agricultural and food models (Deléage & Sabin, 2012; Hervieu & Purseigle, 2015; Albaladejo, 2016; Loring, 2016; Goulet & Giordano, 2017; Cayre et al., 2018; Chia & Dulcire, 2019; van den Berg et al., 2019; Aubert et al., 2020; Gasselin et al., 2020). Also noteworthy is a recent special issue of the Review of Agricultural, Food and

[9] This book does not present a systematic review of the literature on the coexistence and confrontation of agricultural and food models.

[10] Heterogeneity of access to resources and wealth, of actors' practices and projects, of socio-political and economic regulations, etc. (Jollivet & Lepart, 1992; van der Ploeg & Ventura, 2014).

Environmental Studies (Gasselin & Hostiou, 2020), bringing together ten articles on these issues.

In addition to characterising the diversity of agricultural and food models and assessing their relative performance, it is also essential for us to understand their interactions within territories[11]. Indeed, these interactions are intrinsic to territorial development, which we define as an increase in the actors' capacity to control the processes and activities that concern them in their territory, including agriculture and food production (Deffontaines et al., 2001). These activities pertain to all the actors concerned by a territory, across all sectors (Lardon et al., 2015; Torre, 2015). The interactions between models are at the heart of territorial development processes and strategies and open up the field of possibilities. They can take various forms and be combined: copresence, cohabitation, complementarities, synergies, coevolutions, hybridisations and/or confrontations, competitions, marginalisations and exclusions. It is to remove any ambiguity, and to remind us that these interactions are often not peaceful or positive, that this book is entitled 'Coexistence and confrontation of agricultural and food models'.

Our ambition is thus to contribute, through this book, to an improved understanding of the conditions under which the coexistence of agricultural and food models is conducive to sustainable territorial development. We will, however, not undertake a comprehensive, critical and structured discussion on sustainable development of agriculture, food systems or territories (Godard, 1994, 2005; Laganier et al., 2002; Agrimonde, 2009; Esnouf et al., 2011; Zahm et al., 2015; FAO, 2018). We perceive sustainable development in a triple dimension—without claiming to find answers to the political and scientific debates surrounding this concept. First, sustainable development is a capacity to continue to exist whilst taking future generations and uncertainties into account (which refers to the concept of emergent properties of agricultural and food systems, especially resilience). Second, it is an ideological alternative to the industrial agricultural and food model. And, third, it is a combination of ecologically sustainable, economically viable, socially acceptable, and ethically equitable goals. Each of the alternative agricultural or food model embodies a particular vision of this sustainable development.

In this book, we pay attention to controversies and to situations that generate effects that are detrimental to the environment, social justice, the economy or the health of humans, plants or animals (Habte & Krawinkel, 2015; Lindgren et al., 2018). In particular, we examine situations at the margins of the institutionalisation of conventional or alternative models (Bellon & Ollivier, 2018), whether this takes place in science, the market or public policy. In subjecting the coexistence and confrontation of models to scrutiny, we also intend to inform professional and political debates by investigating the sociotechnical controversies through which these models assert themselves and oppose each other.

Thus, this book first of all tests and combines the theoretical frameworks through which agricultural and food models and their coexistence and their confrontations are

[11] In the heritage of French geography, we conceive the territory through the ideal, organisational and material links that are established between societies and their spaces (Lévy & Lussault, 2013).

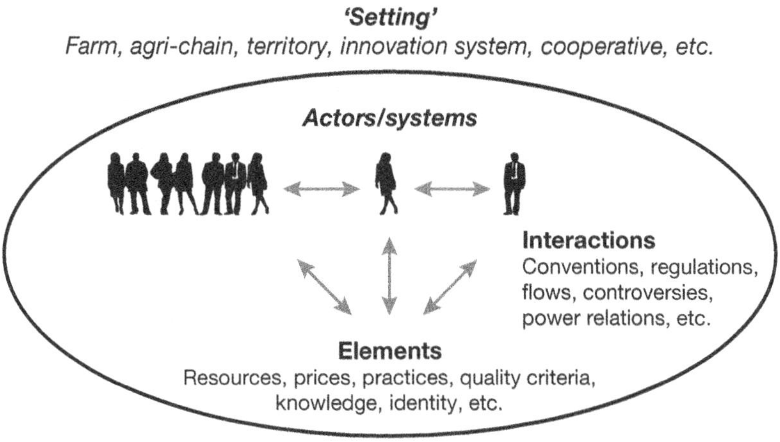

Fig. 1 Situation of coexistence and confrontation of agricultural and food models

constructed and analysed. In doing so, the studies presented highlight the plurality of agricultural and food forms and of their relationships in order to go beyond dual readings. Second, through their investigations, the book's contributors shed light on these models' domains of coherence and major underlying dimensions: technical paradigms, ecological functionalities, relationships with nature, organisation of work, forms of food consumption, etc. We test the hypotheses according to which the diversity of agricultural and food models, and their interactions, confer, under certain conditions, capacities for the diversification, innovation, adaptation or transition of food systems. Finally, we determine the conditions and tools necessary for a territorial governance of the coexistence of agricultural and food models in a perspective of sustainable development of territories and food systems.

Situations of Coexistence

The term coexistence, first used in the 16th century, derives from the Latin *coexistere*, which means 'to exist together'. Dictionaries (Oxford, Websters, Wiktionary, etc.) all list the meaning of coexistence as 'the state of existing together in the same place at the same time'. Its synonyms include contemporaneousness, concomitance, coincidence, simultaneity, copresence and cohabitation. In political vocabulary, coexistence can become 'peaceful' when this adjective is attached to it—which presupposes that it is not necessarily so. In ecology, coexistence refers to various interactions between species in an ecosystem: symbiosis, mutualism, commensalism, neutralism, parasitism, etc. Not all of these interactions are always positive.

Before we can consider a 'situation of coexistence' of agricultural and food models (Fig. 1), we have to specify which actors or systems are interacting (producers, production systems, actors in a territory or agri-chain, etc.) and in which 'settings' or frameworks of interaction (a farm, a cooperative, a territory, an agri-chain, an innovation system, a governance system, etc.). It is also necessary to examine how they interact (conventions, regulations, flows of materials or money, controversies, power relations, etc.) and around which elements (work, technical systems, prices, natural resources, quality criteria, knowledge, identity, etc.).

Situations of coexistence are indeed observed differently at the scale of a farm, a cooperative, a territory or a nation: situations that are 'virtuous' at some scales may not be so at others. Similarly, the issues of coexistence vary according to the problem being addressed. For example, in a given territory, the issue of coexistence may concern the question of competition over resources, but it may also concern the construction of a territorial identity.

A Framework for Analysis and a General Research Issue

This book presents the results of the Format project, funded by the INRA-CIRAD GloFoodS metaprogramme (2015–2017), whose aim was to study combinations of forms of agriculture and food systems at different territorial scales. Some 50 researchers[12] addressed this issue during a series of six seminars that allowed for the analysis of 19 case studies[13]. These seminars concluded with an international symposium (June 2017) and a session of the Living Territories symposium (January 2018). The Format project was part of the 'Coexistence and confrontation of agricultural and food models' Scientific Priority of INRA's Science for Action and Development (SAD) division (2016–2020)[14].

The Format seminars revealed that the coexistence and confrontation of agricultural and food models in a territory both determine and depend on the following four major questions: What are the tensions between specialisation (of production and/or of spaces) and diversification? Is innovation the driving force and/or the product of the coexistence of territorial agricultural and food models? What are the conditions that are suitable for the adaptation of agricultural and food systems in a context of uncertainty? Are the sustainability transition approaches appropriate for designing and supporting situations of coexistence of territorial agricultural and food models? These four dimensions (diversification/specialisation, innovation, adaptation, transition) are addressed in a dynamic way, as processes. They are considered both as

---

[12] These researchers, mainly from the human and social sciences, are affiliated to 13 research and higher education institutions in France (AgroParisTech, CIRAD, CNRS, INRA, IRSTEA, Montpellier SupAgro, VetAgro Sup), Argentina (INTA), Belgium (Catholic University of Louvain), Brazil (Instituto Ambiental do Paraná), Japan (Aichi Gakuin University of Nagoya), Portugal (University of Évora) and Switzerland (University of Neuchâtel).

[13] Seven cases from Europe, six from South America, three from Africa, two from Asia and an international comparative approach concerning seven milksheds.

[14] In January 2020, INRA became INRAE—the French National Research Institute for Agriculture, Food and Environment—and the SAD division became the 'Action and Transitions' (ACT) division.

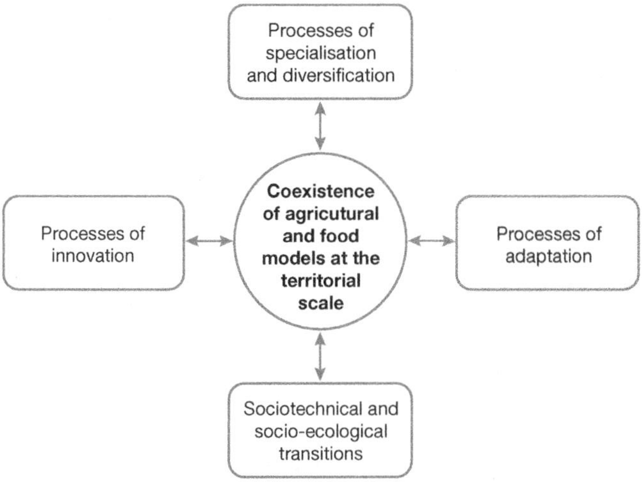

Fig. 2 Framework for analysing the coexistence of agricultural and food models. *Source* Gasselin et al., 2020

factors and as outputs of the dynamics at work in the coexistence and confrontation of territorial agricultural and food models.

These four dimensions (Fig. 2), identified at the end of the Format seminars, echo the focal points of research in the human and social sciences on territorial development during four successive periods (Pike et al., 2006; Jean, 2008; Torre, 2015): specialisation since the 1980s, innovation since the 1990s, adaptation since the 2000s, and transition since the 2010s (Gasselin et al., 2020).

In addition to examining these four dimensions in some detail, this book offers a critical perspective by questioning the relevance of the very notion of the agricultural and food model, by examining how models assert themselves and analysing their coexistence at the territorial scale, and by discussing whether and how these situations of coexistence and confrontation reshape thinking on territorial development.

The Structure of the Book

This book is organised in five parts. The first four parts examine situations of coexistence according to the four dimensions of the analytical framework presented above. Each of these parts is organised as follows:

- An introduction provides a brief overview of the state of the art[15] on the dimension concerned, before proposing some major hypotheses pertaining to the situations of coexistence and confrontation of territorial agricultural and food models. This introductory chapter then presents a summary of the other chapters in the part, followed by a transversal analysis of these contributions;
- Two or three case studies follow, corresponding to specific territories. Each of these case studies formed the topic of a presentation extensively discussed by the Format seminar's participants. An audio or video recording of each session was made, and a report and a transcript of the exchanges were subsequently produced. This material enabled the authors to write their chapter, which was then revised by one or two of the book's scientific editors, an external reviewer and Sylvie Zasser, who was in charge of editorial follow-up;
- Each of the first four parts (except for the second on Innovation) concludes with a 'panoramic' analysis. These chapters of different types (theoretical perspective, comparative analysis, position paper) are original contributions on each of the dimensions considered.

The book's fifth part is divided into two sections. The first gives the floor to three researchers (Jérémie Forney, Kae Sekine and Gilles Allaire) we invited to present situations of coexistence that illustrate contrasting perspectives on territorial development. The second section is divided into two chapters, the first by Ronan Le Velly, the second by Patrick Caron, who were requested to provide a personal reflection and cast a critical look at the entire book. We conclude by examining the title of the book: Do the coexistence and confrontation of agricultural and food models open the way to a new paradigm of territorial development?

With this book, we intend to show that taking the coexistence and confrontation of agricultural and food models into account enriches the conceptual apparatus necessary to analyse and support agricultural and food development in rural and urban territories. These contributions also offer a broad panorama of situations of coexistence around the world, in Europe (France (five cases), Italy and Switzerland), South America (Argentina and Brazil (two cases)), East and South-east Asia (Japan, Vietnam), Burkina Faso and two international comparative approaches. They constitute an analytical framework and a research agenda, the first results of which we present here. We hope readers find this book rewarding and enriching.

[15] The literature review was carried out by querying 4 scientific documentary databases (Web of Science, Agritrop, Prodinra, HAL) using queries adapted to each of them. As an illustration, these were the queries used for Web of Science: TS = ((((intensive or conventional or "high input" or monofunctional or industrial or commercial) near/3 (organic or "low input" or integrated or ecological or familial or multifunctional or sustainable or agroecology))) near/3 (((farm or farming or agricultur* or crop or food or agri$food or livestock) near/3 (system or model)))). The bibliographic material is also based on the expertise of each of the coauthors, who selected the articles which seemed most relevant to them.

References

Agrimonde (Eds.) (2009). Agricultures et alimentation du monde en 2050: Scénarios et défis pour un développement durable, Note de synthèse (February 2009) (p. 34). Inra-Cirad.

Albaladejo, C., (2016). Coexistencia en el territorio de diferentes modelos de desarrollo agropecuario: La teoría de los pactos territoriales aplicada al caso argentino. In D. Nieto, P. Carricart, C. Albaladejo, A. L. de Carvalho Fiúza (Eds.), *Transformaciones territoriales y la actividad agropecuaria. Tendencias globales y emergentes locales* (pp. 27–52). Universidad Nacional de La Plata.

Albaladejo, C., (2020). The impossible and necessary coexistence of agricultural development models in the Pampas: The case of Santa Fe province (Argentina). *Review of Agricultural, Food and Environmental Studies*, March 2020, 1–28.

Allaire, G., & Daviron, B. (Eds.) (2017). *Transformations agricoles et agroalimentaires: Entre écologie et capitalisme* (p. 429). coll. Synthèses, éditions Quæ.

Aubert, P.-M., Ruat, R., Treyer, S., & Rankovic, A. (2020). Holding the ground. Alliances and defiances between scientists, policy-makers and civil society in the development of a voluntary initiative, the "4 per 1000: Soils for food security and climate". *Environmental Science and Policy, 113* (November), 80–87.

Bellon, S., & Ollivier, G. (2018). Institutionalizing agroecology in France: Social circulation changes the meaning of an idea. *Sustainability, 10*(5), 2071–1050.

Bernard de Raymond, A., & Goulet, F. (2014). *Sociologie des grandes cultures: Au cœur du modèle industriel agricola* (p. 224). éditions Quæ.

Beus, C. E., & Dunlap, R. E. (1990). Conventional versus alternative agriculture: The paradigmatic roots of the debate. *Rural Sociology, 55*(4), 590–616.

Bijker, W. E. (1997). *Of bicycles, bakelites, and bulbs: Toward a theory of sociotechnical change*, (p. 360). The MIT Press.

Binder, C. R., Feola, G., & Steinberger, J. K. (2010). Considering the normative, systemic and procedural dimensions in indicator-based sustainability assessments in agriculture. *Environmental Impact Assessment Review, 30*(2), 71–81.

Bosc, P.-M., Sourisseau, J.-M., Bonnal, P., Gasselin, P., Valette, E., & Bélières, J.-F. (Eds.) (2018). *Diversity of family farming around the world. existence, transformations and possible futures of family farms* (p. 341). Springer.

Caron, P. (2016). Climate-smart agriculture: Émergence d'un concept, mise en politique, mise en science et controverses. *Natures Sciences Sociétés, 24*(2), 147–150.

Caron, P., y de Loma-Osorio, G. F., Nabarro, D., Hainzelin, E., Guillou, M., Andersen, I., Arnold, T., Astralaga, M., Beukeboom, M., & Bickersteth, S. (2018). Food systems for sustainable development: Proposals for a profound four-part transformation. *Agronomy for Sustainable Development, 38*(4), 41.

Cayre, P., Michaud, A., Theau, J.-P., & Rigolot, C. (2018). The coexistence of multiple worldviews in livestock farming drives agroecological transition. A case study in french Protected Designation of Origin (PDO) cheese mountain areas. *Sustainability, 10*(4), 1097.

Chayanov, A. V. (2019) [1927]. On differentiation of the peasant economy. *Russian Peasant Studies, 4*(4), 6–21.

Chia, E., & Dulcire, M. (2019). La coexistence de formes de production agricole au prisme des politiques publiques: Le cas de la Guadeloupe. *Études caribéennes* (43–44).

Colson, F. (1986). Le développement agricole face à la diversité de l'agriculture française. *Économie rurale, 172*(1), 3–9.

Darnhofer, I. (2015). Socio-technical transitions in farming: Key concepts. In L.-A. Sutherland, I. Darnhofer, G. A. Wilson, L. Zagata (Eds.), *Transition pathways towards sustainability in agriculture. Case studies from Europe* (pp. 17–31). CAB International.

Daviron, B., (2019). *Biomasse: Une histoire de pouvoir et de richesse* (p. 391). coll. Synthèses, éditions Quæ.

Deffontaines, J.-P., Marcepoil, E., & Moquay, P. (2001). Le développement territorial: Une diversité d'interprétations. In S. Lardon, P. Maurel, V. Piveteau (Eds.), *Représentations spatiales et développement territorial. Bilan d'expériences et perspectives méthodologiques* (pp. 39–56). Hermès Science Publications,.

Deléage, E., & Sabin, G. (2012). Modernité en friche. Cohabitation de pratiques agricoles. *Ethnologie française, 42*(4), 667–676.

Deverre, C., & Lamine, C. (2010). Les systèmes agroalimentaires alternatifs. Une revue de travaux anglophones en sciences sociales. *Économie rurale, 3*, 57–73.

Dumont, A. M., & Baret, P. V. (2017). Why working conditions are a key issue of sustainability in agriculture? A comparison between agroecological, organic and conventional vegetable systems. *Journal of Rural Studies, 56*, 53–64.

Duteurtre, G., Khanh, P. D., & Cesaro, J.-D. (2015). Bassin laitier de Ba Vi (Vietnam). In M. Napoléone, C. Corniaux, B. Leclerc (Eds.), *Voies lactées. Dynamique des bassins laitiers entre globalisation et territorialisation* (pp. 67–87), Inra-SAD-Cardère.

Esnouf, C., Russel, M., & Bricas, N. (Eds.) (2011). *Pour une alimentation durable. Réflexion stratégique duALIne* (p. 288). éditions Quæ.

FAO (2018). *The 10 elements of agroecology: Guiding the transition to sustainable food and agricultural systems* (p. 15). FAO.

Fournier, S., & Touzard, J.-M. (2014). La complexité des systèmes alimentaires: Un atout pour la sécurité alimentaire? *VertigO—La revue électronique en sciences de l'environnement, 14*(1).

Fraser, E. D., Mabee, W., & Figge, F. (2005). A framework for assessing the vulnerability of food systems to future shocks. *Futures, 37*(6), 465–479.

Gaitán-Cremaschi, D., Klerkx, L., Duncan, J., Trienekens, J. H., Huenchuleo, C., Dogliotti, S., Contesse, M. E., & Rossing, W. A. (2019). Characterizing diversity of food systems in view of sustainability transitions. A review. *Agronomy for Sustainable Development, 39*(1), 1–22.

Galliano, D., Lallau, B., & Touzard, J.-M. (2017). Coexistences et transitions dans l'agriculture. *Revue française de socio-économie, 1*, 23–30.

Gasselin, P. (2019). Transformation of French family farming: From diversity study to coexistence analysis of agricultural models (working paper). *The Natural Resource Economics Review*, (March), 61–73.

Gasselin, P., & Hostiou, N. (2020). What do our research friends say about the coexistence and confrontation of agricultural and food models? Introduction to the special issue. *Review of Agricultural, Food and Environmental Studies, 101*(2–3), 173–190.

Gasselin, P., Choisis, J.-P., Petit, S., Purseigle, F., & Zasser, S. (Eds.) (2014). *L'agriculture en famille: Travailler, réinventer, transmettre* (p. 382). EDP Sciences.

Gasselin, P., Lardon, S., Cerdan, C., Loudiyi, S., & Sautier, D. (2020). The coexistence of agricultural and food models at the territorial scale: An analytical framework for a research agenda. *Review of Agricultural, Food and Environmental Studies, 101*(2–3), 339–361.

Geels, F. W. (2010). Ontologies, socio-technical transitions (to sustainability), and the multi-level perspective. *Research Policy, 39*(4), 495–510.

Gisclard, M., & Allaire, G. (2012). L'institutionnalisation de l'agriculture familiale en Argentine: Vers la reformulation d'un référentiel de développement rural. *Autrepart, 3*(62), 201–216.

Godard, O. (1994). Le développement durable. Paysage intellectuel. *Natures Sciences Sociétés, 2*(4), 309–322.

Godard, O., (2005). Le développement-durable, une chimère, une mystification? *Mouvements, 4*, 14–23.

Goodman, D. (2003). The quality 'turn' and alternative food practices: Reflections and agenda. *Journal of Rural Studies, 19*(1), 1–7.

Goulet, F., & Giordano, G. (2017). Searching for family farming in Argentina: Chronicles of a technological innovation between two worlds. *Review of Agricultural, Food and Environmental Studies, 98*(4), 233–253.

Habte, T., & Krawinkel, M. (2015). Nutritional and health implications of conventional agriculture. A review. *Journal of Nutrition and Health Sciences, 2*(1), 1–8.

Hall, P., Schmidt, V., & Thatcher, M. (2015). Cognitive approaches: A French touch? Three Anglo-American perspectives on French policy analysis. In L. Boussaguet, S. Jacquot, P. Ravinet, P. Muller (Eds.), *Une "French touch" dans l'analyse des politiques publiques* (pp. 237–262). Presses de Sciences Po.

Hervieu, B., & Purseigle, F. (2011). Des agricultures avec des agriculteurs, une nécessité pour l'Europe. *Projet, 2*, 60–69.

Hervieu, B., & Purseigle, F. (2013). *Sociologie des mondes agricoles* (p. 320). Armand Colin.

Hervieu, B., & Purseigle F. (2015). The sociology of agricultural worlds: From a sociology of change to a sociology of coexistence. *Review of Agricultural and Environmental Studies*, 96 (1), 59–90.

HLPE (2019). Agroecological and other innovative approaches for sustainable agriculture and food systems that enhance food security and nutrition. Report by The High Level Panel of Experts on Food Security and Nutrition. *HLPE report* 14 (p. 162). FAO.

Hodbod, J., & Eakin, H. (2015). Adapting a social-ecological resilience framework for food systems. *Journal of Environmental Studies and Sciences, 5*, 474–484.

Holling, C. S. (2001). Understanding the complexity of economic, ecological, and social systems. *Ecosystems, 4*, 390–405.

Horlings, L. G., & Marsden, T. K. (2011). Towards the real green revolution? Exploring the conceptual dimensions of a new ecological modernisation of agriculture that could 'feed the world'. *Global Environmental Change, 21*(2), 441–452.

Hubbard, K., & Hassanein, N. (2013). Confronting coexistence in the United States: Organic agriculture, genetic engineering, and the case of Roundup Ready® alfalfa. *Agriculture and Human Values, 30*(3), 325–335.

IPES-Food (2016). *From uniformity to diversity: A paradigm shift from industrial agriculture to diversified agroecological systems* (p. 94). International Panel of Experts on Sustainable Food Systems.

Jank, B., Rath, J., & Gaugitsch, H. (2006). Co-existence of agricultural production systems. *Trends in Biotechnology, 24*(5), 198–200.

Jarosz, L. (2014). Comparing food security and food sovereignty discourses. *Dialogues in Human Geography, 4*(2), 168–181.

Jean, B. (2008). Le développement territorial. Une discipline scientifique émergente. In G. Massicotte (Eds.) *Sciences du territoire. Perspectives québécoises* (pp. 283–313). Presses de l'université du Québec.

Jollivet, M., & Lepart, J. (1992). Hétérogénéité, diversité, complexité: Nuances et convergences. In M. Jollivet (Ed.), *Sciences de la nature, sciences de la société. Les passeurs de frontières* (pp. 373–380), CNRS Éditions.

Kalaitzandonakes, N., Phillips, P. W., Wesseler, J., & Smyth, S. J. (Eds.) (2016). *The coexistence of genetically modified, organic and conventional foods* (p. 426). Springer.

Karlsson, L., Naess, L. O., Nightingale, A., & Thompson, J. (2018). 'Triple wins' or 'triple faults'? Analysing the equity implications of policy discourses on climate-smart agriculture (CSA). *The Journal of Peasant Studies, 45*(1), 150–174.

Laganier, R., Villalba, B., & Zuindeau, B. (2002). Le développement durable face au territoire: Éléments pour une recherche pluridisciplinaire. *Développement durable et territoires, 11*(2).

Lardon, S., Albaladejo, C., Allain, S., Cayre, P., Gasselin, P., Lelli, L., Moity-Maïzi, P., Napoleone, M., & Theau, J.-P. (2015). Dispositifs de recherche-formation-action pour et sur le développement agricole et territorial. In A. Torre, D. Vollet (Eds.), *Partenariats pour le développement territorial* (pp. 47–57). éditions Quæ.

Le Velly, R. (2015). *La promesse de différence: Sociologie des systèmes alimentaires alternatifs*, Application for accreditation to supervise research (HDR) submitted under the guidance of Mrs. Sophie Dubuisson-Quellier (p. 234). Paris Institute of Political Studies.

Lévy, J., & Lussault, M. (2013). *Dictionnaire de géographie et de l'espace des sociétés* (p. 1044). Belin.

Lindgren, E., Harris, F., Dangour, A. D., Gasparatos, A., Hiramatsu, M., Javadi, F., Loken, B., Murakami, T., Scheelbeek, P., & Haines, A. (2018). Sustainable food systems. A health perspective. *Sustainability Science, 13*, 1505–1517.

Loring, P. A. (2016). Toward a theory of coexistence in shared social-ecological systems: The case of cook inlet salmon fisheries. *Human Ecology, 44*(2), 153–165.

Malassis, L. (1997). *Les trois âges de l'alimentaire. Tome 2: L'âge agro-industriel* (p. 376). Éditions Cujas.

Mazoyer, M., & Roudart, L. (1997). *Histoire des agricultures du monde: Du néolithique à la crise contemporaine* (p. 505). Éditions du Seuil.

McMichael, P. (2009). A food regime genealogy. *The Journal of Peasant Studies, 36*(1), 139–169.

Mendras, H. (1967). *La fin des paysans: Innovations et changements dans l'agriculture française* (p. 364). coll. Futuribles, SEDEIS.

Muller, A., Schader, C., El-Hage Scialabba, N., Brüggemann, J., Isensee, A., Erb, K.-H., Smith, P., Klocke, P., Leiber, F., Stolze, M., & Niggli, U. (2017). Strategies for feeding the world more sustainably with organic agriculture. *Nature Communications, 8*(1), 1290.

Muller, P. (1990). *Les politiques publiques* (p. 127). coll. Que sais-je? PUF.

Oui, J., & Touzard, J.-M. (2018). La Climate Smart Agriculture: Un projet politique controversé pour climatiser l'agriculture. In *Colloque SFER «Politiques agricoles et alimentaires: Trajectoires et réformes»* (p. 2).

Pelenc, J., Wallenborn, G., Milanesi, J., Sébastien, L., Vastenaekels, J., Lajarthe, F., Ballet, J., Cervera-Marzal, M., Carimentrand, A., & Merveille, N. (2019). Alternative and resistance movements: The two faces of sustainability transformations? *Ecological Economics, 159*, 373–378.

Petit, S., Hostiou, N., Tallon, H., & Gasselin, P. (2018). Faire recherche sur la coexistence de modèles: Diversité des regards de chercheurs. In *Séminaire permanent «Élevage et développement durable des territoires»: Coexistence et confrontation de modèles d'élevage dans les territoires* (27 June 2018). Inra-Cirad.

Pike, A., Rodríguez-Pose, A., & Tomaney, J. (2006). *Local and regional development* (p. 328). Routledge.

Pimbert, M. P. (2018). Global status of agroecology. *Economic and Political Weekly, 53*(41), 52–57.

Pingali, P. L. (2012). Green revolution: Impacts, limits, and the path ahead. *Proceedings of the National Academy of Sciences, 109*(31), 12302–12308.

Plumecocq, G., Debril, T., Duru, M., Magrini, M.-B., Sarthou, J.-P., & Thérond, O. (2018). Caractérisation socio-économique des formes d'agriculture durable. *Économie rurale, 1*, 99–120.

Purseigle, F., Nguyen, G., & Blanc, P. (Eds.) (2017). *Le nouveau capitalisme agricole. De la ferme à la firme* (p. 305). Presses de Sciences Po.

Regnault, H., De Sartre, X. A., & Regnault-Roger, C. (2012). *Les révolutions agricoles en perspective* (p. 216). Éditions France agricole.

Rosset, P. (2003). Food sovereignty: Global rallying cry of farmer movements. *Food First Backgrounder, 9*(4), 1–4.

Sencébé, Y., Pinton, F., & Cazella, A. A. (2020). On the unequal coexistence of agrifood systems in Brazil. *Review of Agricultural, Food and Environmental Studies*, (February), 1–22.

Seufert, V., Ramankutty, N., & Foley, J. A. (2012). Comparing the yields of organic and conventional agriculture. *Nature, 485*(7397), 229–232.

Sobal, J., Khan, L. K., & Bisogni, C. (1998). A conceptual model of the food and nutrition system. *Social Science and Medicine, 47*(7), 853–863.

Talukder, B., & Blay-Palmer, A. (2017). Comparison of methods to assess agricultural sustainability. In *Sustainable Agriculture Reviews* (pp. 149–168). Springer.

Taylor, M. (2018). Climate-smart agriculture: What is it good for? *The Journal of Peasant Studies, 45*(1), 89–107.

Torre, A. (2015). Théorie du développement territorial. *Géographie, économie, société, 17*(3), 273–288.

Urruty, N., Tailliez-Lefebvre, D., & Huyghe, C. (2016). Stability, robustness, vulnerability and resilience of agricultural systems. A review. *Agronomy for Sustainable Development, 36*(1), 15.

van den Berg, L., Goris, M., Behagel, J., Verschoor, G., Turnhout, E., Botelho, M., & Silva Lopes, I. (2019). Agroecological peasant territories: Resistance and existence in the struggle for emancipation in Brazil. *The Journal of Peasant Studies*, 1–22.

van der Ploeg, J. D. (2010). Farming styles research: The state of the art. In *Keynote lecture for the workshop on 'historicising farming styles'* (pp. 21–23). Melk.

van der Ploeg, J. D. (2012). The genesis and further unfolding of farming styles research. *Historische Anthropologie, 20*(3), 427–439.

van der Ploeg, J. D. (2018). Differentiation: Old controversies, new insights. *The Journal of Peasant Studies, 45*(3), 489–524.

van der Ploeg, J. D., & Ventura F. (2014). Heterogeneity reconsidered. *Current Opinion in Environmental Sustainability, 8*, 23–28.

Wiskerke, J. S., & van der Ploeg, J. D. (2004). *Seeds of transition: Essays on novelty production, niches and regimes in agriculture* (p. 356). Assen, Van Gorcum.

Zahm, F., Alonso Ugaglia, A., Boureau H., Del'homme, B., Barbier, J.-M., Gasselin, P., Gafsi, M., Guichard, L., Loyce, C., Manneville, V., Menet, A., & Redlingshofer, B. (2015). Agriculture et exploitation agricole durables: État de l'art et proposition de définitions revisitées à l'aune des valeurs, des propriétés et des frontières de la durabilité en agriculture. *Innovations agronomiques, 46*, 105–125.

Part I
Specialisation/diversification

Productive and Territorial Specialisation: A Hindrance or a Resource? Introduction to Part I

Pierre Gasselin and Denis Sautier

The first part of this book examines specialisation and diversification, concepts commonly used to qualify the economic or ecological processes that differentiate agricultural and food systems. These concepts can also apply to transformations of a territory, depending on whether the territory evolves towards the domination by a productive activity or, on the contrary, towards a greater plurality of functions. We define a territory as specialised when the vast majority of systems of production and of derivation of value from products are part of the same development model. Conversely, the diversification of activities in a territory implies the presence of a plurality of systems which are part of development models with different orientations.

In this introduction to the first part of the book, we first recall the historical trajectory of the 'specialisation vs diversification' controversy, followed by a discussion of the positive and negative effects of specialisation and of diversification at the socio-economic and socio-ecological levels. This review of the debate leads us to conclude that the processes of specialisation or diversification are still too little studied through the prism of the situations of coexistence of agricultural and food models, thus raising new questions in the field. We then introduce the three chapters of this book that deal with specialisation and diversification, and we conclude with a comparative reading according to the analytical framework of situations of coexistence and the general hypotheses proposed.

Overview of Analytical Frameworks and Hypotheses

Modernising Paradigm Versus Sustainable Development

The process of productive specialisation was one of the drivers of the modernising paradigm of agriculture in the latter half of the twentieth century. It was also bolstered in the countries of the Global North by price stabilisation mechanisms (Allaire & Daviron, 2019). The debate between specialisation and diversification first emerged

P. Gasselin et al. (eds.), *Coexistence and Confrontation of Agricultural and Food Models*,
https://doi.org/10.1007/978-94-024-2178-1

in the 1970s with a strong focus on social issues of dependence and of the loss of autonomy of farms and rural spaces (Kayser, 1992). However, observing and evaluating specialisation requires precise definitions and methodological precautions, given the different possible scales of analysis: farms, rural territories and sectoral organisation (Mathieu, 1985). In the present century, the 'specialisation vs diversification' controversy has intensified with the rise of the concept of sustainable development, focusing on a key question: How can agricultural production be reconciled with the preservation of natural spaces and land use in a global context of pressure on resources and accelerating biodiversity loss?

Economies of Scale Versus Autonomy and Resilience

At the socio-economic level, the respective merits of the specialisation and diversification processes oppose each other (Mathieu, 1984). The specialisation of rural spaces or productive activities has been advocated in pursuit of economies of scale, economic integration, low food prices or even agglomeration externalities (Antoine, 2016). For its part, diversification is frequently defended for the sake of the advantages of complementarity and the autonomy or resilience that are associated with it (Suryanata, 2002; de Roest et al., 2018). Some observers advocate the specialisation of spaces and forms of production in order to intensify, while correcting and compensating for the possible negative effects of such a specialisation (Pingali, 2012). Others prefer instead the integration of agriculture's multiple functions with a diversification of farms and landscapes (IPES-Food, 2016). This debate can be extended to the organisation of food production, the organisation of rural spaces, ecological intensification, and the conservation of the environment.

Land Sparing Versus Land Sharing

At the socio-ecological level, two ideal types can be considered to manage a territory and its resources: 'We can distinguish a first model, qualified as "segregationist", separating what can be cultivated from what should not be from the point of view of environmental protection, in which "natural" processes will nevertheless have to be managed. […] From another point of view, a model that can be described as "integrationist" combines ecological and productive functions of agro-ecosystems in the same territory' (Agrimonde, 2009, p. 31). The first approach divides the territory into spaces dedicated to intensive agriculture, on the one hand, and to the preservation of natural environments, on the other. The second proposes the conception of a diversity and complementarity of forms of agriculture arranged to create ecological mosaics producing various ecosystem services. This debate on the best way to protect nature has been structured around the notions of 'land sparing' and 'land sharing' (Green et al., 2005; Byerlee et al., 2014). The proponents of land sparing hypothesise that high-yield agriculture, based on advanced technology and industrial inputs, is the best way to produce on limited areas and therefore can best preserve large uninhabited nature reserves sustainably (Green et al., 2005). Conversely, the advocates of land sharing do not believe in isolated protected spaces surrounded by regions inhospitable to biodiversity. According to them, 'to avoid ecosystem collapse, we must integrate biodiversity conservation throughout the landscape we use' (Kremen & Merenlender 2018). A tension definitely exists between these two types of intervention. However, a compromise was subsequently reached between these streams of thought, according to which the choice of approach would depend in part on the scale (Fischer et al., 2014). Proponents of land sparing agree that 'land sparing and land sharing describe two ends of a continuum of intentional spatial organisation of food production and biodiversity conservation: whether separated or integrated' (Phalan, 2018). For their part, the advocates of land sharing recognise the importance of the traditional protected area approach, but they argue that the range of tools available to maintain biodiversity in more or less anthropised areas must include an agroecological approach to cultivated spaces (Kremen & Merenlender, 2018).

Hypotheses and Major Questions

This brief overview of the debate shows that the issues are not framed in terms of the coexistence or confrontation of territorial agricultural and food models, which nevertheless drive the processes of domination or diversity, and aggregation or dispersion (of actors, of productive and/or spatial arrangements, etc.). It should be remembered that a situation of coexistence or confrontation of agricultural and food models has to be examined in terms of the interactions between actors or systems around particular objects in a given setting. Analysis of these interactions sheds new light on specialisation and diversification processes, which leads us to propose three hypotheses to test in future research and to raise associated questions:

Hypothesis 1: *Specialisation and diversification are not always in opposition, but can follow one another, combine together or be nested, depending on the scale (spatial, temporal, social) under consideration.* Indeed, the issues of specialisation and diversification can be expressed differently at the scale of the farm, the territory, the sector, the production basin, etc. The articulation of these scales is essential. In order to endure sustainably, how can specialised systems be made to take into account and maintain the diversity of a territory's resources? How do successful diversification trajectories borrow elements of innovation from the specialisation model? Is the diversification of productive spaces not based on a certain level of farm specialisation, for example, by relying on acquired professional skills to expand farms towards new activities or new markets?

Hypothesis 2: *Diversification and specialisation are differentiated by different power relationships.* What are the forms of power in a territory around agricultural and food issues? Is specialisation a process of internal organisation or is it a progressive subordination to an ordered pattern of decision-making and functioning? Is diversification underpinned by a plurality of powers in the territory concerned or by the absence of a common project? Does it necessarily lead to a multifunctional balance between activities?

Hypothesis 3: *Both for specialisation as well as for diversification, there are intended evolutions, of course, but territorial actors also undergo evolutions.* The determinants are not necessarily found at the territorial scale, because the dynamics of markets and of macro-economic actors (firms, States) shape these processes to a large extent (van der Ploeg et al., 2008). Diversification, for example, can either result from an active and deliberate strategy or, on the contrary, passively reflect the lack of structuring opportunities. While the coexistence of models in a territory can result in conflicts and synergies, it can also lead to ignorance or mutual tolerance. How does competition over resources (land, water, labour) crystallise tensions between these models? What roles do the representations that actors have of specialisation and diversification play? In which cases is the territory the sole determinant of the dynamics of specialisation or diversification?

The multiple interactions between combinations of specialisation and diversification and their territorial effects deserve to be better examined. Thus, taking the coexistence and confrontations of agricultural and food models into account makes it possible to revisit the debate on the opposition between specialisation and diversification at the territorial scale. It shows dialectical relationships between these two processes and, above all, it leads to the necessity of better articulating different temporal, spatial and actor-organisation scales.

Testing in the Field

Following this short review of the scientific literature and the central hypotheses that arise from it, we present the three chapters that form this part of the book. They analyse the tensions between specialisation and diversification in situations of coexistence of agricultural and food models at the territorial level. The first chapter is a case study of pig and poultry production in southern Brazil, while the second compares five milksheds in South America and France. The last chapter is a panoramic reflection of territorial economics on the specialisation of agricultural and rural Europe. The synthesis of these studies leads us to undertake a comparative analysis that highlights their common points and divergences.

Three Illustrations

In the first chapter, Claire Cerdan analyses the processes of specialisation and diversification of food systems in Santa Catarina state, in southern Brazil. In the 1970s, integrated pig and poultry for meat production was organised in a win-win relationship between agro-industries and multi-crop-livestock family farming with the support of public authorities and the training of workers by industry. In the 1980s and 1990s, the international-level agro-industries consolidated through a process of intensification, increased specialisation and concentration of activities. Farmers lost their autonomy and more than two-thirds of pig farmers disappeared in a period of 15 years. From the 1990s onwards, agro-industries implemented new diversification strategies within and outside the territory to make their industrial units profitable, while diversifying their range of offerings (frozen products and ready meals). With the support of public authorities, trade unions and social movements promoted the diversification projects of farmers who were excluded from the agro-industrial integration model or of those who refused to be part of it. These artisanal projects benefited not only from traditional knowledge but also from the know-how acquired during years of specialisation in farms and industries. The exclusion of a large number of livestock farmers called into question the idea of a territory specialised by an integrated agro-industrial model. Producers developed small-scale processing, while continuing to raise animals for industry. At the end of the 2000s, a 'passive (or peaceful) coexistence between the agro-industrial model and the on-farm and artisanal production model' was organised. Unlike other Brazilian regions, this region has been able to retain its working population due to the presence of the food industries. This population is now the main market for family agribusinesses. Claire Cerdan considers that this case study 'confirms the hypothesis that diversification and specialisation are part of the same process of adaptation of productive spaces to the global system'.

In the second chapter, Martine Napoléone, Marie Houdart and Guillaume Duteurtre discuss three archetypal development pathways of dairy activities in five contrasting rural territories in South America and France. The authors attempt to identify the main factors impacting the dynamics of specialisation and diversification in these territories. They identify and characterise three types of milksheds: those dominated by industrialised forms of development, which are part of a globalised development model (Salto in Uruguay); those dominated by territorialised forms, which are part of a territorialised development model (Brasil Novo in Brazil); and those in which different forms of development coexist, which belong to both the aforementioned development models (Livradois-Forez, Cévennes and Drôme Provençale in France). The authors consider that the processes of specialisation and diversification are sometimes driven more by food demand and influenced by the territorial resources available than by adaptation to global change. Thus, specialisation can apply both to globalised models (Salto) as well as to territorialised ones (Brasil Novo). Moreover, the coexistence of models may result from different dynamics of specialisation: in France, the different models are spatially distributed over the territory, whereas in Uruguay, an agro-industrial model and a territorialised model are superimposed on the same space. The authors also show that diversification and specialisation trajectories can be analysed as a combination of market dynamics and the action of strategic actors (firms, States, professional organisations or civil society). In line with DuPuis & Block (2008), the authors confirm that specialisation and diversification are not solely an adaptation to the globalised market. Finally, they categorise three situations: agro-industrial development geared towards the production of commodities; dynamics that are conducive to the commercial positioning of agro-industrial processors in the long supply chains of national or regional mass distribution entities; and finally, a territorial logic with little connection to world markets or supermarket chains, in which the territory's dairy supply meets the demand for products arising from local relationships. However, they believe that 'while some territories are favourable to the expression of several forms of development, others are not', depending on the characteristics of territorial resources and food demands.

In the third chapter, Frederic Wallet offers a panoramic reflection on the evolution and coexistence of agricultural production models in the light of European knowledge and policies on regional specialisation. He notes that low energy prices, the search for economies of scale and the criteria for granting aid help explain productive specialisation in France. This is marked by a geographical dissociation of crop and livestock production, a massive decrease in the number of farms and a concentration of agri-food industries that capture most of the added value. Frederic Wallet emphasises that the many alternatives (short supply chains, organic farming, etc.) to the dominant model 'make up a complex array of production and distribution models for food products and services', which makes uniform or sectoral policies irrelevant. On the basis of this observation, European territorial development policies with a 2025 horizon have adopted a 'place-based' logic. These 'smart specialisation and growth' policies have the goal of offering a differentiated development of territories depending on their resources, their technological capacities and their modes of organisation. It is a matter of promoting sectors of activity in which regions have a competitive advantage over others, without, however, losing the 'related variety' (i.e. interconnected and complementary activities). This consists of combining specialisation with a coherent diversity of sectors in order to take advantage of the processes of production and knowledge dissemination and thus stimulate innovation between various value chains. However, these 'smart specialisation' strategies are more favourable to metropolitan areas than to rural territories, which struggle to benefit from scale effects. The conditions for the success of smart growth policies in rural spaces include the fight against land competition and specialisation, on the one hand, and well-structured agricultural and food innovation systems, on the other. Finally, Frederic Wallet discusses the processes of coexistence marked by the mechanisms for allocating aid and land, as also by oppositions between the value systems underpinning the different agricultural models. He calls for the modification of regulatory mechanisms to reduce the asymmetries of resources, market configurations and public intervention mechanisms, which still clearly favour a dominant form of agriculture.

In this way, he argues that smart specialisation 'will only prove beneficial to agricultural value chains and the rural economy if it allows the full expression of the transformative potential of the dynamic co-evolution of the various agricultural and food models'.

Comparative Reading

Combined Specialisation and Diversification

These three chapters mobilise analytical frameworks from various disciplines, in particular rural geography, agro-economics and territorial economics. They also situate productive and territorial specialisation in a historical movement: since 1970 for Claire Cerdan; at the earliest since the 1930s for Martine Napoléone, Marie Houdart and Guillaume Duteurtre; and since the 2000s for Frederic Wallet. To do so, the authors rely on various diachronic approaches: a historical trajectory of the region, with particular attention to the power relations between actors and the interactions between different development models (Cerdan); specialisation pathways divided into a sequence of periods with a view to chronicling the transformation of dairy activities within a territory and identifying explanatory factors (Napoléone et al.); and the analysis of the emergence of alternative agricultural models and the mutations of territorial development policies (Wallet). These temporal perspectives lead these authors to highlight, in each of their respective chapters, that there are different types of specialisation: 'dual process of specialisation/diversification' and 'flexible specialisation' (Cerdan), 'agro-industrial specialisation' and 'territorialised specialisation' (Napoléone et al.), 'open specialisation' and 'smart specialisation' (Wallet). In so doing, they confirm our first hypothesis which is that *specialisation and diversification are not always in opposition, but can follow one another, combine together or be nested, depending on the scale (spatial, temporal, social) under consideration.*

Choice or Submission?

The historical trajectories also allow these authors to identify the main determinants of the process of specialisation. Their analyses converge to a large extent and point to the deregulation of markets and the opening up of borders; public aid directed to agro-industries favouring the concentration of value chains and integration; land-use planning facilitating commodity flows; certain networks of actors (particularly between agro-industries); and foreign private investment. In addition, Napoléone et al. emphasise the extent to which the characteristics of demand are also determinants of specialisation or diversification (commodities for international trade; generic products for distribution through national or regional long supply chains;

products with a quality label and/or originating from a particular area for distribution in regional, or even national, long supply chains; food produced in proximity (geographical or organisational) to consumers). The underlying or explicit rationales for specialisation are the search for better productivity of production factors through economies of scale and cost reductions (in industries, land structures, seed markets, distribution, etc.).

These three chapters also highlight the importance of the State's role in productive and territorial specialisation and diversification. In Brazil's Santa Catarina state (chapter by Claire Cerdan), public actors first encouraged specialisation by setting up a research centre specialising in agro-industrial meat production in 1975. In 1996, they began hosting an international food industry fair, which showcases regional industrial know-how and a network of service providers. 'This fair reflects the dual process of specialisation/diversification underway in the region'. Later on, public programmes and the research centre for family farming encouraged the professionalisation of family farming and the growth in a number of small-scale processing units, some of which were associated with the agro-industry network. In the Salto milkshed in Uruguay (chapter by Martine Napoléone et al.), specialisation towards an agro-industrial model is taking place in a landlocked region with no dairy tradition thanks to the support of the State, which invested between the 1930s and 1980s in infrastructure and dairy industries oriented towards supplying the city of Salto with standard products. In contrast, the State only intervened to support a 'horizontal' territorial specialisation already undertaken by the actors in the Brasil Novo milkshed in Brazil. In Europe (chapter by Frederic Wallet), the criteria for allocating aid have largely contributed to territorial specialisation and concentration. Moreover, the mechanisms for uniform intervention in all territories have shown their limitations. As a result, the new regional development policies propose an alternative based on the recognition of the importance of regional particularities according to a principle of 'smart specialisation'. Thus, these three chapters confirm our third hypothesis which states that *both for specialisation and for diversification, there are intended evolutions, of course, but territorial actors also undergo evolutions.*

Towards Moderate Specialisations?

These three studies are also unanimous in criticising the disadvantages of excessive specialisation, in particular the sociotechnical lock-in by systems of standards and networks of actors established during specialisation (which leaves little room for alternatives and innovations), the increase in farm size (which generates social and economic exclusion), the disappearance of small and medium-sized agrifood businesses, and the capture of a growing share of added value along the entire lengths of value chains. But it is all a matter of extent. Indeed, all three studies are in favour of moderate specialisation. Claire Cerdan praises the synergies between specialised industries, which represent a source of employment, and numerous alternative micro-activities. These complementarities have stabilised the working population in rural areas as well as in urban and peri-urban areas, which is now the main market for family agro-industries. Martine Napoléone et al. note the benefits of a territorial specialisation of activities in Brasil Novo, Brazil. This is achieved thanks to 'horizontal relations between different actors who undertake their activities following the same logic of territorial anchoring'. Frederic Wallet, for his part, considers that the 'related variety', linking specialisation to a diversity of interconnected and complementary activities, is a favourable path for territorial development in Europe, even though he underscores the many conditions necessary for this 'smart specialisation' to succeed. The three studies call for a deepening of our second hypothesis, which states that *diversification and specialisation are differentiated by different power relations.* Indeed, these three chapters show that specialisation and diversification can be both a process of internal organisation and subordination to an ordered pattern of decision-making. Thus, specialisation can apply to both globalised and territorialised models.

The Coexistence of Models is Not Self-evident

These three studies describe the interactions between agricultural and food models in their territories. In the Brazilian state of Santa Catarina, Claire Cerdan notes that while there were 'win-win' relationships between family farmers and agribusinesses in the 1970s and 1980s, the advantage shifted in the 1990s to equipment suppliers, agribusinesses and maize distributors. The losers were the farmers who had to bear the costs of the intensive production model. At the end of the 2000s, solidarity started being established, with ethical and sustainable development values coexisting peacefully between the agro-industrial model and the family farming and artisanal production model. Martine Napoléone et al. show that specialisation can lead to a spatial distribution of various agricultural models within a territory (French case) or, as in Salto (Uruguay), to the exclusion of certain actors from the agro-industrial model and to coexistence in the same territory. Frederic Wallet focuses his attention on the hybridisation of practices and resources, the competition in the mechanisms for allocation of aid or land and the opposition between value systems. He invites us to consider the processes of coexistence from a transition perspective and to set up appropriate governance mechanisms to support initiatives that respond to local issues.

Conclusion

A diversity of agricultural and food models in a territory can appear to be propitious to increases in the capacities for innovation and for taking initiatives. However, the coexistence of models does not guarantee sustainable development. As Frederic Wallet points out, 'Some niches are oriented more towards a posture of resistance or a rationale of subsistence than they are towards unseating the industrialised and globalised agricultural model'. Thus, moderate and horizontal specialisation, whether 'smart' or 'territorial', implies that the coexistence and confrontation of agricultural and food models, and therefore of their interactions, have to be managed.

The chapters in this part of the book help to inform the three major hypotheses we are proposing, which encourages us to put them on our research agenda. Research on the coexistence and confrontation of agricultural and food models is renewing our understanding of the forms, determinants and impacts of processes of productive and territorial specialisation and diversification. However, these studies leave the door open to new research on topics that have not been studied much in this book, for example the analysis of specialisation and diversification of socio-ecological systems and the examination of their resilience.

References

Agrimonde (Eds.) (2009). *Agricultures et alimentation du monde en 2050: scénarios et défis pour un développement durable*, Report of the working group (February 2009) (p. 194). Inra-Cirad.

Allaire, G., & Daviron, B. (2019). *Ecology, capitalism and the new agricultural economy: The second great transformation* (p. 194). Routledge.

Antoine, A. (Ed.) (2016). *Agricultural specialisation and rural patterns of development* (p. 304). Brepols Publishers.

Byerlee, D., Stevenson, J., & Villoria, N. (2014). Does intensification slow crop land expansion or encourage deforestation? *Global Food Security, 3*(2), 92–98.

DuPuis, E. M., & Block, D. (2008). Sustainability and scale: US milk-market orders as relocalization policy. *Environment and Planning A, 40*(8), 1987–2005.

Fischer, J., Abson, D. J., Butsic, V., Chappell, M. J., Ekroos, J., Hanspach, J., Kuemmerle, T., Smith, H. G., & von Wehrden, H. (2014). Land sparing *versus* land sharing: Moving forward. *Conservation Letters, 7*(3), 149–157.

Green, R. E., Cornell, S. J., Scharlemann, J. P., & Balmford, A. (2005). Farming and the fate of wild nature. *Science, 307*(5709), 550–555.

IPES-Food (2016). *De l'uniformité à la diversité: Changer de paradigme pour passer de l'agriculture industrielle à des systèmes agroécologiques diversifiés* (p. 110). International Panel of Experts on Sustainable Food Systems.

Kayser, B. (Eds.) (1992). *Naissance de nouvelles campagnes*, (p. 175). coll. Monde en cours, Éditions de l'Aube/Datar.

Kremen, C., & Merenlender, A. (2018). Landscapes that work for biodiversity and people. *Science, 362*(6412), eaau6020.

Mathieu, N. (1984). Mécanismes et limites des processus de spécialisation, diversification de l'espace rural. *Économie rurale, 162*(1), 31–32.

Mathieu, N. (1985). Un nouveau modèle d'analyse des transformations en cours: La diversification-spécialisation de l'espace rural français. *Économie rurale, 166*(1), 38–44.

Phalan, B. T. (2018). What have we learned from the land sparing-sharing model? *Sustainability, 10*(6), 1760.

Pingali, P. L. (2012). Green revolution: Impacts, limits, and the path ahead. *Proceedings of the National Academy of Sciences, 109*(31), 12302–12308.

Roest, K., Ferrari, P., & Knickel, K. (2018). Specialisation and economies of scale or diversification and economies of scope? Assessing different agricultural development pathways. *Journal of Rural Studies, 59*, 222–231.

Suryanata, K. (2002). Diversified agriculture, land use, and agrofood networks in Hawaii. *Economic Geography, 78*(1), 71–86.

van der Ploeg, J. D., van Broekhuizen, R., Brunori, G., Sonnino, R., Knickel, K., Tisenkopfs, T., & Oostindië, H. (2008). Towards a framework for understanding regional rural development. In J. D. van der Ploeg, & T. K. Marsden (Eds.), *Unfolding webs: The dynamics of regional rural development* (pp. 1–28). Koninklijke Van Gorcum.

Chapter 1
From Agro-industrial Specialisation to a Plurality of Models in Southern Brazil

Claire Cerdan

In this chapter, we analyse the processes of specialisation and diversification of food systems in the southern Brazilian state of Santa Catarina. For a long time, this state was known for its unique dynamics of development based on a balanced distribution of consumption centres (secondary cities) and local production systems spread throughout the territory. From the 1950s onwards, it was an example of diffuse industrialisation based on a flexible, skilled, low-cost and entrepreneurial workforce (Vieira et al., 2009). The economist Raud (1997) identified six specialised production clusters in Santa Catarina: metalworking/mechanics (Joinville), textiles/clothing (Blumenau), ceramics (Criciúma), furniture (São Bento), paper/cellulose (Lages) and agrifood (Chapecó). These clusters are based on a process of specialisation with the development of small and medium-sized companies, alongside companies with national and international visibility and reach.

The success of this development model, based on a balanced distribution of local productive systems, is well known (Storper, 1997; Vieira, 2002). However, since the late 1980s, the gradual loss of competitiveness of these different systems has raised questions about this model's sustainability (Vieira et al., 2009). What do diffuse industrialisation and sectoral specialisation contribute to these different regions? What are the elements that have contributed to the processes of specialisation and diversification? How do recent developments call into question the modalities of coexistence of agricultural models?

To answer these questions, we draw on research conducted in the Chapecó region among small family agro-industries (on-farm and artisanal processing) and larger agro-industries (poultry and pork) between 2004 and 2011 (Mior, 2004; Andion, 2006; da Silva, 2009; Vitrolles, 2011).

C. Cerdan (✉)
UMR Innovation, Cirad, Saint-Pierre, Réunion, France
e-mail: claire.cerdan@cirad.fr

P. Gasselin et al. (eds.), *Coexistence and Confrontation of Agricultural and Food Models*,
https://doi.org/10.1007/978-94-024-2178-1_1

The Chapecó agrifood cluster, located in Santa Catarina's westernmost region, is specialised in the production and processing of meat products (pork and poultry). A study of the region's historical trajectory helps us understand the conditions of emergence of this local productive system organised around agro-industries, which are now leaders on the world market, and small on-farm processing units, based on know-how and on the production of traditional products resulting from European colonisation at the end of the nineteenth century (cheeses, cold meats, breads, fruit-based preparations and desserts). An analysis of the dynamics of specialisation and diversification makes it possible to identify the power relations between actors and the interactions between different development models.

1 The 1970s: The Integrated Production Model

Santa Catarina state[1] originally developed through a process of settlement of several waves of European migrants.[2] On arrival, migrants were given a plot of land of about 20 ha. With their agricultural and artisanal skills, they established a subsistence agricultural development model based on multicrop farming and livestock husbandry on small farms. This model persisted until the 1960s.

The federal government's agricultural modernisation programme in the 1970s marked a turning point by supporting the establishment of agro-industries and off-farm livestock production (intensive systems). Some families in the region already involved in the processing of animal products expanded their facilities. A few became international leaders in the processed meat sector (Sadia, Perdigão).[3]

One of the most significant challenges for these companies was to increase production volumes. This was achieved by implementing integrated production throughout the state. Production was organised in a manner originally observed in the 1950s, associating family producers with large industries. Industry signed contracts with producers and supplied animals (chicks, piglets) and inputs, and provided technical advice. For their part, the producers invested in the construction of buildings and raised the animals. The industry then took back the finished animals. In addition, spouses and children of farmers were often employed at the slaughtering and cutting units, which were located in urban areas.

---

[1] Santa Catarina is one of the smaller states in Brazil. It accounts for 1.12% of the country's surface area, is home to 3.39% of the Brazilian population and contributes 4.2% of the gross domestic product (GDP).

[2] The state experienced several waves of settlement between the sixteenth and twentieth centuries. The study area was mainly settled after 1870 by descendants of Italians and Germans who came from the country's south in search of available land.

[3] In 2019, these two food industry giants merged with 28 other brands to form Brasil Foods S.A. This company now represents over 13,000 integrated producers, 30,000 suppliers and 200,000 customers in over 140 countries.

This system of integration was to make a decisive contribution to the development of pig and meat poultry production in the region. It allowed for the rapid dissemination of new production technologies and flexible financing of breeding activity. It also ensured significant economies of scale, optimisation of equipment and available resources, a decrease in production and transaction costs, and reduction in commercial risks for breeders. Access to national markets (São Paulo, Rio de Janeiro) and export markets motivated all the actors of the sector and of the territory to invest in the production of meat and meat products.

At the time, the industries were concerned about the qualification and training of their workers and producers. They set up training programmes that helped in the flexible specialisation of the workforce. Thus, for example, when working on the butchering of animals, the worker was trained to intervene at all stages of the production process: slaughtering, cutting, packaging and/or quality control. We will see later how this training policy had an important impact on the evolution of activities in this territory.

In this way, the system of integration made it possible to maintain small farms with multicrop-livestock farming. Family labour was used to raise animals. Chicken feed was produced on the farm (maize and, to a lesser extent, soya, squash and manioc). Only feed concentrate was purchased. Thus, when the chicks were received, the family was assured of a financial income as well as indirect benefits (self-consumption and use of the litter from the chicken houses to fertilise fields). These were the reasons behind the enthusiasm of the farmers for this system. They also admitted that the presence of one or more processing units on their farms was a distinguishing sign, a social recognition: 'There are those who have been able to jump on the modernisation bandwagon, but the others…,' one producer said (our interviews, 2009).

This new economic process motivated public actors to invest in research. In 1975, Embrapa[4] built a research centre specialising in agro-industrial meat production in the region. Eight years later, Santa Catarina state inaugurated a family farming research and development centre (Research Centre for Family Farming—Agricultural Research and Rural Extension Company, Epagri/Cepaf).

The 1970–1980 period was marked by strong economic growth and a common vision of the actors on the future of their territory, i.e. to make it a specialised cluster for the production and processing of pigs and poultry. This led to the emergence of a localised agrifood system with positive externalities such as the availability of a skilled workforce, the emergence of a large number of service providers and suppliers (equipment manufacturers, suppliers of ingredients, additives and condiments, transport), a good reputation and remunerative markets.

---

[4] Embrapa: *Empresa Brasileira de Pesquisa Agropucària* (Brazilian Agricultural Research Corporation).

2 The 1980s and 1990s: Concentrating Production and Specialising

The move towards specialisation that began in the 1970s continued over the next two decades. Agro-industries consolidated through a process of intensification, specialisation and concentration of activities. This specialisation strategy enabled the region to gain international recognition. The territory's ability to produce quality meat at a low cost thanks to the existence of a qualified workforce both in production and in processing forged its reputation. New businesses sprang up. Some of them were created by the producers themselves, who formed their own cooperatives. Integrated production remained the reference.

Signs of increased specialisation appeared within farms. For example, studies have pointed to a certain disconnect between maize production and livestock activities (Testa, 2004). Increasingly, animal feed was made from ingredients produced off-farm. Land formerly used for maize cultivation was repurposed for the construction of new livestock buildings to accommodate larger numbers of animals. Among pig farmers, this process resulted in specialisation depending on the animals' growth phase: pig breeder, farrow-to-finish farmers, and fattening farmers.

These different developments jeopardised the economic viability of farms. Producers gradually lost their autonomy, with the industry supplying piglets, feed and medication. This led to the exclusion of a large number of pig farmers,[5] who were unable to meet the additional costs of feed, infrastructure expansion, and waste management (previously recycled on the farm).

Tensions crystallised between the agro-industries and the farmers, and, at the end of the 1990s, led to strong social mobilisation and the emergence of trade unions to defend farmer interests (Mior, 2004).

Later, and to a lesser extent, the meat poultry agri-chain went through the same process. As contracts became more and more demanding, farmers had to invest in modernising their buildings.[6]

While the 1970s and 1980s had been marked by a 'win–win' relationship between family farmers and agro-industries, the late 1990s saw them being divided into winners and losers. The winners were the equipment suppliers, the agro-industries and those who marketed maize. The losers were the breeders who had to bear the costs of the intensive production model. These developments led to the emergence of a process of diversification at the level both of industries as well as of the producers.

---

[5] There were 67,000 pig farmers in the region in the early 1980s. Their numbers steadily declined, to 45,000 in 1985, 30,000 in 1990 and 20,000 in 1995.

[6] Later, in 2006, measures against bird flu resulted in the disappearance of many farms, unable to bear the additional costs of protecting chickens. Regulations required the installation of a disinfectant gate at the entrance to the farm and the fencing of the animal facilities to prevent the possible contamination of flocks by infected animals and birds.

3 The 1990–2000 Period: A Dual Process of Diversification

The 1990s were marked by the opening up of the Brazilian economy, a movement that threatened the viability of the economic models of many regions (production costs that were too high). The agro-industries adopted new diversification strategies within and outside the territory to keep the industrial units profitable. These strategies had an impact on farms and the territory.

3.1 Diversifying Production Areas to Supply Agro-industries

To begin with, the industries introduced selection criteria based on the capacity to produce a greater number of animals per farm, their proximity to processing units (to reduce transport costs), and a sufficient 'safety' distance from water courses to avoid 'trouble' with environmental authorities. For this reason, the industries had to cast a wider geographical net to find eligible suppliers.

The agro-industries started looking at other regions of Santa Catarina and elsewhere in Brazil. They provided subsidies to businessmen and investors to set up new livestock units in the country's centre-west, a region specialising in large-scale grain production (maize, soya). The proximity of the feed production locations and the size of the units allowed a significant reduction in production costs.[7] In the 2010s, several agro-industries expanded to the tobacco growing regions of northern Santa Catarina. One industrialist told us: 'We receive aid[8] to set up, tobacco producers already have experience with integrated agri-chains, they are also pluriactive farmers… and the low concentration of livestock units in their region eliminates the risk of bird flu' (interview with a cooperative manager, 2009).

This migration of industries to other agricultural regions worried farmers and local communities. Several press articles announced the end of pig and poultry production in Santa Catarina (da Silva, 2009). However, these fears were proven to be unfounded; the industries remained and continued to source from Santa Catarina. Despite the higher cost of production, the industrialists recognised the advantages of production in western Santa Catarina: the skills of the breeders, their ability to respect commitments and specifications, the family workforce and the pluriactivity of the breeders, all of which facilitated negotiations on the animal procurement prices paid by the industry.

---

[7] In June 2006, the total cost of chicken delivered to the factory was 1.176 reais/kg in Goiás (a central-western state) based on an average flock size of 25,000 finished birds, with air-conditioned livestock facilities, while it was 1.263 in Santa Catarina based on flocks of 12,000 chickens (Embrapa, 2006).

[8] In 2006, Article 17 of Decree 5.658.2006 of the World Health Organisation's Framework Convention on Tobacco Control (WHO FCTC) provided for support for the development of viable alternative economic activities for tobacco workers and growers. Support was thus provided for diversification projects in tobacco-producing regions.

In addition, the capacities and qualifications of the workers in the Santa Catarina factories were widely recognised and appreciated by the agro-industries. For chicken processing, the workers in the industrial units of Chapecó had the best carcass butchering performance. As the manager of an agro-industry told us, 'When a chicken passes through the hands of a worker from the western region of Santa Catarina,[9] not a gram of flesh remains on the bones and this in a very short time. They are very good workers!'

3.2 Developing New Processed Products and Leveraging Agro-industrial Know-How

Another way in which the agro-industries diversified was to offer new frozen products and ready-made meals in order to expand their range not only for the national market (cuts, salami, sausages, ready-made meals), but also for export. Today, western Santa Catarina is known as one of the main production clusters for chicken nuggets for fast-food restaurants in many countries.

The 30 years of specialisation have contributed to a large extent to the emergence of significant industrial know-how and the consolidation of a network of actors and service providers: equipment manufacturers and input suppliers in the region. In 1996, these actors decided to leverage their skills in the field of agro-industrial production. With the support of public authorities, they organised the first international food industry fair, Mercoagro. The first edition brought together 148 exhibitors. The 12th edition of this event was held in 2018 with 250 exhibitors and more than 15,000 visitors.

This fair reflects the dual process of specialisation/diversification underway in the region. Specialisation has indeed contributed to the emergence of specific capacities: industrial know-how, equipment, food ingredients, etc. It is therefore leading to the creation of activities that are complementary to food production, mainly the sale of equipment and inputs to food industries, and training. Chapecó's production system is thus both recognised for its production of meat products and more generally as a centre of excellence for the food industry.

4 The Return to a Diversified Agricultural Model for Family Farms

In the 1980s, the contracts that bound different entities were the result of negotiations and compromises. Tensions over the purchase price of inputs or the sale price of animals contributed to the structuring and collective organisation of farmers. Several agricultural unions originate from this region, including the Federation of

[9] These workers are often simply called the 'catarinas'.

Family Farmers (Fetraf Sul) and the Agricultural Workers' Union. Non-governmental organisations (NGOs) such as the Apaco association are also active (Box 1).

Box 1 Association of Small Producers of Western Santa Catarina (Apaco)
Apaco is an NGO created in 1989 following a seminar on agricultural cooperation. It brings together 76 groups of producers who wished to participate in collective projects. Apaco provides technical assistance and promotes appropriate technologies. In the 1990s, it helped set up solidarity credits to help these farmers invest in small-scale processing units (fruit, meat, fish). These artisanal units were very successful and it soon became necessary to organise the sale of their production. The association then helped the farmers to set up cooperatives to sell their products, to sell directly at open-air markets and to sell some of their produce to the region's supermarkets. In the late 1990s, it helped conduct a fair-trade experiment with several groups of producers.

These trade union organisations and social movements have promoted diversification projects for farmers who have been excluded from the integrated system and those who have simply refused to participate in the integration model of the agro-industries. They are also supported by the Research Centre for Family Farming (Epagri/Cepaf). For them, development projects must 'think of another way of working with family farmers by creating innovative organisational structures' (interview, Apaco director, 2009). According to Apaco officials, the challenge is indeed not only to improve farmers' incomes and living conditions, but also to discuss collectively a 'new development model, based on cooperation between family production units, on solidarity, on the independence of farmers with regard to the unit, on the use of technologies that reduce production costs and preserve natural resources' (interview, Apaco director, 2009). With support from their organisations and NGOs, rural families have gradually turned towards individual or collective diversification projects: dairy production, on-farm processing of traditional products (fruit compotes, cheese, meat products). New development models and projects for the territory have emerged and have been supported by a large number of farmers. These initiatives have been facilitated by new government measures aimed at the professionalisation of family farming.

The trajectory of two small-scale agro-industries illustrates how family farmers have implemented these diversification activities (Box 2).

Box 2 Trajectory of two family-owned agro-industries: the Santa Rosa de Modelo agro-industry and the Pinhalzinho-Santa Catarina slaughter unit

In 2001, all the families in the commune of Modelo produced their own sugarcane and sugar. With the support of the mayor and the Epagri centre, three families came together to set up a sugarcane processing unit. They received financial support and technical advice. The construction of the building took four years, the time it took to gather the necessary funds; the equipment was acquired with the three families' own resources. By 2009, five other families had joined them and the facility was producing a diverse range of sugarcane products (8000 L of eau-de-vie (fruit brandy) per year and more than 10 tonnes of sugar). Some of their customers buy products at the site. The rest of the production is sold directly or in retail shops in a nearby area. However, one of the partners has maintained his contract to raise industrial chickens for Sadia, a leading company in the field.

The second example is a pig slaughtering and processing unit run by two producers from Pinhalzinho. They own 17 ha of farmland, 6 of which are reserved for maize and soya bean cultivation. These producers work for a cooperative that is integrated with an agro-industry. Following the pork crisis in 1992, in order to 'survive', they decided to add value to the sale of pigs by producing meat and processed products themselves. The first trials were satisfactory and motivated them to formalise their activity. The farmers received State aid to construct the building. The equipment was offered to them under the 'Development and added value' programme of the Ministry of Agriculture's Family Farming Secretariat. The slaughterhouse is now associated with agro-industry networks and affiliated to Apaco, which allows them to use the latter's umbrella brand, Sabor Colonial, while also using with their own brand.

Source Interviews (Vitrolles, 2011).

These illustrations highlight the fact that the know-how mobilised in the artisanal units is based not only on traditional knowledge, but also on more recent skills acquired during the years of specialisation on farms and in industries. We refer here to the observation made above on the willingness of industrialists to train their staff. The knowledge and skills acquired while working in industry stand them in good stead for their artisanal processing unit, and it must be noted that compliance with standards and procedures is satisfactory on the whole.

Thanks to public programmes for supporting family farming, the number of artisanal units has grown. The presence of many intermediate towns has facilitated the sale of products through short supply chains. While agro-industrial products are exported, artisanal products are mainly sold and consumed locally (direct sales or in supermarkets), including in the canteens of agro-industries. In 2010, there were more than 1000 small-scale processing units in the western region of the state, the majority of them involved in fruit processing, sugarcane processing, bread-making (corn bread) and milk derivatives. They represented more than 45% of the artisanal food units in Santa Catarina. Apaco's umbrella brand, Sabor Colonial, is helping build the reputation of the region's products and is raising the reputation of the farming profession.

5 Identifying the Interplay of Actors in the Processes of Specialisation/Diversification of Activities

The dynamics of specialisation and diversification are at the origin—and are also the product—of power relations and the actors' representations of the future of the territory and the development models they advocate.

5.1 The Interplay of Actors Between 1970 and 1990: 'What is Good for Agro-industry is Good for the Territory!'

Figure 1 shows the system of actors involved in meat production: breeders, agricultural support services, and representatives of the public authorities at the local and federal levels. During the 1970–1990 period, a shared vision of the territory prevailed, and all stakeholders collaborated in the development of this specialised cluster of pig and poultry production and processing.

Research was oriented towards improving industrial processes, animal breeds and the production of animal feed. Banks financed new livestock projects, and municipalities facilitated land acquisition for new industrial units and the development of new urban neighbourhoods for workers' families. The meat production sector was organised around associations between meat industries, integrated producers and a few cooperatives.

This convergence of vision can be explained not only by the economic dynamism of the specialisation process, but also by the involvement of some industry leaders in the political sphere, such as Attilio Fontana,[10] who co-founded the Sadia company in 1944. However, given the companies' paternalistic relationships with the other actors, there was little opportunity during this period for the integrated producers, the employees or the company managers to challenge the existing arrangements (Fig. 1).

[10] For many years, A. Fontana (1925–2000), while being head of his company, took on many public functions. He was responsible for improving road infrastructure, investment in workforce training (creation of schools, specialised technician courses, adult literacy initiatives, continuing education for workers), and subsidised credit programmes for the modernisation of agro-industrial units.

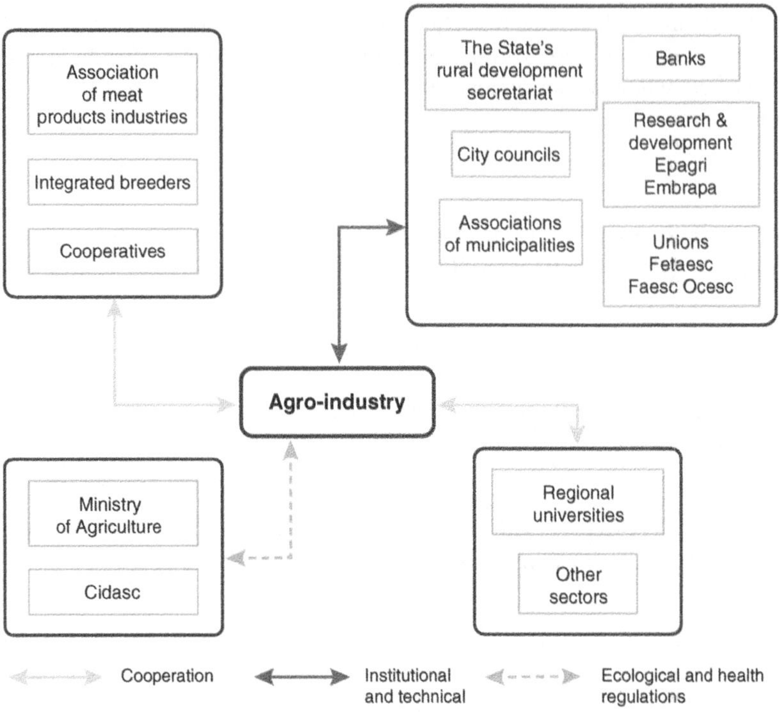

Fig. 1 Mapping of actors and their relationships in the Santa Catarina meat production territory, 1970–1990. Epagri: Agricultural Research Company; Embrapa: Brazilian Agricultural Research Corporation; Fetaesc: *Federação dos Trabalhadores na Agricultura do Estado de Santa Catarina*; Faesc Ocesc: *Federação da Agricultura e Pecuária do Estado de Santa Catarina Organização das Cooperativas do Estado de Santa Catarina*; Cidasc: *Companhia Integrada de Desenvolvimento Agrícola de Santa Catarina*. *Source* Cerdan

5.2 From the 1990s Onwards, a System of Actors Spanning a Plurality of Territorial Projects

From the 1990s onwards, the system of actors became more complex and diversified. The exclusion of a large number of livestock farmers from the agro-industrial model called into question the idea of a specialised hub for the production and processing of meat products. A number of initiatives were taken and programmes created to offer alternatives to the existing model (Fig. 2).

These developments were part of the democratic change process of 1988. Armed with political and fiscal autonomy, the municipalities involved all their inhabitants and union representatives in the development of local policies. This led to the deconstruction of the integrated production model and the emergence of new development models, based on forms of solidarity and incorporating values of ethical and sustainable development.

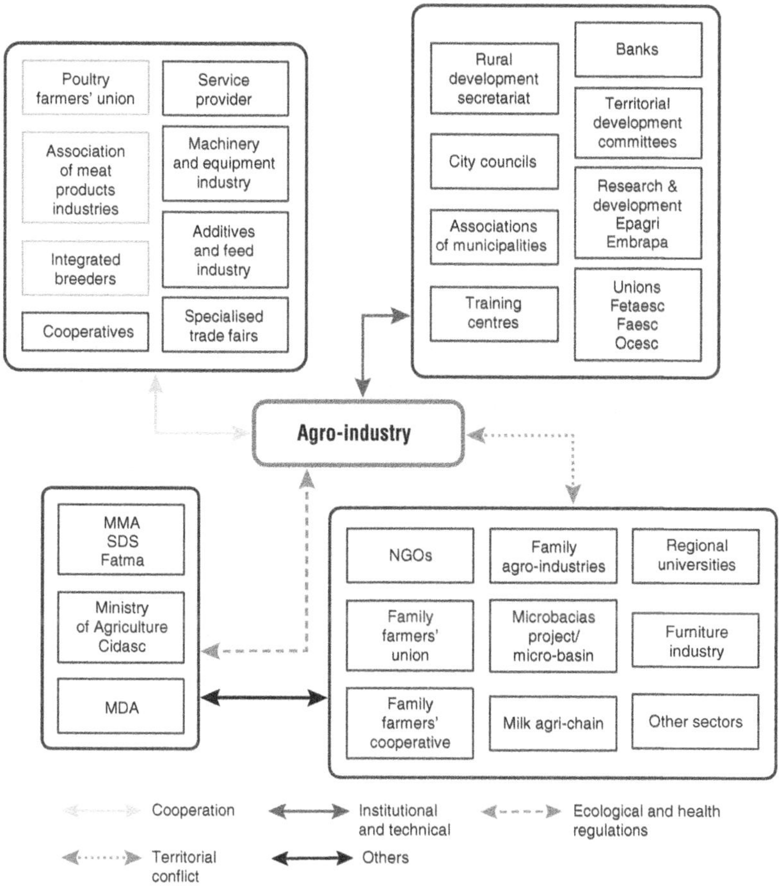

Fig. 2 Mapping of actors and their relationships in the Santa Catarina meat production territory, 1990–2019. Epagri: Agricultural Research Company; Embrapa: Brazilian Agricultural Research Corporation; Fetaesc: *Federação dos Trabalhadores na Agricultura do Estado de Santa Catarina*; Faesc Ocesc: *Federação da Agricultura e Pecuária do Estado de Santa Catarina Organização das Cooperativas do Estado de Santa Catarina*; MMA: Ministry of Environment; SDS: Sustainable Development Secretariat; Fatma: *Fundação do Meio Ambiente*; MDA: Ministry of Agrarian Development; Cidasc: *Companhia Integrada de Desenvolvimento Agrícola de Santa Catarina*. *Source* C. Cerdan

However, the interplay of actors at the end of the 2000s highlighted a passive (or peaceful) coexistence between the agro-industrial model and the on-farm and artisanal production model. Producers developed small-scale processing capabilities, while continuing to raise and supply animals for industry. Provided that artisanal production does not increase health risks, the industry is not opposed to a diversification of household incomes. Quite the contrary! Indeed, we noted that agro-industrial companies offered local and artisanal products in their company restaurants canteens.

Finally, the development of the milk sector, the emergence of family processing units, fruit growing and organic production have also led to new development models for the territory, without however challenging the agro-industrial model.

6 Conclusion

This study confirms the hypothesis that diversification and specialisation are part of the same process of adaptation of productive spaces to the global system. The processes oriented towards specialisation are responses to market challenges and societal demands. They are also the product of the interplay of the actors involved and their ability to coordinate collectively to influence territorial dynamics. We can thus assert that these specialisation/diversification processes are based on effective coordination and on support from public authorities.

The economic development of western Santa Catarina is currently based on the presence of specialised industries that represent a source of employment and of numerous alternative micro-activities. Unlike other rural regions of Brazil, which have seen an exodus of their populations, this agricultural region has benefited from the presence of food industries to retain the working population (especially young people) in rural, urban and peri-urban areas. This population is now the main market for family agro-industries.

A historical look at processes of specialisation/diversification helps us better understand the determinants of these changes. It also shows that development models are dynamic in nature and take shape gradually according to the objectives of local actors and the opportunities they have. In our case study, the agro-industrial model is the result of a process of flexible specialisation driven by a productive organisation that was original at the beginning, but which has also subsequently been strongly called into question. Faced with these new challenges, the actors aspire for other development models. These actors can either be against the agro-industrial model or position themselves on the basis of a passive coexistence of different models.

References

Andion, C. (2006). Développement territorial durable en milieu rural, gouvernance et rôle des organisations non gouvernementales: l'État du Santa Catarina au Brésil. *Mondes en Développement, 136*(4), 85–100.

Da Silva, E. I. (2009). O papel da avicultura na construçao do territorio e na reproducao social da agricultura familiar: O caso de Chapeco e Quilombo no oeste catarinense. Mémoire de mestrado Agroecossistemas, Universidade Federal de Santa Catarina, Centro de Ciências Agrárias, 1, Florianópolis, Brazil, UFSC, 114 p.

Embrapa. (2006). Relatório anual de atividades 2006 [da] Embrapa Suínos e Aves, Concórdia, Embrapa Suínos e Aves.

Mior, L. C. (2004). *Agricultores familiares, agroindustrias, e territorio: a dinâmica das redes de desenvolvimento rural no Oeste Catarinense*. Doctoral thesis in Humanities, Society and Environment, supervised by Julia Silvia Guivant and Terry Mardsen, UFSC, 2003, 316 p. Federal University of Santa Catarina.

Raud, C. (1997). Potentiel et modalités d'une industrialisation décentralisée au Brésil. *Espaces et Sociétés, 88*(88/89), 161–183.

Storper, M. (1997). *The Regional World* (p. 338). Guilford Press.

Testa, V. M. (2004). Desenvolvimento sustentável e suinocultura do Oeste Catarinense: Desafios éconômicos, sociais e ambientais. In *Desafios para o desenvolvimento sustentável da suinocultura*, Argos, Chapeco, Brazil, pp. 23–72.

Vieira, P. (2002). *A pequena produção e o modelo catarinense dde desenvolvimento (p. 377)*, APED, Florianópolis, Brazil.

Vieira, P., Cazella, A. A., Cerdan, C., & Andion, C. (2009). Potencialidades e obstáculos à construção de territórios sustentáveis no estado de Santa Catarina. *Política e Sociedade, 8*(14), 335–380.

Vitrolles, D. (2011). *La promotion de l'origine au Brésil*. Geography thesis, Université Lumière de Lyon-2, 325 p. + Annexes.

Chapter 2
The Construction of Specialisation and Diversification Pathways in Selected Milksheds: Understanding the Plurality of Agricultural Development Models

Martine Napoléone, Marie Houdart, and Guillaume Duteurtre

Agriculture and food are at the core of current societal debates. For more than half a century, agricultural spaces have been—and continue to be—profoundly transformed by the combined evolution of agriculture and the agrifood sector. With the modernisation of agriculture and the rise of agrifood firms, a profit-oriented logic has gradually taken hold, profoundly changing our relationship with the land and living organisms, with local society, with work and with food (Hervieu & Purseigle, 2013). However, under certain conditions, traditional activities driven by other rationales based on social and territorial balances continue to persist (Rieutort, 2009). With peasant principles, as also the terroir and the local, once again finding recognition and generating value in alternative food systems (Tregear, 2011), new links between societies, rural activities and consumers are opening up. Are we going to witness the end of the peasant, as Mendras (1967) predicted, and the inexorable growth of corporate agriculture (Purseigle et al., 2017)? Is a new horizon opening up for peasant forms of farming in a fragile balance with other forms of farming within agricultural territories? Our aim in this chapter is to analyse the processes of specialisation and diversification at the territorial level in all their complexity. How and why does a diversity of models exist in some territories and not in others? How are these processes of specialisation and diversification constructed within the same territory? How do they evolve together? What are the factors and conditions that favour or hinder a plurality of development models within a territory?

M. Napoléone (✉)
INRAE, UMR Selmet, Montpellier, France
e-mail: martine.napoleone@inrae.fr

M. Houdart
INRAE, UMR Territoires, Clermont-Ferrand, France
e-mail: marie.houdart@inrae.fr

G. Duteurtre
Cirad, UMR Selmet, Montpellier, France
e-mail: guillaume.duteurtre@cirad.fr

P. Gasselin et al. (eds.), *Coexistence and Confrontation of Agricultural and Food Models*,
https://doi.org/10.1007/978-94-024-2178-1_2

To find answers to these questions, we focus on the dairy sector, which is especially subject to the interplay of forces between territorialisation and globalisation (Napoléone et al., 2015). Dairy farming involves both powerful agro-industrial companies and dynamic local actors. It concerns a product, milk, which can be transported over long distances, processed into standard products or according to ancient know-how. Milk can be sold on generic-product markets, on 'top-of-the-range' niche markets or even through local channels for processing into traditional products.

By analysing the pathways of recomposition of dairy activities in five contrasting rural territories (Salto in Uruguay, Brasil Novo in Brazil, and Livradois-Forez, Cévennes and Drôme Provençale in France), we examine the underlying hypothesis of this part of the book: 'Diversification and specialisation are part of the same process of adaptation of productive spaces to the global system.' We first present the method we used to understand these processes in their complexity. We then describe three archetypal development pathways, identified from the analysis of recompositions in the five territories, in order to isolate the main factors influencing the dynamics of specialisation and/or diversification at the scale of these territories. In the discussion, we return to the links between development models and specialisation/diversification processes at the territorial level, as well as to the key factors favouring these processes.

1 Understanding the Processes of Specialisation and Diversification at a Territorial Scale by Analysing Development Models

Given that our interest is in studying specialisation and diversification at the territorial scale, we start by considering that it is the processes that take place over time as well as a state at a given moment that, taken together, characterise a situation of specialisation or diversification. We define a territory as specialised when a large majority of production and processing systems in it are part of the same development model. In contrast, a territory is considered diversified when it is host to a plurality of systems that are part of different development models. We consider that a development model is an aim, an orientation, and an ethic in which an individual or collective project is embedded. The development model thus mobilises not only conceptions and technical and economic solutions, but also political choices (Duteurtre, 2014). It brings into play a system's rationality and coherence, which condition or orient the form of development, i.e. the strategic choices made, the way of determining what is a resource or not, and the methods used to mobilise them, set up governance modalities and evaluate the result. We characterise the form of development by the specific way in which attributes of the territory, the farming systems and the agri-chain are linked and translated into particular modes of production (Napoléone & Boutonnet, 2015b). In a territory, there may be a single or several forms of development, all of which may evolve. We will use the term 'development pathway' to describe, at the territorial scale, the types of succession involving one or more forms of development over time.

The method we have implemented is intended to account for the processes of transformation of dairy activities within a territory and to identify underlying factors. Our analysis is based on the proposition that reconfigurations in milksheds are driven by two types of processes: one of globalisation, the other of territorialisation. We are interested in the forms of development of dairy activities, through the interrelations between the transformations of agri-chains, of animal husbandry systems and of the territory in question. Using this analytical framework, we represent the processes of globalisation and territorialisation through two ideal-types (Box 1). The changes in the milksheds were identified from comprehensive analyses, including of archival material, and then recorded on a historical timeline (50 years on average). This chronicle allowed us to analyse the forms of interrelations present in each territory at a given time (synchronic analysis) and their evolution (diachronic analysis) (Napoléone & Corniaux, 2015).

Box 1 Two contrasting ideal-types

Globalisation. Process 'driven' by an agro-industrial and sectoral dynamic: concentration of companies, lengthening of commercial supply chains, standardisation of products; concentration of activities in the most favourable geographical areas; in livestock farming, intensification, expansion.

Territorialisation. Process 'driven' by local and territorial dynamics: artisanal production and processing units, collective project(s) involving a diversity of local activities, local distribution channels, territorial anchoring of products, development of identity dynamics. Farmers adapt the size and scope of their activity and their practices to the resources available to them.

2 Construction of Specialisation and/or Diversification Pathways

Based on the analysis of the development pathways of the five rural territories, we identify three types of milksheds. In the first type, industrialised forms of development dominate, belonging to a globalised development model. The second type is one in which territorialised forms of development dominate, belonging to a territorialised development model. And, finally, the third type is one in which different forms of development can be found alongside each other, which can belong to either of the two development models.

2.1 An Agro-industrial Specialisation Pathway

The example of the Salto milkshed in Uruguay (Correa et al., 2015) helps us understand how a territorial specialisation pathway towards an 'agro-industrial' model develops. This milkshed developed in a landlocked region which did not have any dairy tradition. During a first period (from the 1930s to the 1980s), the State encouraged the production of milk, investing in constructing infrastructure and in the development of a cooperative dairy industry to supply the city of Salto with standard products. A second period was characterised by the extension of milk markets to neighbouring countries within the framework of protected trade agreements (Mercosur). Agro-industry modernised, benefiting from private foreign investment, production intensified and farms grew in size. Cereal farmers turned to dairy production. In the current period, the milkshed's development is marked by the conquest of the (unprotected) world market. The largest dairy farms continue to grow. This growth is based on a mechanised, capital-intensive farming model. Recently, the increased demands of the dairy industry have led it to stop collections from farms smaller than 50 ha. Some of these farms, close to urban areas, have thus been compelled to reorient their system in a territorial logic by associating dairy farming and market gardening, with products being sold via short urban channels.

During this journey of development, this system expanded by adjusting the products produced to the chosen market (type and volume): primary production to meet industrial demand, and territorial development to suit the development of this production (increase in surface area, intensification of land, etc.). The aim and rationale have remained unchanged, oriented towards the increased productivity of production factors. They are shared by all the actors in the system, who manage their activities according to this logic at their own levels. The physical characteristics (large surface areas available, land suitable for intensification, etc.) have facilitated this process. This pathway is oriented by a development process driven by a globalisation model. However, while this is the current trend, the dynamics reveal an orientation towards two forms of development: a form of development articulated around a powerful agro-industry, collecting milk from large farms, oriented towards the export market and long urban supply chains, and localised forms, concerning small farms selling via niche channels oriented towards meeting local demand for diversified agricultural products.

To summarise, the process of a territory's agro-industrial specialisation is characterised by a top-down pathway, with the dissemination of a State-approved development model in which each actor shares the same objective. The model's development and success create sociotechnical barriers that reinforce its development. At the same time, this model leads to the exclusion of activities that do not correspond to its standards: new forms of development are then forced to be created, leading to a diversification of development models in this specialised territory.

2.2 A 'Territorialised' Specialisation Pathway

The case of the Brasil Novo milkshed in Brazil (Poccard & Carvalho, 2015) illustrates a development pathway towards a 'territorialised' model, based on the local 'reinvention' of know-how and products. The origins of this landlocked milkshed along the Trans-Amazonian highway date back to the time migrants from dairy regions settled on this agricultural frontier, bringing with them their dietary habits and their cheese processing know-how. Dairy activity in this territory developed on local bases (economic, social and cultural) in three main stages. First, dairy production with on-farm processing emerged from the migrants' traditional know-how, as a complement to meat production. During a second period, artisanal dairies were created to meet the demand of Brasil Novo, a small town in this enclave. Production methods were similar to traditional ones. Finally, at present, dairy activity is consolidating to meet the growing demand of consumers in Brasil Novo. Regional investors are facilitating the modernisation of processing facilities. The State is relaxing cheese production standards so that artisanal processing becomes compliant.

This milkshed's dynamics are tied to its territory in many ways: through its isolation, which has limited trade with the outside world and maintained a local urban demand; through its farmland, whose size and quality have allowed production to develop; through the presence of farmers who are ready to orient part of their activities towards dairy production; through the products and know-how originally brought by migrants and now adopted by artisanal dairies; and through the demand of local consumers for these artisanal products.

In summary, the pathway here is characterised by a bottom-up process, initiated by the pioneers, with the State intervening only at a later stage to ensure the sustainability of these activities by adapting the relevant standards. The milkshed's development is based on processing know-how shared by producers and processors, shared dietary habits and knowledgeable consumers. All the resources necessary for this 'territorialised' pathway's development are present within the territory (common culture, fodder resources, livestock farming, local demand).

2.3 Pathways that Lead to a Diversity of Models in a Territory

Three French milksheds have followed pathways that have led to a diversity of development models: Cévennes (Napoléone & Boutonnet, 2015a), Drôme (Napoléone & Boutonnet, 2011), Livradois-Forez (Houdart et al., 2015; Houdart, 2018). All three are mountain or semi-mountain territories (Box 2).

Box 2 The characteristics of these territories

These are semi-mountainous territories with a rich heritage value: landscapes recognised in the form of protected natural parks and traditional products rooted in the local food culture. They exhibit specific spatial characteristics: a core surrounded by a massif, with difficult land, and more open, cultivable parts of the territory, close to major roads. These milksheds border urbanised plains that represent major consumption basins (Montpellier, Clermont-Ferrand, Marseille, etc.).

Cow milk is produced in Livradois-Forez and almost all the farms in this territory are dairy farms. Most of the utilised agricultural area (UAA) is arable. In contrast, the farms in Cévennes and Drôme mainly undertake goat breeding. These territories are wooded and arable land is scarce. More than 80% of the farms are managed by tenant farmers.

These pathways were constructed in four main stages.

Throughout the first half of the twentieth century, the peasant model of the household economy prevailed in all of these territories. Surplus production was sold locally. Artisanal dairies and maturing units were set up, relying on traditional know-how and selling the products locally.

After the Second World War, production models started diversifying under the impetus of public policies. The State encouraged agricultural modernisation, the creation of robust agri-chains, and the development of mass distribution channels. Consumption patterns began changing. In the three milkshed studied, a diversity of models, spatially distributed, emerged: intensification of production and industrialisation of processing in the areas most favourable to this type of development; and maintenance of territorialised activities in isolated areas, thanks to the continued demand for local products (Fourme d'Ambert, Bleu d'Auvergne, Pélardon, Picodon).

The 1980–2000 period was marked by major difficulties. The success of the productivist model led to overproduction, especially as major investments were required to bring dairies up to standard and increases in production followed to make these investments viable. The 'mad cow crisis' in 1986 undermined consumer confidence. In order to gain market share, agribusinesses segmented their product ranges with labels of differentiation (name of famous place or product, organic label, cheese with a 'terroir' identity), which created confusion between local cheeses and industrial cheeses that also claimed a 'terroir' identity. In order to differentiate their cheese from industrial cheeses and to protect the name and reputation of their products, the actors of traditional production sought the recognition of their cheeses as Protected Designations of Origin (PDOs).[1] Once PDOs were granted, companies could no longer produce cheeses outside the area bearing the name of traditional products. In the 1990s, mergers and commercial agreements between industrial groups from outside the area and local dairies multiplied. In this way, these industrial groups

[1] PDOs: Fourme d'Ambert, 1972; Bleu d'Auvergne, 1975; Picodon, 1983; Pélardon, 2000.

could maintain PDO products in their offerings, and the small dairies could reduce the logistical costs of accessing long supply chains. The diversity of models struggled to be maintained during this period. The coherence of the territorial model was undermined.

More recently, the diversification of food demands is encouraging the return to a diversity of models in these territories. The criteria pertaining to proximity are becoming more important for consumers. At the same time, demand from emerging countries is a growth opportunity for the dairy industry (Idèle, 2016). In the three milksheds studied, we are witnessing both a return to territorial dynamics for some of the actors and activities, and an increase in industrialisation for others. The increase in demand for direct sales and the diversification of marketing methods allow the artisanal dairies that have remained independent and the farmers who do on-farm processing (or farmer producers) to take advantage of their artisanal character and the product's local image. Most of them are refocusing on channels based on proximity (geographical or organisational) or on niche distribution. Political actors are supporting these changes. The industrial dairies continue on their path to concentration. Industrial groups can adopt one of two strategic orientations: to position themselves on the world market and maintain a place in the domestic market, or to remain the leader of a regional market by emphasising traditional products.

Ultimately, territorial and sectoral actors were able to mobilise the diversity of territorial resources according to the orientation of the development model to which they belonged. These developments were driven by forms of consumption. The strategies of processing companies also played an important role. While in some cases, artisanal enterprises were absorbed by industrial ones, in others cooperation between globalised industrial firms and artisanal enterprises enabled the latter to survive in difficult times. The protection of typical products by official designations of specific quality has favoured the diversity of forms of development, some focused on sectoral development (often mobilising a diversity ranging from standards to 'terroir' products), others anchored in a logic of territorialisation. Industry has adapted its processing chains to accommodate both generic products and more typical products in order to reach 'connoisseur' consumers who are accessible through long regional urban distribution channels. Artisans and farmer producers have taken advantage of their small-scale character via specialised distribution channels. Products have been delivered to consumers through a variety of channels (short or long, generic niche or local), which has helped to develop product awareness. It is certain that the current situation, with demand for generic products (export) and local products (short distribution channels), facilitates the deployment of distinct forms of development.

3 Discussion

Studying the processes of specialisation and diversification at the territorial level by analysing the development pathways being followed allows us to understand the ways in which these processes are tied to development models. This approach sheds light on the main factors of diversification and specialisation at the territorial level: these processes are more than an adaptation to global changes, indeed they are driven by food demand and influenced by the territorial resources available.

3.1 Complex Linkages

Our analysis shows that specialisation can take place in both globalised and territorialised models. It is not always a process dependent on a single decision-making or power centre. In some cases, specialisation is the result of the predominance of agro-industrial models functioning in a context of globalisation. In such cases, the goal is oriented by public policies. Even though there are several decision-making and power centres, all the actors follow the same logic, at their different levels (policymaker, agro-industry manager, breeder and farmer). All the activities are organised—and segmented—along a value chain ranging from production to consumption. In other cases, the specialisation is that of activities in a territory-centric logic. The milkshed is then characterised by the predominance of horizontal relations between different actors who undertake their activities following a same logic of territorial anchoring.

As for processes of diversification, we highlight the fact that these processes challenge the coexistence, or the concomitancy, of different development models at the territorial scale. In some cases, as in the three French ones, the coexistence of models may result from the dynamics of specialisation under different models, spatially distributed over the territory or it may result from the adaptation of the models present to different territorial and global constraints and opportunities. In other cases, such as of the Salto milkshed in Uruguay, diversification results from the effects of the specialisation of the agro-industrial model, which leads to the exclusion of certain actors from the system, and thus to the emergence of a territorialised model that then coexists in the same territory.

3.2 Between Food Demands and Territorial Resources

Trajectories of diversification and specialisation can be analysed as the consequence of market dynamics and the actions of strategic actors (firms, the State, professional organisations or civil society) (van der Ploeg et al., 2008). In the case of animal husbandry agri-chains, researchers have highlighted the numerous economic advantages that accrue from specialisation through concentration, such as gains of productivity or economies of scale (Roguet et al., 2015). However, several authors are less convinced about the importance of the 'global' factor in specialisation and diversification processes, especially in the case of dairy production, and note the fact that specialisation or diversification is never solely an adaptation to the globalised market

(DuPuis & Block, 2008). This is what our work confirms by highlighting the influence of the diversity of food demands and territorial resources in both diversification and specialisation processes.

Food demand influences the orientation of development models. We note that four main types of demand have influenced, to varying degrees, the evolution of development models in the milksheds studied: the demand for commodities for international trade; the demand for generic products for distribution through national or regional long supply chains; the demand for products with a quality label and/or originating from a particular area for distribution in regional, or even national, long supply chains; and, finally, the demand by consumers for food produced in proximity (geographical or organisational).

Not all of these demands are compatible with all forms of production, processing and the 'milkshed role' of the territory. Territorial characteristics can lead to different developments in various parts of the milkshed: the conditions necessary to satisfy each type of demand require different resources (Houdart & Poccard, 2015). So while some territories are favourable to the expression of several forms of development, others are not.

Finally, combinations of food demand and territorial resources lead to three possible situations. In the first, agro-industrial development is geared towards the production of commodities (standard products) in order to gain market share in countries in which consumption is growing. This type of export-oriented industrial development takes place only in those geographical, social, technical and economic situations that meet its requirements. The second situation is characterised by dynamics that are conducive to the commercial positioning of agro-industrial processors in the long supply chains of national or regional mass distribution entities. This can apply both to generic products as well as to products sold under quality labels. In all cases, the supply of milk must be compatible with industrial processing and distribution via long supply chains (regularity and homogeneity of supply, high volumes, density of livestock in the territory). In this situation, the size of production systems tends to increase, relying, if necessary, on the purchase of inputs if the territory cannot satisfy the herd's food requirements. This agro-industrial process can accommodate certain requirements and constraints (e.g. specifications) if compliance with them will allow the industrial operator to differentiate itself in commercial segments. This is the case of the growth in collections by agro-industries in PDO territories. These products offer a competitive advantage, allowing the industry and distributor to segment their ranges. Finally, the third situation is that of processes tied to a territorial logic that is not very connected to world markets or to mass distribution, in which the territory's milk offer corresponds to a demand for geographical, relational or organisational proximity (Rallet and Torre, 2007). In these situations, supply is composite; it is driven by various exchanges and relationships at the territorial or agri-food system levels. These interpersonal relationships contribute to the construction of common norms and values between the people involved. The horizontal dynamics in play synergise elements other than purely commercial ones between the agri-chain's actors (Pecqueur, 2014).

4 Conclusion

The major issue concerning the future of milksheds is the tension between a selective evolution of corporate forms of agriculture and the diversification of models leaving room for forms that are very much rooted in their territories. The diversity of forms of agriculture are then strongly tied to the political regulations that are put in place and to the way in which each form fits in and weaves links with an economic, social and territorial environment. Ultimately, specialisation and diversification at a territorial scale are processes of adaptation controlled by actors with very different registers of legitimacy, which go beyond the sole productivist aspect. Some are strongly driven by policies and encouraged by the agro-industrial sector, while others are supported by a civil society in search of different values. The diversity of models in a territory allows for a diversity of market access and a plurality of forms of conducting farming activity. While diversification can be considered to be a richness that increases the capacity for initiative at the territorial scale, the sustainability of the coexistence of models cannot be taken for granted. For the sustainable development of territories and agri-chains, the challenge seems to us to be to recognise these forms of development, through a territorial approach, to analyse them with regard to their own interest and, if necessary, to put in place regulatory measures to encourage their coexistence.

References

Correa, P., Arbeletche, P., Piedrabuena, L., Bataburu, D., Tourrand, J. -F., & Morales, H. (2015). L'expansion d'un bassin laitier basé sur le développement de l'agro-industrie et de l'exportation. In M. Napoléone, C. Corniaux, & B. Leclerc (Eds.), *Voies lactées. Dynamiques des bassins laitiers, entre globalisation et territorialisation* (pp. 39–65). Cardère.

DuPuis, E.-M., & Block, D. (2008). Sustainability and scale: US milk-market orders as relocalization policy. *Environment and Planning A: Economy and Space, 40*(8), 1987–2005.

Duteurtre, G. (2014). Les minilaiteries, «modèle» d'avenir pour les filières élevage en Afrique? In C. Corniaux, G. Duteurtre, & C. Broutin (Eds.), *Filières laitières et développement de l'élevage en Afrique de l'Ouest: L'essor des minilaiteries* (pp. 66–91). Karthala.

Hervieu, B., & Purseigle, F. (2013). *Sociologie des mondes agricoles* (p. 318). Armand Colin.

Houdart, M. (2018). Le bassin laitier du Livradois-Forez, de l'analyse historique aux modèles de dynamique. *Mappemonde*, (123).

Houdart, M., Baritaux, V., & Cournut, S. (2015). Bassin laitier du Livradois-Forez: dés-ancrage/ré-ancrage de la production laitière dans le territoire: quelle influence des acteurs de l'aval? In M. Napoléone, C. Corniaux, & B. Leclerc (Eds.), *Voies lactées. Dynamiques des bassins laitiers, entre globalisation et territorialisation* (pp. 76–97). Cardère.

Houdart, M., & Poccard, R. (2015). Les conditions d'interaction entre dynamique de bassins laitiers et dynamique territoriale? In M. Napoléone, C. Corniaux, & B. Leclerc (Eds.), *Voies lactées. Dynamiques des bassins laitiers, entre globalisation et territorialisation* (pp. 169–185). Cardère.

Idèle. (2016). Conférence sur les marchés mondiaux du lait. Risques et opportunités pour les filières françaises et européennes, 8 June 2016, Paris.

Mendras, H. (1967). *La fin des paysans, innovations et changement dans l'agriculture française*, SEDEIS.

Napoléone, M., & Boutonnet, J. -P. (2011). Lecture diachronique de l'évolution des systèmes de production et des stratégies de firmes, en élevage caprin. *Options méditerranéennes, «Economic, social and environmental sustainability in sheep and goat production systems», série A*(100), 91–100.

Napoléone, M., & Boutonnet, J. -P. (2015a). Bassin laitier «Pélardon en Cévennes méridionales»: une filière localisée qui reste affranchie des dynamiques industrielles. In M. Napoléone, C. Corniaux, & B. Leclerc (Eds.), *Voies lactées. Dynamiques des bassins laitiers, entre globalisation et territorialisation* (pp. 157–184). Cardère.

Napoléone, M., & Boutonnet, J. -P. (2015b). Entre local et global: quelles reconfigurations à l'œuvre dans les bassins laitiers? Analyse comparative dans les bassins laitiers au Nord au Sud. In M. Napoléone, C. Corniaux, & B. Leclerc (Eds.), *Voies lactées. Dynamiques des bassins laitiers, entre globalisation et territorialisation* (pp. 249–277). Cardère.

Napoléone, M., & Corniaux, C. (2015). De la trajectoire singulière aux processus communs. In M. Napoléone, C. Corniaux, & B. Leclerc (Eds.), *Voies lactées. Dynamiques des bassins laitiers, entre globalisation et territorialisation* (pp. 21–36). Cardère.

Napoléone, M., Corniaux, C., & Leclerc, B. (Eds). (2015). *Voies lactées. Dynamiques des bassins laitiers, entre globalisation et territorialisation* (313 p). Cardère.

Pecqueur, B. (2014). Esquisse d'une géographie économique territoriale. *L'espace Géographique, 43*(3), 198–214.

Poccard, R., & Carvalho, S. (2015). Bassin laitier de Brasil Novo: l'émergence d'un bassin laitier localisé sur un front pionnier. In M. Napoléone, C. Corniaux, & B. Leclerc (Eds.), *Voies lactées. Dynamiques des bassins laitiers, entre globalisation et territorialisation* (pp. 185–205). Cardère.

Purseigle, F., Nguyen, G., & Blanc, P. (2017). *Le nouveau capitalisme agricole, de la ferme à la firme* (p. 305). Les Presses de Sciences Po.

Rallet, A., & Torre A. (coord.). (2007). *Quelles proximités pour innover?* (221 p). L'Harmattan.

Rieutort, L. (2009). Dynamiques rurales françaises et re-territorialisation de l'agriculture. *L'information Géographique, 73*(1), 30–48.

Roguet, C., Gaigné, C., Chatellier, V., Cariou, S., Carlier, M., Chenut, R., Daniel, K., & Perrot, C. (2015). Spécialisation territoriale et concentration des productions animales européennes: État des lieux et facteurs explicatifs. *INRA Prod Anim, 28*, 5–22.

Tregear, A. (2011). Progressing knowledge in alternative and local food networks: Critical reflections and a research agenda. *Journal of Rural Studies, 27*(4), 419–430.

van der Ploeg, J. D., van Broekhuizen, R., Brunori, G., Sonnino, R., Knickel, K., Tisenkopfs, T., & Oostindië, H. (2008). Towards a framework for understanding regional rural development. In J. D. van der Ploeg, & T. K. Marsden (Eds.), *Unfolding webs: The dynamics of regional rural development* (pp. 1–28). Royal Van Gorcum.

Chapter 3
Does the Evolution of Agricultural Production Models Allow for Their Coexistence in a Territory?

Frédéric Wallet

A growing number of studies have now highlighted the limitations of the agro-industrial model and the need to reform it because of the negative externalities it generates and because of its inability to meet the objective of providing everyone with access to quality food. Based on a logic of cost reduction and economies of scale, this agricultural model has led to a process of concentration along the entire lengths of value chains, starting from land structures and the seed market all the way up to distribution systems. This process is also reflected geographically in the increasing regional specialisation of forms of agricultural. And yet, the need for a transition to more sustainable agricultural models has led to dynamics of innovation that take the form of both an internal reconfiguration of the agro-industrial model and multiple alternative models. New actors in the agricultural and food sector are the main drivers of this innovation, as also existing farmers who are beginning to view their profession differently.

Initially considered marginal, these new or rediscovered agricultural models are now becoming more firmly rooted in the landscape, calling for an examination of the forms of coexistence they maintain, ranging from opposition to hybridisation, with the agro-industrial model. Thus, agricultural and food systems, as they are deployed in a territory, create multiple configurations, far from a binary competition between a dominant model and alternatives that naturally tend to converge towards a common horizon and are expected to overturn the established sociotechnical regime.

These developments are taking shape in a context of changing regional policies where the idea of identical intervention mechanisms and development principles for all territories has shown its limits. Thus, the recognition of the importance of regional and territorial specificities, and the search for higher performance in terms of innovation and sustainable development now form the basis of the so-called 'smart specialisation' principle. It is thus necessary to not only examine the contribution

F. Wallet (✉)
INRAE, UMR Sadapt, Paris-Saclay University, Paris, France
e-mail: frederic.wallet@agroparistech.fr

P. Gasselin et al. (eds.), *Coexistence and Confrontation of Agricultural and Food Models*,
https://doi.org/10.1007/978-94-024-2178-1_3

made by agricultural and food systems to this strategy in each region, but also the relevance of applying this policy to agricultural activities and rural areas: which productive and organisational structure should be promoted to make this approach work for agriculture and be coherent within a territorial strategy?

In the first part of this chapter, we start by revisiting the characteristics and foundations of the regional specialisation of agricultural models. We then discuss the variety of initiatives that currently constitute avenues for diversification, and the way in which these different models make up territorial food systems. In the second part, we situate these changes in the broader context of the evolution of regional development policies in Europe and their impact on rural spaces and agricultural value chains. In the third and final part, we discuss the notion of coexistence and the questions it raises for researchers and decision-makers in order to move towards a desired transition in agriculture and development models in rural territories.

1 Specialisation of Agricultural Models and Emerging Alternatives

The process of increasing regional specialisation in forms of agriculture and of the increased concentration of resources at all levels of value chains are calling into question the capacity of this agricultural regime to undertake the transition required to meet the challenges of sustainability. This is reflected in the emergence of alternative solutions that call into question the modes of articulation between sociotechnical models.

1.1 Regional Specialisation and Concentration in Value Chains

An analysis of French agriculture reveals a strong movement towards regional specialisation, marked by a geographical dissociation of crop and livestock production, a replacement of grasslands by field crops (oilseeds, cereals), a decline in perennial crops (vines, arboriculture) and the geographical concentration of livestock production (Gaigné, 2012).

The cheap price of energy and the search for economies of scale are behind this productive specialisation. These effects of scale are coupled with the expected benefits of agglomeration economies, encouraging co-location and an increasing recourse to integrative logic (Chatellier & Gaigné, 2012). Finally, the criteria for allocating aid have spatial implications, which also encourage specialisation and concentration.

The growth strategies of farms bring with them a corollary of an increase in economic and health risks. The result is a massive decrease in the number of farms, especially small and medium-sized ones, to the benefit of large farms,[1] especially in the poultry, pig and dairy sectors. This movement of concentration and integration also applies to the processing and marketing functions of agrifood industries, leading to the disappearance of small and medium-sized enterprises (SMEs) from the agrifood sector. This results in the capture of an increasing share of added value in the value chains and the imposition of an industrial model that leaves little room for alternatives.

1.2 A Host of Initiatives Proposing Alternatives … Which Still Remain Relatively Marginal

Faced with the limits of the dominant model, many alternatives are emerging in the agricultural and food sectors, including short supply chains, local supply, organic farming and agroecology, as well as products of origin with quality labels. These alternative movements have been adopted in various forms by both consumers and producers, as expressions of freedom in farming and food choices. While these movements are enjoying an increasing exposure in the media, their place in the agricultural and food sector remains marginal compared to the large agribusiness and distribution companies.

In several cases, these models are economically fragile, thus requiring public policy support (Brand et al., 2019) at different scales with a view to re-territorialise agriculture, sustain employment and preserve the environment and biodiversity. What is being debated through these approaches not only concerns the importance and support granted to these alternative models, but also the way in which value chains based on distinct principles can coexist in the same territory, given that they will most likely be competing for resources, public subsidies and potential outlets.

Thus, irrespective of the scale considered, food systems display a profound diversity. Far from being in a face-off between dominant agro-industrial systems and alternative ones (Goodman, 2003), they make up a complex array of production and distribution models for food products and services built on a diversity of organisational principles, technical and production standards, and quality conventions. Each of these models—including agro-industrial ones—finds it difficult or even impossible to feed on its own large populations with multiple socio-economic realities. This invites us to reflect on these models' potentials for and conditions of complementarity, and the modalities of their co-evolution. As a result, 'one size fits all' policies are proving to be ineffective in addressing major food challenges (environmental

[1] Classification of farms: distribution of farms according to their specialisation and economic size. Since the 2010 agricultural census, classification of farms is based on the concept of standard gross production (SGP). On the basis of the SGP coefficient, farms are divided into three size-classes: small (SGP less than 25,000 euros), medium (SGP between 25,000 and 100,000 euros) and large (SGP greater than 100,000 euros) (*Source* Insee, TEF, 2019, https://www.insee.fr/fr/statistiques/3696937).

impacts, food security, etc.), or at least are not able to exploit the levers presented by the diversity of models (Fournier & Touzard, 2014).

This observation echoes the shifts that have characterised regional and rural development policies over the last ten years.

2 Place-Based Strategies: An Interpretation of Open Specialisation

In a context of the evolution of regional development policies, these changes in agricultural models are taking shape by proposing an alternative based on the recognition of the importance of regional and territorial particularities, and the search for higher performance in terms of innovation and sustainable development. Based on the principle of 'smart specialisation', the relevance of this strategy's implementation needs examination when it comes to agricultural activities and rural spaces.

2.1 European Territorial Development Policies with a 2020 Horizon: Choosing a Place-Based Logic

The failure of sectoral approach policies has highlighted the deterioration in the competitiveness of European regions in comparison to their American and Asian counterparts, in particular due to insufficient innovation. The Barca Report (2009) notes three main reasons for these failures:

- the excessive uniformity of regional policies and sectoral orientations in favour of high technology at the European level, even though not all regions are equally capable of taking on international competition;
- insufficient specialisation of regions, which consequently tends to spread their resources over too wide a range of sectors and technologies;
- the lack of interest in public policies about the way in which spatial dimensions are integrated into corporate localisation and coordination strategies.

This is the framework for the EU's new growth strategy for 2020, now extended for another five years, which is based on the objective of becoming a 'smart, sustainable and inclusive' economy (European Commission, 2010). This objective is founded not only on the identification, in a context of global competition, of the comparative advantages of regions and their coherent integration into global value chains, but also on innovation processes within pre-existing sectors, thus allowing the definition of a particular regional development model.

The policies of smart specialisation and growth thus aim to offer the possibility of a differentiated development of territories depending on their resources, their technological capacities and their modes of organisation. These policies represent a shift from an identical approach for all regions to a recognition of places and development policies based on local knowledge (Foray, 2015).

The smart specialisation strategy thus put in place is distinctly different from those that preceded it, in that it incorporates a greater consideration of knowledge networks and spatial dimensions, and also challenges current modes of governance and institutional arrangements (McCann & Ortega-Argilés, 2015). The place-based logic (Rodríguez-Pose & Wilkie, 2017) thus constitutes a shift in favour of giving primacy to the territory as a crucible for development models and innovation, in which previously the logics of spatial deployment of sectoral forms prevailed. Introduced in Europe, these strategies are now seeing their rationale spread across the globe, especially in Latin America (Barroeta et al., 2017).

2.2 The Principles of Regional Smart Specialisation Strategies

Smart specialisation strategies are based on a combination of several structuring principles.

To begin with, it is matter of identifying the sectors of activity in which the region has a competitive advantage at a European or even global scale, while avoiding mimicking what is being done elsewhere. In other words, the objective is to find the field of activity, the organisational structure and the range of functions within the value and technology chains to be established in the territory that can ensure a sustainable competitiveness of the region and its enterprises.

To this end, targeted choices must be made in areas with sufficient critical mass, which means prioritising a relatively small number of sectors and technologies. Does this mean that each region should opt for a narrow and strict specialisation? It is here that the originality of the smart specialisation approach is to be found: it stresses the importance of a 'related variety', linking the specialisation approach with the coherent diversity of technologies and sectors in order to take advantage of the processes of production and dissemination of knowledge externalities. In this way, the smart specialisation approach stimulates the dynamics of innovation within different sectoral value chains as well as in the linkages between them.

Research in geographical economics has shown that one of the conditions for the success of this model is its degree of regional integration, i.e. the strong regional or local connections with certain industries, in terms of input–output links of flows (material, informational and monetary) and labour, along the entire lengths of the value chains. Furthermore, there is also the matter of the links between regional companies and the outside world. Connectivity thus plays an important role through its capacity to promote network externalities, both via clusters structured in spatial proximity and via longer distance networks. Finally, the smart specialisation model accords a central role to entrepreneurial discovery as a vector of innovation, resituated in a context of dynamic interactions between firms, research laboratories, public actors and citizen-consumers, as described in the quadruple or even 'quintuple helix' models (Carayannis et al., 2012) (Fig. 1).

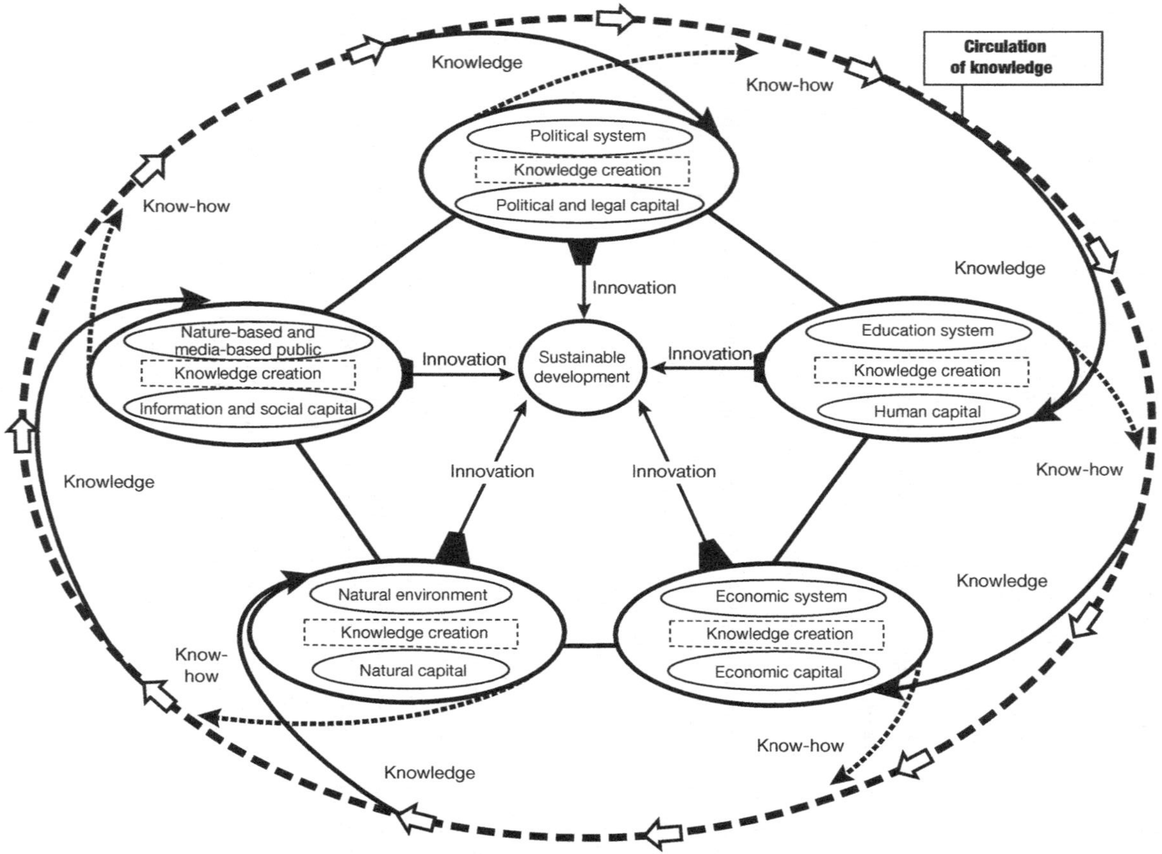

Fig. 1 The 'quintuple helix' model and its functions. *Source* Carayannis et al. (2012) (figure under Creative Commons licence)

In concrete terms, the European Union has thus invited each region to establish programmes on the basis of an explicitly developed strategy (McCann, 2015), and to choose a few priority areas, activities or key technologies, according to three criteria: the insertion of the activity in a value chain at the regional level; a specialisation in specific fields of activity that have a competitive advantage; and a coherent diversification through related varieties (interconnected and complementary fields of activity).

While the overarching objective of the smart specialisation strategy is clearly to improve the long-term competitiveness of European regions and enterprises, the regional programmes have been fine-tuned over time to better respond to the challenges of transition to more sustainable industrial models. Thus, innovation in this framework is seen as a vector of both competitiveness and sustainability.

2.3 *The Requirements for Smart Rural Growth*

Smart growth policies in rural areas makes sense because, unlike other European economic policies, these approaches explicitly take into account the differences between territories, and are meant to adapt to the particular characteristics of different types of regions in Europe. At the same time, however, they are based on structuring principles (see 'The principles of regional smart specialisation strategies' above) that are liable to be applied imperfectly to rural regions. Indeed, the latter are often known to suffer from a number of limitations, which correspond precisely to the weakness of the entrepreneurial fabric. This leads not only to the absence of a critical mass, which then reduces the possibilities of connectivity, but also to an inability to establish mechanisms of integration and related variety at a significant scale. In other words, the less dense network of interactions between the different components of value chains and the different sectors often condemns these territories to a less sustained or incomplete development.

The research carried out in particular in the framework of the European TASTE[2] project partly puts the relevance of these new regional and innovation policy orientations into perspective. Smart growth strategies seem to be suitable for well-developed or intermediate regions, both urban and rural, if their population is large enough. However, they offer only very limited possibilities for outlying regions due to their difficulties in exploiting effects of scale, which leads to the following problems of:

- low density, limiting the number of strong relationships;
- the lack of diversification, limiting the potential application of technological and productive relationships;
- the lack of 'innovation broker' organisations that act as facilitating intermediaries within innovation networks.

[2] Towards A Smart Rural Europe, EraNet-Ruragri (2013–2017).

Nevertheless, it is necessary to take into account the great diversity of rural areas, which leads to a very different relationship with smart growth principles and policies. As a simplification, it can be said that:

- rural areas close to cities are good candidates for smart growth policies in the sense the European Union ascribes to them: this group includes areas that are more or less integrated with cities and intermediate regions combining urban and rural areas;
- the more outlying rural regions are less suitable, precisely because of the lack of dimensions recognised as favourable for smart growth policies;
- however, some of them could potentially engage in smart specialisation by exploiting local facilities and other resources (such as tourism, natural resources or the service economy for the elderly) (Torre et al., 2020).

Furthermore, two additional dimensions need to be considered with regard to smart specialisation principles in rural or peri-urban zones: land use and agricultural activities.

Land use and its evolution play a crucial role in the capacities and policies of development of European rural regions, as they determine the new activities to be embarked upon and which existing ones should be replaced. Thus, the use of land by competing activities can lead to an increase in land value or even to conflicting relationships detrimental to the dynamics of territorial development. The processes of soil artificialisation have demonstrated often enough that this competition occurs in particular between agricultural activities and other potential land uses. But it has also been observed between different types of production and agricultural production models, in a context marked by a scarcity of arable land, both at the European and global scales (Le Mouël et al., 2018). On the other hand, excessive specialisation on a single type of land use can entail high vulnerability in the event of an economic crisis or a climate shock, for example.

Thus, the basic principles of smart growth in land use should be based on two principles that correspond to those established at the industrial level (Darly et al., 2020):

- avoiding homogenous and rigid regional land use with insufficient variety that may result in vulnerability in case of drastic changes (climate change, economic crisis, policy changes, etc.);
- avoiding major fragmentation between rival land uses, which can lead to uncontrolled competition and even conflict, and thus appear as an obstacle to smart growth processes.

Moreover, if environmental issues are taken into account, these principles get reinforced through the consideration of ecosystem services, soil quality and biodiversity preservation. To draw a conceptual parallel with smart specialisation principles, we can consider that questions arise concerning the modalities of integration of different land use practices, the critical mass necessary for these uses with regard to the territorial development orientations decided upon and sustainability issues, and finally the connections between the different types of spaces—especially in the urban–rural relationship—, which are underpinned by changes in the forms of land use and the resulting flows.

The second dimension pertains to the opportunity to launch a smart agriculture process, given the major role played by agricultural activities in land use in rural areas and their crucial role in supplying food to the European population. The limits of the conventional agricultural model have now been clearly identified and call for the search for alternative solutions and resilience in production systems. The prospects offered by new technologies and digital applications certainly offer interesting avenues for adaptation, but they cannot be the only answer to the challenges of agroecological transition.

Smart agriculture is polymorphous, whatever the scale considered. When considered at the regional level and understood on the basis of synthetic and thematic indicators, it reveals performances that vary from one region to another, and which, depending on the case, may be based on economic, environmental or social dimensions (Corsi et al., 2020). At the scale of sub-regional territories, it is the proliferation of initiatives and their diversity that are the distinguishing feature of the dynamics underway (Duvernoy & Soulard, 2020). These approaches can be considered to be niches of innovation with disparate development potentials, which combine more or less harmoniously in a territory, often within the framework of an agricultural and food project supported by local authorities, to steer the system towards greater sustainability. If they appear simultaneously in sparsely populated rural zones and close to urban ones, this latter configuration seems to be able to benefit, despite the pressure on agricultural land, from the consumption basin and the infrastructure needed for their sustainable development. Agriculture practised further away from towns can also rely on product quality and a diversification towards derivation of value from local facilities (especially through tourism).

More broadly, a large set of agrifood systems appears as a key strategic component in many European regions. Indeed, the Eye@RIS3[3] database identifies it among the priorities in three out of four regions; the areas concerned mainly pertain to new agrifood technologies (23%), agrifood and tourism (20%), and high value-added food (13%) (Ciampi and Cavicchi, 2019).

Nevertheless, the smart specialisation strategy remains essentially an industrial and innovation policy at the regional scale. And a precise and systematic examination of the coherence between regional orientations, their territorial deployment and local initiatives remains to be undertaken.

From this point of view, food innovation initiatives can be considered to be drivers of smart regional growth. These initiatives can take a wide range of forms since they are an expression of entrepreneurial discovery in the broadest sense, and thus their characteristics vary in terms of the number and forms of interaction between the actors involved, their geographical coverage and the stage of maturity of innovations.

---

[3] Eye@RIS3 is an online database created by the European Union as a tool to assist in the development of regional strategies for investment in resources in order to stimulate knowledge-led growth.

3 Discussion: Is It Enough to Coexist?

3.1 The Need to Go Beyond the Simple Notion of Coexistence: What Kind of Compatibility of Practices?

The results presented above encourage us to think about the types of coexistence possible between agricultural models: between the utopic expectation of convergence of the whole towards a sustainable food system and the danger that the industrial model with its devastating effects could ultimately prevail by absorbing the alternatives using its resilience and power.

A condition for the overall evolution of coexistence towards a more equitable balance is to recognise the differentiated contributions models make. However, can the multi-functionality of agroecosystems and the alternative and family systems coexist with land concentration and productive specialisation?

The processes of coexistence are not only marked by the phenomena of hybridisation of practices and resources, but also by the rationales of competition, resistance, and even conflict. The latter concern not only different dimensions, specially mechanisms for allocating aid or land, but also opposition between the value systems represented by the principles of regulation of these agricultural models. Characteristics such as the stretching of the link to nature, industrialisation, cornering of resources, production and value addition, etc. are thus often evoked to justify the rejection of agro-industry, and to encourage a resistance to it.

However, the current context is marked by the asymmetry of resources and configurations in terms of market regulation and public intervention mechanisms, which are very clearly oriented towards supporting a dominant form of agriculture. Regulatory mechanisms may need to be adjusted to take into account the contributions and impacts on the territory of the various agricultural models: services, value addition and employment, externalities (pollution, landscape, culture, etc.), etc.

At the territorial level, it is necessary to put in place appropriate governance mechanisms to support innovative initiatives that address local issues, while meeting the diversity of expectations, and to consider the adaptation of consumer behaviour as a powerful lever for change (expectations of local food, social link, food justice, decrease in health risks, gustatory expectations, etc.). Such governance mechanisms should make it possible to remove the stumbling blocks to a transition to a more sustainable agriculture, and to facilitate the compatibility of practices belonging to different sociotechnical models.

3.2 *Coexistence Processes to Be Considered in the Context of the Transition of Agricultural Models*

Faced with the limitations of the dominant industrial agricultural models, the search for solutions for a transition towards greater sustainability is becoming increasingly important. Alternative models, in all their diversity, thus appear to be niches in which agrifood initiatives can emerge and which can help make changes in the conventional model (Touzard et al., 2014). The possibility of a comprehensive adaptation of food systems therefore depends on this complex interplay between innovation processes based on the confrontation between agrifood models. In this sense, the processes of coexistence must be considered in a transition perspective (Gasselin et al., 2020). They are a means for removing lock-in effects to promote the resilience of agricultural models in territories, but to do so they must be able to deploy their potential (Chiffoleau et al., 2020). Promoting alternative forms and innovative solutions is a major challenge for interventions that address the issue of coexistence. This calls for a reflection on the criteria and modalities not only of public intervention, especially in terms of financing and investment support, but also of access to certain strategic resources, most importantly agricultural land and the conditions for processing and marketing.

Furthermore, we no doubt need to re-examine the notion of niche as it is often presented in transition theory, which has borrowed the concept from Geels's (2002) multi-level perspective. Given the proliferation of experiments in sociotechnical niches, certain successful innovations in these models are expected to weaken and manipulate the dominant sociotechnical regime, and eventually replace it. Reality, however, is undoubtedly more complex in more ways than one, as it reveals the issues surrounding the coexistence of models. In a context of permanent market segmentation, coexistence can also be perceived between different forms of alternative models, e.g. organic farming, products with official quality signs and short supply chains. The complementarity between these models is not self-evident, nor do they possess a capacity to combine efficiently to accelerate the transition process.

Thus, we can ask ourselves whether the diversity of niches contributes to a dispersion that is favourable in sustaining the industrial model, which is working to reinvent itself in order to preserve its dominant position, or whether they work together to contribute, over the long term, to a shift towards an alternative proposal. This is all the more true since some niches are oriented more towards a posture of resistance or a rationale of subsistence than they are towards unseating the industrialised and globalised agricultural model. More broadly, the perspective of coexistence in the long term and the transformative capacity of the whole to respond to the global challenges of transition have thus been brought to the fore.

4 Conclusion

Finally, several observations can be made regarding the coexistence of agricultural models in territories and their insertion into sustainable development processes.

First, while the spatial inscription of the agro-industrial model has led to a productive specialisation of agricultural territories, a myriad of initiatives stemming from alternative forms of agriculture are now emerging and gaining traction. They are helping create a complex patchwork of systems in which a dominant agro-industrial model in the process of recomposition and a diversity of innovative niches with sometimes contradictory orientations are linked in conflicting or complementary ways, or even hybridising. Recognising the usefulness of the solutions proposed by these different models, especially in terms of multi-functionality for a more sustainable agriculture, appears to be a necessity for adapting support and regulatory mechanisms to prevent their disappearance or their absorption into a recomposed agro-industrial model operating on a single principle. The structuring of adapted territorial food governance systems plays an essential role in this process of recognising the respective strengths of different models and their ability to address priority local challenges (Viljoen and Wiskerke, 2012). Thus, some approaches do not get the attention they deserve within territorialised food systems even though they constitute important pathways towards a more global transition. Still emerging initiatives to develop social innovations, eco-innovations or innovations linking health and food lack prominence in territorial food projects.

Second, these changes in food systems are taking place in a wider movement to recognise territorial diversity in the formulation of regional development policies. This principle, embodied today in particular by smart growth strategies, aims to promote a diversity of innovation models, based on support for a limited number of sectors of activity and technologies that offer opportunities for sustainable and inclusive knowledge intensive growth. Since this strategy is, a priori, less favourable to the development of rural territories than of urban areas, it will only prove beneficial to agricultural value chains and the rural economy if it allows the full expression of the transformative potential of the dynamic co-evolution of the various agricultural and food models. Furthermore, the challenge is also to make these models coherent within integrated approaches to innovation in territories, encouraging dialogue between, on the one hand, agricultural and food initiatives, and, on the other, the social and solidarity economy, industrial and territorial ecology, or even forms of collaborative and functional economy.

References

Barca, F. (2009). An agenda for the reformed cohesion policy. Report to the Commissioner for Regional Policy, Brussels, April.

Barroeta, B., Gómez Prieto, J., Paton, J., Palazuelos, M., & Cabrera Giráldez, M. (2017). Innovation and regional specialisation in Latin America. JRC Technical Report, EUR 28511 EN.

Brand, C., Bricas, N., Conaré, D., Daviron, B., Debru, J., Michel, L., & Soulard, C. -T. (Eds.). (2019). *Designing urban food policies: Concepts and approaches* (142 p). Springer International Publishing.

Carayannis, E. G., Barth, T. D., & Campbel, D. F. J. (2012). The Quintuple Helix innovation model: Global warming as a challenge and driver for innovation. *Journal of Innovation and Entrepreneurship, 1*, 2.

Chatellier, V., & Gaigné, C. (2012). Les logiques économiques de la spécialisation productive du territoire agricole français. *Innovations Agronomiques, 22*, 185–203.

Chiffoleau, Y., Brit, A. -C., Monnier, M., & Akermann, G., Lenormand, M., & Saucède, F. (2020). Coexistence of supply chains in a city's food supply: A factor for resilience? *Review of Agricultural, Food and Environmental Studies,*, September, 1–24.

Ciampi, S. K., & Cavicchi, A. (2019). *Smart specialisation and the agri-food system* (p. 90). Palgrave Pivot.

Corsi, S., Wallet, F., Cahuzac, E., Maigné, E., & Filippini, R. (2020). Smart Agriculture and empirical evaluation of smartness using composite indicator. In Torre, et al. (Eds.) *Smart development for rural areas*. Routledge.

Darly, S., Torre, A., & Olivier, C. (2020). Smart land use for smart rural development. In Torre, et al. (Eds.), *Smart development for rural areas*. Routledge.

Duvernoy, I., & Soulard, C. (2020). Smart agriculture, grassroots initiatives and urban food governance. In Torre, et al. (Eds.), *Smart development for rural areas*. Routledge.

European Commission. (2010). *Europe 2020. A European strategy for smart, sustainable and inclusive growth, Brussels.*

Foray, D. (2015). *Smart specialisation: Opportunities and challenges for regional innovation policy* (p. 104). Routledge.

Fournier, S., & Touzard, J. -M. (2014). La complexité des systèmes alimentaires: Un atout pour la sécurité alimentaire? *VertigO—La Revue Electronique en Sciences de l'environnement, 14*(1).

Gaigné, C. (2012). Organisation des filières animales et environnement. Vingt ans après la directive Nitrates. *INRA Productions Animales, 25*(4), 275–388.

Gasselin, P., Lardon, S., Cerdan, C., Loudiyi, S., & Sautier, D. (2020). The coexistence of agricultural and food models at the territorial scale: an analytical framework for a research agenda. *Review of Agricultural, Food and Environmental Studies*, July, 1–23.

Geels, F. W. (2002). Technological transitions as evolutionary reconfiguration processes: A multi-level perspective and a case-study. *Research Policy, 31*(8–9), 1257–1274.

Goodman, D. (2003). The quality "turn" and alternative food practices: Reflections and agenda. *Journal of Rural Studies, 19*(1), 1–7.

Le Mouël, C., de Lattre-Gasquet, M., & Mora, O. (Eds.). (2018). *Land use and food security in 2050: A narrow road* (398 p). Éditions Quæ.

McCann, P. (2015). *The regional and urban policy of the European Union: Cohesion* (p. 304). Edward Elgar Publishing.

McCann, P., & Ortega-Argilés, R. (2015). Smart specialization, regional growth and applications to European Union cohesion policy. *Regional Studies, 49*(8), 1291–1302.

Rodríguez-Pose, A., & Wilkie, C. (2017). Revamping local and regional development through place-based strategies. *Cityscape, 19*(1), 151–170.

Torre, A., Corsi, S., Steiner, M., Wallet, F., & Westlund, H. (Eds.). (2020). *Smart development for rural areas* (232 p). Routledge.

Touzard, J. -M., Temple, L., Faure, G., & Triomphe, B. (2014). Systèmes d'innovation et communautés de connaissances dans le secteur agricole et agroalimentaire. *Innovations—Cahiers d'économie et de Management de l'innovation, 43*, 13–38.

Viljoen, A., & Wiskerke, J. S. C. (Eds.). (2012). *Sustainable food planning: Evolving theory and practice* (600 p). Wageningen Academic Publishers.

Part II
Innovation

Innovation: Driver and Outcome of the Coexistence of and Confrontation Between Agricultural and Food Models. Introduction to Part II

Pierre Gasselin

The second part of this book examines innovation (Faure et al., 2018), as found in situations of coexistence of agricultural and food models in territories. The concept of innovation, first conceptualised in the twentieth century, primarily refers to a technical and/or organisational process through which something new is being thought about, invented, and developed, which actors then appropriate and tailor to their requirements. This then leads to questions on the determinants and trajectories of this process (Temple et al., 2018). At the same time, innovation is the result of a system of actors and knowledge with a more or less explicit 'project' (Hall et al., 2003; Klerkx et al., 2010; Touzard et al., 2015). Finally, innovation is also the driving force behind the transformation of territories and of sociotechnical and socio-ecological systems (Allaire & Daviron, 2017).

Innovation marks the boundaries between different agricultural or food models and in doing so often helps to define them. These models are thus qualified by a set of technical and organisational innovations that serves as their qualifying banner: organic farming, conservation agriculture, digital farming, corporate farming, fast food, etc. In addition, innovation transforms the conditions of coexistence and confrontation of agricultural and food models, by influencing the four elements that make up the 'situations of coexistence': the actors or systems; their interactions; the specific objects concerned; and the 'setting' under consideration (refer to this book's General Introduction). Indeed, innovation in agricultural and food systems, just as in other sectors, is never neutral in terms of its justifications (productivity, well-being, environment, fairness, justice, etc.), its forms and its impacts on territorial development (Torre & Wallet, 2013). As a result, innovation gives rise to sociotechnical controversies and leads to differential impacts and arrangements between actors who advocate different development models (Akrich et al., 2002). Through the paths innovation adopts (e.g. who are the actors involved? what knowledge? what sociotechnical and socio-ecological paradigms? which challenges are targeted?), it helps determine a desired future and the principles of action (Coudel et al., 2013). Innovations in agriculture and food help reflect specific relationships actors have not only with nature, with space and with societies (Touzard, 2018), but also with the knowledge,

P. Gasselin et al. (eds.), *Coexistence and Confrontation of Agricultural and Food Models*,
https://doi.org/10.1007/978-94-024-2178-1

technology, the State, the markets and territories (Albaladejo, 2020). Thus, innovations not only depend on, but also determine a specific development model, for example, a high-tech agricultural trajectory or an agroecological goal (Bonny, 2017), and configure situations of coexistence.

In this introduction to Part II of the book, we first undertake a quick review of the existing literature on innovation, which leads us to propose hypotheses that are little addressed in existing research, in order to analyse the relationships between innovation and the coexistence of agricultural and food models in territories. We then present four chapters that pertain to innovation and then offer a comparative reading in the light of the analytical framework for situations of coexistence and of the general hypotheses proposed.

Overview of Analytical Frameworks and Hypotheses

The three main acceptations of the concept of innovation (process, object of a system of actors and knowledge, driver of transformations) can be broken down into different theoretical streams. We focus here on sociotechnical regimes and innovation niches, the conventionalisation and hybridisation of innovations, and finally territorial and social innovations. This brief state of the art leads us to formulate two general hypotheses for orienting a research agenda on innovation in situations of coexistence.

Sociotechnical Regimes and Innovation Niches

Multi-level perspective, applied in different economic sectors (energy, agriculture, etc.), is one of the major theoretical frameworks used for understanding and managing transitions. It is presented in detail in the introduction to Part IV of the book. We recall here that it distinguishes between three sociotechnical levels (Geels & Schot, 2007; Duru & Therond, 2015; Dumont et al., 2020): the landscape (macro-context of the sociotechnical system); the regime (meso-level that imparts stability to the dominant sociotechnical regime); and the niches (sociotechnical spaces in which innovations emerge at the margins of the dominant regime). Questions then naturally arise about several dimensions of innovation: the nature of the innovations, their aims, the principle challenges they are supposed to address, the target actors, the conditions for the emergence of, and support for, the innovation, and finally the modalities of scaling up (Wigboldus et al., 2016). As an illustration, we can refer to two innovation regimes in plant genetics that have emerged since the Second World War, and which are closely tied to contrasting agricultural and food models (Bonneuil et al., 2006). The first is sectoral, industrial, oligopolistic, linear, integrated and based on the paradigm of fixed varieties. It is designed to serve a 'modernising' and productivist development model that is largely based on the use of machinery, fertilisers and phytosanitary products and on a standardisation of food. The second is territorialised, polycentric and participatory. It attaches importance to intra-varietal genetic heterogeneity, in line with local ecological and social conditions. It is part of an economy of quality (Allaire & Daviron, 2017) based on product differentiation and on leveraging quality (organic farming, geographical indications, reserved industrial varieties, 'peasant seeds', etc.). The opposition between these two varietal innovation regimes results in the structuring of actor communities on the basis of affinities or oppositions around agricultural and food development models embodying a particular set of shared values and principles of action.

Duality and Hybridisation

The failures of and crises of confidence in the agro-industrial model have made way for a positive re-qualification of various innovations that are considered to be solutions, albeit partial ones. We thus see, since the 1990s, the development of alternative food systems around innovations that define them and which differentiate them from other food systems: geographical indications, organic farming, fair trade, short supply chains, urban agriculture, etc. (Le Velly, 2017). They are opposed to the conventional food system, but the binary oppositions between alternative and conventional food systems quickly reveal these oppositions' limits. Indeed, systems are often hybrid, for example, as a result of the 'conventionalisation' of organic farming or the mainstreaming of fair trade. There exists a diversity of implementations of organic farming, fair trade or short supply chains, which implies that these forms have to be studied in detail, especially when attempting to assess their impacts. Indeed, while the aim of innovations is often 'problem solving', i.e. addressing certain environmental, social, health or economic limitations of the systems in place, they can, just as often, induce negative 'secondary effects' (social and economic exclusion, harm to the environment or health, etc.). The impacts of innovations therefore imply politicising their implementations with regards to the risks they may pose (Beck, 2001), to their purposes (in particular, to examine the notions of growth, progress and the trade-offs often necessary between economic and environmental issues) and to the choice of target actors (solvent *vs* non-solvent, entrepreneurs *vs* vulnerable populations, etc.) (Leach et al., 2012).

Territorial and Social Innovations

Thus, two approaches open up to help reflect on innovation in situations of coexistence of agricultural and food models. The first considers that agricultural and food innovations take place in sociotechnical transitions by examining various scales of analysis. They are in particular that of the niche, in which an innovation occurs, and that of the sociotechnical system, in which a sector's operating regime is institutionalised, with or without integrating these innovations (an approach is presented in Part

IV of the book). The second approach looks at innovation as an intrinsic part of the territory, which is regarded as a geographical space that is appropriated, organised, managed, lived in and represented by a social group. This second approach can itself be sub-divided into two: concerning territorial innovation and concerning social innovation. Territorial innovation encompasses both innovative territories as well as the territorialisation of innovations (Giraut, 2009; Soulard et al., 2018). It concerns new spatial forms (e.g. with regard to relationships between the centre and the periphery or interstitial areas), the emergence of coordination efforts between heterogeneous actors, and new material and symbolic relationships with the local places (e.g. in the activation of territorial resources and localised production systems). For its part, social innovation is viewed in the existing literature in three different ways. It can be defined as a process of social change (a revamping of the established order), as novel social practices, or as innovations that concern categories of actors that are very much in the minority and are often marginalised (Chiffoleau & Paturel, 2018). As a result, territorial and social innovations raise questions on the respective places accorded to the various agricultural and food models, and thus of their coexistence in society and in a territory, in a criticism of technological and productivist progress, and with particular emphasis on actors in social, economic, political and spatial situations of marginality.

Hypotheses and Important Questions

This state of knowledge on innovation in situations of coexistence of agricultural and food models in territories leads us to propose two hypotheses, which are little studied in the existing literature, and to ask some related questions.

Hypothesis 1: *Innovation modifies the conditions for the coexistence of agricultural and food models in territories.* Innovation, whether radical, systemic and/or disruptive, presents distinct continuities with the old, in a tension between continuity and discontinuity[1]. Consequently, how do tradition and innovation coexist, for example, with respect to appellations of origin and protected geographical indications? Innovation normally takes place quietly, whether 'discreet' (Albaladejo, 2004) with a certain conformity with the dominant regime, or 'ordinary' (Alter, 2000) and thus transgressive of social norms. In this way, innovation gives rise to linkages, emancipation and empowerment, for example, in short supply chains or peri-urban land tenure arrangements, as well as to exclusion and eviction. So does innovation cause a rift between different development models or, on the contrary, does it reconcile opposing views that were believed to be incompatible?

[1] Innovations that are radical, systemic or disruptive necessarily maintain continuity with the old, not only with regard to historically inherited structures (concerning land, organisations, markets, public policies, etc.), but also because of the sociotechnical and socio-ecological interactions that are established between the innovation that breaks with the past and its social, economic and technical environment.

Hypothesis 2: *The different configurations of coexistence of agricultural and food models influence innovation.* Can innovation be driven by either economic competition or cooperation? If yes, then under what conditions? Aren't conflicting interactions between actors, based on the different models they advocate, sometimes favourable to innovation? How can we unlock innovations in situations where the dominant model imposes its sociotechnical systems (Plumecocq et al., 2018)? How to prevent certain innovations from becoming 'conventionalised' during changes in scale and from losing their 'promise of making a difference' (Le Velly, 2017) by forsaking certain features that existed in the initial proposal?

Innovations Tested in the field

Following this brief review of the scientific literature and the elaboration of the central hypotheses of our research agenda, we present four case studies that explicitly analyse innovation in situations of coexistence of agricultural and food models in territories. This part of the book includes three chapters that are devoted to innovation (Baritaux and Houdart; Lucas and Gasselin; Toillier, Bancé and Faure). A fourth chapter, which also deals with adaptive processes, can be found in Part III of the book (Iceri). After these presentations, we conduct a transversal analysis highlighting their common points, their divergences and their grey areas.

Four Illustrations

In Chap. 4, Virginie Baritaux and Marie Houdart analyse hybridisation and innovation processes in the context of a tripartite partnership between a multinational company and local actors in milk production and cheese processing in Auvergne (France). In 2005, the Carrefour supermarket chain entered into a partnership with a family dairy in the Livradois-Forez region and 19 medium-sized farms to produce and market, under its private-label brand name, two protected designation of origin (PDO) cheeses (Bleu d'Auvergne and Fourme d'Ambert) made from cow's milk. The specifications of this 'quality line' are stricter than those of the PDO, in particular as it imposes an 'all hay' cattle feed and the use of raw milk[2]. The technical innovations[3] and organisational innovations[4] derive value from local resources, especially

[2] Grass and dry fodder-based silage-free feed, cheese from raw milk, maturing for three to six months longer than required by the PDO specifications, commitment to collect milk from the dairy for seven years, higher price for milk producers, and sale in all Carrefour outlets in France.

[3] Changes in livestock farmers' practices: innovation through withdrawal in the form of abandoning silage, development of barn drying of hay, etc.

[4] Changes in contractual relationships: reorganisation of the collection and production chain by the dairy; for producers, revision of commitments within the cooperatives for the use of agricultural silage equipment, exchange networks for barn drying, etc.

the grasslands. Some livestock farmers are reluctant to adopt these innovations as they run counter to their own strategy for increasing milk productivity. It may therefore seem paradoxical that these innovations were proposed by Carrefour, a typical actor of the 'conventional' model, which is characterised by 'a high concentration of downstream actors, financialisation and an intensification of production systems' (Rastoin, 2008). These reservations on the part of farmers were overcome by an incentivizing strategy of the dairy and the creation of a local professional group. Basing themselves on the work of Bloom and Hinrichs (2011), Virginie Baritaux and Marie Houdart explain the hybridity of the agri-chain by the fact that 'actors often have to mobilise resources and practices associated with both the conventional and the alternative models'. This hybridity allows Carrefour to consolidate its legitimacy in a context of changing expectations regarding food quality, while the farmers and the dairy diversify their sources of revenue in a context of increasing competitive pressure. Hybridisation has favoured technical and organisational innovation and the activation of territorial resources. However, the innovation was driven by a multinational entity, which raises questions about the 'conventionalisation' of the alternative model, on the one hand, and its capacity for innovation, on the other, 'to ensure a transition towards more sustainable food systems on a larger scale'.

In Chap. 5, Véronique Lucas and Pierre Gasselin report on modalities of local cooperation between heterogeneous farmers involved in agroecological innovations in France (soil conservation agriculture and the development of fodder legumes). These modalities include place-based coordination to co-construct references among peers or to design systems that optimise ecological processes. The authors report that formal cooperation first emerged among small groups of homogeneous peers with similar production systems. These groups were mainly part of farm machinery cooperatives (CUMA[5]), whose aim is to modernise farms by increasing labour productivity through a sharing of equipment. This common strategy and the homogeneity of the production systems facilitate the coordination and the conditions for sharing pooled resources, as well as the reciprocity of the exchange of services and material between farms. This type of cooperation also helps determine the 'right action' through fruitful dialogues between peers who share the same professional standards. However, the categories of farmers (new entrants, pluriactive, etc.) and their practices (product qualification, ecologisation, development of digital capability, etc.) have seen a rapid diversification. In such a context, how do farmers belonging to a CUMA manage their interactions with different colleagues? The authors show that the farmers in each of the five CUMAs studied, while different (organic and conventional agriculture, conservation agriculture or not, different productions and marketing methods, etc.), establish functional complementarities and find it beneficial to share material and cognitive resources, requiring however specific means of coordination. An agroecological orientation reconfigures social positions among farmers, leading to the adoption of new tactics for cooperating among peers in a reciprocal manner, without however completely eliminating the risks of splits in

[5] French: *Coopératives d'Utilisation de Matériel Agricole (CUMA).*

groups. Tacit silence is also a way of managing differences and preserving the conditions for technical cooperation. These results call for a qualification of observations that consider heterogeneity an obstacle to cooperation between farmers. The authors call for a deeper analysis of the mechanisms farmers use to manage heterogeneity and mutual interdependence among themselves. The mobilisation of ecological processes, anchored in territories, is likely to generate new favourable interactions between farmers who are *a priori* different, linked by cycles of reciprocity, new values and objectives of creating knowledge and shared management of work and tangible resources (materials, seeds, natural resources, etc.).

In Chap. 6, Aurélie Toillier, Saydou Bancé and Guy Faure analyse the coexistence of three agricultural advisory sub-systems for particular paths of ecological intensification in Burkina Faso. The first is aimed at solving the problems of sustainable intensification in conventional agriculture, the second encourages conversion to organic farming, and the third raises awareness about and provides training in agroecology. The authors study these three advisory sub-systems by characterising the networks of actors, their zones of intervention, and their registers of action and justification (transferring techniques, participatory problem solving, and education to help build capacity). They show that each of these three sub-systems brings together specific actors in close interactions (funding and injunctions, transfer of techniques and knowledge, training, co-production of technical and knowledge references). Moreover, each of these sub-systems operates in a particular zone: agroecology is mainly promoted in the country's Sahelian zone, where commercial agriculture is not very developed; the sustainable intensification of conventional agriculture can be found in the southern zones (especially the cotton basin); and organic farming is promoted around large cities with organised agri-chains, or in more rural areas for export crops (cotton, shea, etc.). However, these advisory sub-systems 'are relatively compartmentalised due to the type of the actors they involve and by their geographical areas of operations'. The advisory actors who promote the sustainable intensification of conventional agriculture are mainly public officials, while the other two sub-systems (organic farming and agroecology) generally involve actors from the private and associative sectors. The authors note that 'these compartmentalised configurations may be the root cause of hold-ups in innovation in the domain of ecological intensification'. They highlight in particular the 'rigidity' of advisory sub-systems (lack of spaces and time for debates on diagnoses and solutions) and their tendency to rely on 'turnkey' technical packages. 'As a consequence, certain agricultural production practices persist, even if they are not the most effective in preserving the environment'. Furthermore, the authors point out that the governance, knowledge and financing structures of all three sub-systems are tied to the influence of foreign entities (NGOs, research centres, donors) operating in Burkina Faso. As a result, the authors call for a more in-depth characterisation of the geographical, technical, institutional, political and financial divides between agricultural advisory sub-systems. They consider that the level of development and the dissemination of innovation rely on linkages between agricultural advisory sub-systems, especially through actors who play a bridging role between them and who facilitate knowledge and experience sharing.

We also refer to Chap. 9, which belongs to Part III of the book, to shed light on problems related to innovation. In this chapter, Vanessa Iceri reports on the complementarities between tradition and innovation in a Brazilian farming community in the centre-south of Paraná state. The *Faxinal Emboque* community, consisting of 68 families, mainly of Polish descent (who migrated here in the nineteenth century), is heir to a collective management of natural resources (water, land, forest) and of agricultural production, in particular the extensive rearing of free-range pigs. In 2011, the municipality culled these animals as they were suspected of carrying diseases, a move that triggered a mobilisation of the community. Unlike other communities that are engaged in political activism to counter the modernisation of Brazilian agriculture, *Faxinal Emboque* has launched a development project funded by an oil company. The project promotes various technical innovations (new productions, new pig breeds, rice hulling, new animal feed, seed selection, vegan recipes, etc.) and organisational innovations (pastry and bakery production units, seed processing unit, etc.). These innovations leverage an identity (*faxinalense*), products (dry-cured ham known for the quality of its breeding, ice cream made from a fruit previously reserved for animal feed, etc.), know-how (biodiversity management, seed exchanges, extensive pig breeding, etc.) and symbols (collectively managed pig pens, forest-grown mate tea, free-range pigs) that are claimed by the inhabitants as being traditional to *Faxinal Emboque*. This tradition values and draws on Polish and Brazilian cultures, especially through culinary recipes, and asserts itself as a way of life. This project is upsetting the farmers' normal relationship with work and with the market, with the dual requirement of preserving the traditional character of the production system and food practices, while meeting market-imposed health and commercial standards. The marketing of *faxinalense* products not only helps achieve the economic goal of increasing community income but also sheds light on this agricultural model and the families' way of life. Finally, the author examines how this hibridisation and this permeability between tradition and innovation are being used as a form of resistance by a weakened group.

A Comparative Reading

These studies cover four sites in France and elsewhere (Burkina Faso and Brazil) at different socio-spatial scales: a dairy and some of its suppliers in partnership with a multinational food retailer; small agricultural machinery cooperatives; an agricultural advisory system at the country level; and a farmer community with links to the market. Each of these socio-spatial scales is a 'setting' for coexistence, as defined in this book's general introduction, i.e. an arena or framework in which different models coexist. The innovations studied pertain mainly to environmental issues (all four case studies), economic issues (Baritaux and Houdart, Iceri) and health issues (Iceri). These case studies cover the three acceptations of innovation mentioned in this chapter's introductory paragraph. Innovation is considered a process in a territory and a value chain (Baritaux and Houdart) or in an organisation (Lucas and Gasselin).

It is regarded as a system of actors and knowledge (Toillier, Bancé and Faure). And it is viewed in terms of the transformations induced in a community (Iceri). While these studies mobilise several human and social science disciplines (economics, sociology, management sciences, geography), each of them pays particular attention to the technical aspect, which represents an essential marker of agricultural and food models. In this way, they confirm the intrinsic link that exists between technical and organisational innovations.

While avoiding a binary vision of agricultural models, two of these studies are based on a dual analysis (conventional versus alternative, Baritaux and Houdart; traditional versus modern, Iceri) to qualify hybridisation processes. Aurélie Toillier, Saydou Bancé and Guy Faure offer an analysis of ecological intensification based on three agricultural models. Véronique Lucas and Pierre Gasselin recognise the heterogeneity in technical systems and actors, without however associating a model with them, even though they note the divides between organic agriculture, conservation agriculture and conventional agriculture. These four case studies show firstly that innovations at the interface of two or more agricultural and food models are, or can be, capable of helping meet the challenges confronting the actors concerned. They emphasise that coexistence allows functional complementarities (sharing of resources, knowledge, work, market niches, value, etc.) and hybridisation. Vanessa Iceri argues that marketing products with a strong traditional aspect (symbolic and technical) can even constitute a path of resistance for a community wanting to claim a traditional way of life. However, only two of the case studies explicitly report on the mechanisms through which innovation creates conflict between actors (Lucas and Gasselin; Toillier, Bancé and Faure) because of the models they defend or criticise. This may involve the compartmentalisation of actor networks as a result of, among other factors, the support of international institutions (Toillier, Bancé and Faure), or splits in farmer groups (Lucas and Gasselin).

Conclusion

Research on the coexistence and confrontation of agricultural and food models helps our understanding of the drivers, methods, targets, challenges and impacts of innovations as well as of the barriers to them. These four case studies confirm the relevance of the two major hypotheses we propose, which encourages us to include them in our research agenda. Indeed, innovations contribute to defining agricultural models, often in terms of technical modalities, but also in terms of the specific relationships that actors (producers, consumers, citizens) establish with markets, the territory, the State, work, capital, nature and knowledge. In doing so, innovations modify the conditions, and act as catalysts, for the coexistence of agricultural and food models in territories. But innovations are also the result of the situations of coexistence that give rise to new potentials for innovation, for example, through functional complementarities between different agricultural and food models or through their hybridisation. However, it seems to us that certain currents of research, in pursuit of innovations that promote a positive vision of development, pay too little attention to controversies, power relations between actors and asymmetries (economic, institutional, spatial, power, symbolic, etc.). We argue for a critical analysis of innovation and for a debate on the possible goals of territorial development in situations of coexistence and confrontations of agricultural and food models.

References

Akrich, M., Callon, M., Latour, B., & Monaghan, A. (2002). The key to success in innovation, part I: The art of interessement. *International Journal of Innovation Management*, *6*(2), 187–206.

Albaladejo, C. (2004). Innovations discrètes et re-territorialisation de l'activité agricole en Argentine, au Brésil et en France. In C. Albaladejo & B. Cara (Eds.), *Développement local et multifonctionnalité des territoires ruraux en Argentine* (pp. 413–456), UNS Departamento de Geografía/IRD UR102/Inra-SAD/Université Toulouse Le Mirail, UMR Dynamiques rurales.

Albaladejo, C. (2020). The impossible and necessary coexistence of agricultural development models in the Pampas: The case of Santa Fe province (Argentina). *Review of Agricultural, Food and Environmental Studies*, (March), 1–28.

Allaire, G., & Daviron, B. (Eds.) (2017). *Transformations agricoles et agroalimentaires: Entre écologie et capitalisme* (p. 429). coll. Synthèses, éditions Quæ.

Alter, N. (2000). *L'innovation ordinaire* (p. 278). coll. Sociologies, PUF.

Beck, U. (2001). *La société du risque. Sur la voie d'une autre modernité* (p. 521). Aubier.

Bloom, J. D., & Hinrichs, C. C. (2011). Informal and formal mechanisms of coordination in hybrid food value chains. *Journal of Agriculture, Food Systems, and Community Development*, *1*(4), 143–156.

Bonneuil, C., Demeulenaere, E., Thomas, F., Joly, P. B., Allaire, G., & Goldringer, I. (2006). Innover autrement? La recherche face à l'avènement d'un nouveau régime de production et de régulation des savoirs en génétique végétale. In P. Gasselin & O. Clément (Eds.), *Dossier de l'environnement de l'INRA*, n° 30, «Quelles variétés et semences pour des agricultures paysannes durables?» (pp. 29–53), Inra.

Bonny, S. (2017). High-tech agriculture or agroecology for tomorrow's agriculture? *Harvard College Review of Environment and Society*, *4*(Spring 2017), 28–34.

Chiffoleau, Y., & Paturel D. (2018). Social innovation through short food supply chains: Between networks and individualities. In G. Faure, Y. Chiffoleau, F. Goulet, L. Temple, & J.-M. Touzard (Eds.), *Innovation and development in agricultural and food systems*. éditions Quæ (Chapter 5).

Coudel, É., Devautour, H., Soulard, C.-T., Faure, G., & Hubert, B. (2013). *Apprendre à innover dans un monde incertain: Concevoir les futurs de l'agriculture et de l'alimentation* (p. 246). éditions Quæ.

Dumont, A., Gasselin, P., & Baret, P. V. (2020). Transitions in agriculture: Three frameworks highlighting coexistence between a new agroecological configuration and an old, organic and conventional configuration of vegetable production in Wallonia (Belgium). *Geoforum*, *108*, 98–109.

Duru, M., & Therond, O. (2015). Designing agroecological transitions. A review. *Agronomy for Sustainable Development*, *35*(4), 1237–1257.

Faure, G., Chiffoleau, Y., Goulet, F., Temple, L., & Touzard, J.-M. (2018). *Innovation and development in agricultural and food systems*. éditions Quæ.

Geels, F. W., & Schot, J. (2007). Typology of sociotechnical transition pathways. *Research Policy*, *36*(3), 399–417.

Giraut, F. (2009). Innovation et territoires: Les effets contradictoires de la marginalité. *Journal of Alpine Research/Revue de Géographie Alpine*, *97*(1), 6–10.

Hall, A., Sulaiman, V. R., Clark, N., & Yoganand, B. (2003). From measuring impact to learning institutional lessons: An innovation systems perspective on improving the management of international agricultural research. *Agricultural Systems, 78*(2), 213–241.

Klerkx, L., Aarts, N., & Leeuwis, C. (2010). Adaptive management in agricultural innovation systems: The interactions between innovation networks and their environment. *Agricultural Systems, 103*(6), 390–400.

Leach, M., Rockström, J., Raskin, P., Scoones, I., Stirling, A. C., Smith, A., Thompson, J., Millstone, E., Ely, A., & Around, E. (2012). Transforming innovation for sustainability. *Ecology and Society, 17*(2), 11.

Le Velly, R. (2017). *Sociologie des systèmes alimentaires alternatifs. Une promesse de différence* (p. 200). coll. Sciences Sociales, Presses des Mines.

Plumecocq, G., Debril, T., Duru, M., Magrini, M.-B., Sarthou, J.-P., & Therond, O. (2018). Caractérisation socio-économique des formes d'agriculture durable. *Économie Rurale, 1*, 99–120.

Rastoin, J.-L. (2008). Les multinationales dans le système alimentaire. *Projet, 6*(307), 61–69.

Soulard, C.-T., Perrin, C., Scheromm, P., Chia, E., Clément, C., Michel, L., Hasnaoui Amri, N., Duffaud-Prévost, M.-L., & Ubilla-Bravo, G. (2018). Territorial innovation in the relationships between agriculture and the city. In G. Faure, Y. Chiffoleau, F. Goulet, L. Temple, & J.-M. Touzard (Eds.), *Innovation and development in agricultural and food systems*. éditions Quæ (Chapter 7).

Temple, L., Chiffoleau, Y., & Touzard, J.-M. (2018). A history of innovation and its uses in agriculture (Chapter 1). In (G. Faure, Y. Chiffoleau, F. Goulet, L. Temple, & J.-M. Touzard (Eds.), *Innovation and development in agricultural and food systems*. éditions Quæ.

Torre, A., & Wallet, F. (2013). Innovation and governance of rural territories. In E. Coudel, H. Devautour, C.-T. Soulard, G. Faure, & B. Hubert (Eds.), *Renewing innovation systems in agriculture and food* (pp. 147–164). Wageningen Academic Publishers.

Touzard, J.-M. (2018). Agricultural and agrifood innovation in the 21st century: Maintaining, erasing or reshaping its specificities? In G. Faure, Y. Chiffoleau, F. Goulet, L. Temple, & J.-M. Touzard (Eds.), *Innovation and development in agricultural and food systems*. éditions Quæ (Chapter 2).

Touzard, J.-M., Temple, L., Faure, G., & Triomphe, B. (2015). Innovation systems and knowledge communities in the agriculture and agrifood sector: A literature review. *Journal of Innovation Economics and Management, 2*, 117–142.

Wigboldus, S., Klerkx, L., Leeuwis, C., Schut, M., Muilerman, S., & Jochemsen, H. (2016). Systemic perspectives on scaling agricultural innovations. A review. *Agronomy for Sustainable Development, 36*(3), 46.

Chapter 4
Supermarket Chains as Drivers of Hybridisation and Innovation in Territorial Food Systems

Virginie Baritaux and Marie Houdart

Some authors are now calling for going beyond approaches that are based on a fundamental opposition between the so-called 'conventional' and 'alternative' agricultural and food models (Winter, 2003; Sonnino & Marsden, 2006; Le Velly, 2017). Seen as an archetype of an observed reality, the conventional model refers to 'the global food system, where relations between producers and consumers are distant, anonymous, and motivated by profits' (Bowen & Mutersbaugh, 2014, pp. 204–205). This model is characterised by a large concentration of downstream actors, and financialisation and intensification of farming systems (Rastoin, 2008). For its part, the so-called 'alternative' model is constructed as an opposition. It is associated with approaches aimed, in particular, at rekindling a closeness between producers and consumers (Milestad et al., 2010), and at instituting a social and territorial 're-anchoring' of food (Marsden et al., 2000; Renting et al., 2003). This process implies a re-localisation of food systems, and is reflected in an assertive stance of eliminating intermediaries between agricultural producers and consumers (Bloom & Hinrichs, 2011). For Mount (2012), this dichotomous vision prevents the diversity of forms of the localisation of food systems from being taken into account. In particular, it excludes initiatives that involve more than one intermediary, despite the possible role of intermediaries in creating a kind of proximity (Praly et al., 2014). For Bloom and Hinrichs (2011), the notion of hybridity must then be recognised to account for the fact that actors often have to mobilise resources and practices associated with both the conventional and the alternative models.

The strategies adopted by French supermarkets, as part of the re-localisation of food production, appear to be conducive to hybridisation. The development of a

V. Baritaux (✉)
VetAgro Sup, UMR Territoires, Clermont-Ferrand, France
e-mail: virginie.baritaux@vetagro-sup.fr

M. Houdart
INRAE, UMR Territoires, Clermont-Ferrand, France
e-mail: marie.houdart@inrae.fr

P. Gasselin et al. (eds.), *Coexistence and Confrontation of Agricultural and Food Models*,
https://doi.org/10.1007/978-94-024-2178-1_4

range of 'terroir' products sold under their private-label brands, the implementation of territory-specific brands and, more recently, the re-localisation of a part of their procurement or the development of direct relationships with local producers (Baritaux & Billion, 2016) are, in fact, based on the mobilisation of local resources and the appropriation by these 'global conventional' actors of objectives and values associated with alternative models (quality, territorial anchoring, proximity, etc.) (Maye & Kirwan, 2010). For the supermarket chains, this hybridisation represents a way of finding a new legitimacy in a context of changing consumer expectations concerning food quality: healthy, organoleptic, environmental, and ethical (Beylier et al., 2011). This hybridisation process involves changes and technical adjustments on the part of actors who are in partnerships with retailers, as well as changes in the coordination of relationships within the agri-chains concerned (Mazé, 2002). Furthermore, because of the mobilisation of local resources, these innovations are also likely to have impacts on territorial construction (Lamara, 2009).

In this chapter, we analyse the way in which this hybridisation, driven by a supermarket chain, is expressed, and its impact in terms of innovation, at the level of the agri-chain's actors and of the territory. What are the dimensions that make up the 'hybridity' of the studied value chain? Where can the innovation that is linked to this hybridisation process be found?

We use the example of supermarket chain Carrefour's 'Quality Commitment' product line in the Livradois-Forez territory in France (Fig. 1) to find answers to these questions. Launched in the early 2000s, this quality line is sourcing cheese, since 2012, through an agri-chain made up of a tripartite partnership between 19 milk producers, their dairy[1] (the Société fromagère du Livradois, SFL) and the supermarket's product line. The aim is to produce and market, under the supermarket's private-label brand, two kinds of cow milk cheeses under their protected designation of origin (PDO Bleu d'Auvergne and Fourme d'Ambert). This partnership is based on a threefold commitment: the milk producers and the dairy are to produce milk and cheese that comply with the retailer's requirements; the milk is to be procured by the dairy for a period of seven years, at a premium procurement price; and, finally, the retailer must buy an agreed quantity of cheese at a higher price, and sell the cheese (in the form of cut cheese) in all its outlets in France. The specifications concerning cattle rearing practices and cheese processing defined by the retailer are more stringent than those of the PDO: the cows must be fed on grass (direct grazing) or on hay harvested in the PDO's farming region, with silage (hay or maize) being strictly forbidden, and the cheeses must be made with raw milk, and be matured for longer than required by the relevant PDO specifications (3–6 months longer).

We used interactionist approaches to understand the way in which the relationship of cooperation was created between the retailer and its suppliers. These approaches regard supplier-retailer relationships as processes, with time representing a key dimension to understand and characterise the nature of these relationships (Ford & Håkansson, 2006; Nogatchewsky & Donada, 2005). Using a literature search and 22

[1] 19 producers out of the 190 who supply milk to the dairy are part of this agri-chain, and 10% of the total volume of milk the dairy processes annually is destined for the agri-chain.

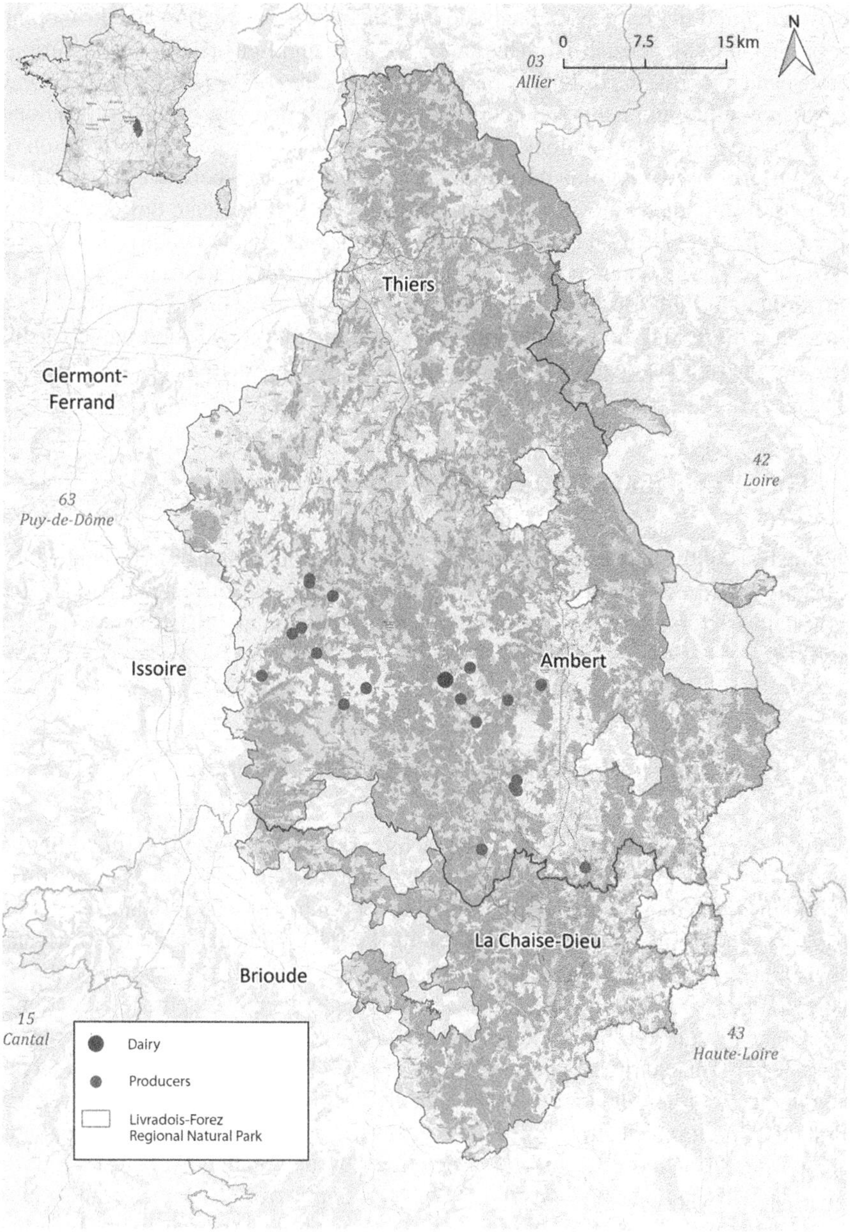

Fig. 1 The SFL dairy and its suppliers in the Livradois-Forez territory. *Sources* M. Houdart/IGN/OSM, 2019; execution: F. Johany, UMR Territoires, 2019

semi-structured interviews, we analysed the process of constructing and developing the partnership (Mendez, 2010). This analysis helped highlight the roles of the various actors in the process, the different phases in the trajectory of the relationship, the way in which conflict and cooperation are articulated and play out over this trajectory, and the elements of the relationship's environment that contribute to the creation of the approach.[2] Without going into the details of this trajectory here, we describe how the hybridisation and innovation processes came together to create this quality line. To this end, we present, in the first part, the dimensions around which hybridisation takes shape. The second part describes the process of innovation through withdrawal, the result of this hybridisation. In conclusion, we examine the link between these two processes, discussing how the hybridisation of agricultural and food models helps turn the supermarket chain into an actor of territorial construction.

1 The Dimensions of the Agri-chain's 'Hybridity'

The analysis of the agri-chain's trajectory reveals the four dimensions around which hybridity emerges: the different scales of action, the objectives pursued by the agri-chain's actors, the coordination mechanisms, and the mobilisation of territorial resources.

1.1 Actor Interactions at Different Levels of Action

The tripartite partnership brings together actors who undertake activities at different levels, from the global to the very local. The multinational company Carrefour is a well-known brand in the French supermarket food retailing sector (Daumas, 2006). It is also known in France as a pioneer in the development of 'terroir' private-label brands and in the organisation of agri-chains. As for SFL, it is a 'hybrid'-scale company exhibiting a strong local anchorage, even as it pursues a national development strategy and markets its products in France and overseas (Corniaux et al., 2015). This family business was created in 1949 in the Livradois region by the grandfather of the current owners. Although it is deeply anchored to its territory, it has adopted a development strategy for the past several years that is based on horizontal integration, through the acquisition of other dairies, and on the diversification of its activities. This has led to the creation of a holding company that owns SFL and which is managed by the family. It groups together several production sites in PDO regions in France, as well as a cheese cutting and wholesale marketing unit. Finally, at the local level, the 19 dairy producers participating in the agri-chain, located within a 40-km radius around SFL (Fig. 1), are largely representative of the agricultural

[2] For more details on the case and the methodology used to study it, see Baritaux and Houdart (2015).

structures present in the territory, i.e. medium-sized farms with extensive farming systems, self-sufficient in fodder, and with moderate milk production.

1.2 Quality and Territorial Anchoring Are Essential for Economic Value-Addition

Most of the actors in the partnership recognise the terms of the agri-chain's specifications as a means of guaranteeing specific qualities of the final product linked to a particular territory. These specifications can thus be seen as meeting the expectations the consumers have of territorial anchorage and product quality, often considered characteristic of alternative initiatives (Goodman, 2003; Watts et al., 2005). Nevertheless, the actors' engagement in the process also seems to be strongly motivated by economic objectives and profitability (maintaining market share, creating value, etc.), reflecting a form of 'instrumentalism' in the sense of Block (1990) (Hinrichs, 2000).[3] Thus, although the retailer includes this partnership in its social responsibility policy (support for farmers and small processors), this initiative is also part of a broader strategy, launched in the 1990s, of developing an offering of high-quality food products (authentic, good taste), that aims to increase market share by catering to growing consumer demand for more authentic and ethical products. For the dairy, the partnership with the retailer is an opportunity to secure an outlet and pursue its diversification strategy, after an unsuccessful attempt to develop a venture for certified organic cheeses. This partnership achieved the objective of creating a new commercial outlet and of obtaining better value from its products. The implementation of this quality approach is thus seen as the logical continuation of the production and commercial choices that were made earlier, especially in terms of re-appropriating raw milk production and benefitting from certain production methods. As for the farmers, their participation seems to be guided more by economic interests than by ecological concerns. The first farmers involved in the initiative were already producing 'all hay' milk and already had an organic farming certification. On the other hand, the farmers who were 'recruited' subsequently to construct the partnership were selected more for their superior milk quality and their willingness to modify their practices in order to join the partnership than for their environmentally friendly practices. In addition, while the dairy made an economic commitment (price premium and duration of collection), it did not put in place any particular mechanism to encourage farmers to use and manage their pasture and grassland resources in a way that could ensure real positive ecological impacts.

---

[3] A strong instrumentalism stands in contradiction to the objective of developing alternative systems that are deeply socially anchored, in which choices, even economic ones, are guided by 'social' objectives (friendship, moral values, etc.) (Hinrichs, 2000).

1.3 The Link Between Formal and Informal Coordination Mechanisms

The contrast between conventional and alternative model is also evident in how economic relationships are governed in these models (Forsell & Lankoski, 2015). Informal coordination based on interpersonal trust, developed through proximity and the social embeddedness of relationships, is often perceived as a method of good governance since it allows objectives other than purely commercial ones to be considered (Bloom & Hinrichs, 2011). In contrast, formal coordination mechanisms such as contracts or labelling schemes are often associated with a conventional model. The development of these mechanisms reflects the rise in power of supermarket chains (Burch & Lawrence, 2005; Filser & Paché, 2008), and are often seen as a way for the latter to strengthen their sway over other agri-chain actors (Filser et al., 2001; Berges-Sennou & Caprice, 2003). The agri-chain under study links these two disparate mechanisms: although governed by contractual mechanisms, long-term relationships of trust between different actors have played a major role in promoting coordination, resulting over the years in a relatively rebalanced power relations, despite the presence of an actor that remains dominant in its negotiating capacity (Baritaux & Houdart, 2015). For example, before the agri-chain's official launch, the dairy's production manager relied on his long-standing good relationships with producers to encourage them to join the agri-chain. When it was officially launched in 2005, coordination was based on a contract between SFL and the retailer, and on a moral commitment for milk collection between the SFL and the producers, and the payment of a premium of €30 per 1000 L of all-hay milk. Subsequently, the dairy continued to leverage its close relationships with producers to encourage new memberships (meetings, farm visits). However, faced with a slowdown in membership, and finding itself in a difficult economic situation,[4] the retailer and SFL coordinated more closely to develop an incentivising strategy for producers. These interactions helped grow the trust between the retailer and the producers. The result was that contracts were formalised by the dairy that guaranteed producers a premium of €60 per 1000 L of milk, and a commitment to collect their milk for a period of seven years, which made it possible to guarantee a return on the investments necessary to convert to barn drying.

1.4 The Mobilisation of Territorial Resources by a 'Global' Actor

In hybrid value chains, the retailer mobilises local resources that helps support an activity that it carries out at a national or even international scale (Bloom &

[4] The year 2008 was marked in particular by a sharp fall in the price of dairy products resulting from the cumulative effects of a fall in consumption and in exports, and an increase in milk quotas.

Hinrichs, 2011). The agri-chain being studied fits into this scenario. Indeed, production methods based on traditional know-how (use of grass and hay, processing of raw milk), which were abandoned in the wake of the industrialisation of agriculture and processing, are being implemented by an actor that is emblematic of the agro-industrial system (huge volumes, standardised products, low production costs). By using the PDOs and the grass resource, and by mobilizing local actors, their earlier relationships (dairy-producer links) and their know-how, Carrefour brought into play both material and immaterial territorial resources (Gumuchian & Pecqueur, 2007). The very way in which the agri-chain was initiated illustrates this significant mobilisation of territorial resources. In the early 2000s, Carrefour executives involved in developing private-label quality cheese lines saw the use of grass and hay for animal feed and the use of raw milk as indicators of an adherence to the values promoted by the chain's brand. Auvergne was therefore identified as a production region that met these expectations and that offered products that could be added to the 'cheese platter' sold in the cut cheese section of Carrefour outlets. Indeed, in addition to the existence of several PDOs, this region is characterised by a large number of livestock farms using grass and hay as feed. This choice was also reinforced by the existence of a similar partnership, already established with a cooperative for the production of Cantal all-hay cheese, which served as a firm basis to define the basics of the approach (the specifications in particular). To meet its objective of developing an agri-chain for the two 'Blue' PDOs, the retailer implemented a traditional sourcing system enabling it to identify dairies likely to meet its requirements in the Auvergne territory. This resulted in the selection of two companies on the basis of various criteria, including the ability of the dairies to mobilise producers who were already producing all-hay milk, their experience in making raw-milk cheese, and their presence in the production territory of the PDOs concerned. SFL was selected at the end of this process due to four major assets. It was already procuring milk from producers whose forage systems are solely based on grass and dry forage. It is characterised by its experience in producing raw-milk cheese. It is an independent, medium-sized company, which is in line with the retailer's stated philosophy of participating in a territory's economic development. Finally, the company already had business relationships with the retailer for several years.

2 Innovation: From the Agri-chain to the Territory

The establishment and development of this agri-chain appears to be a process of organisational and technical innovation for the various actors involved. These innovations were conceived not only at an individual scale, but also at the scale of a group of actors and the Livradois-Forez territory.

2.1 Innovations at the Scale of the Agri-chain

In a context of increasing competitive pressure and the expression of new consumer expectations in terms of the quality of food products and production methods linked in particular to various health crises, the brand implemented a strategy clearly oriented towards the development of a 'socially responsible' product range and created its first quality line in 1992. This innovative strategy provided the initial impetus (Baker & Mehmood, 2015), reflected in the years that followed by the establishment of other quality agri-chains, including the quality agri-chain for the two Blue cheese PDOs from Auvergne in 2005.

The creation and development of this agri-chain was based on the gradual establishment of a tripartite partnership that was new for all concerned. This partnership was gradually standardised mainly because of the special relationships between the dairy and the farmers. In addition to the implementation of this new coordination mechanism in the territory, the organisational innovation is also evident in the reorganisation of the collection and the production chain by the dairy in order to satisfy the new specifications. It has put in place a routine for a daily collection from the producers and a rigorous system to ensure traceability of the milk and cheese meant for the agri-chain. In addition, it made investments in unloading and storage facilities to prevent mixing with other types of milk. For farmers who were not on an all-hay feed routine, joining the chain entailed an abandonment of the practice of silage. This 'innovation through withdrawal' (Goulet & Vinck, 2012) led to a series of technical and organisational adaptations on the farms on relatively large scales depending on their original production system (especially concerning fodder management) and on investments (financial and learning) in the barn drying technique.

This innovation by withdrawal also has had an impact at the scale of the Livradois-Forez territory. It influenced the dynamics of innovation because of the links existing between the farmers involved in the agri-chain and the other actors in the territory.

2.2 Innovations as Part of a Territorial Approach

At the territorial level, this process of abandoning silage was not without conflict, and it triggered a complex process of legitimisation of all-hay practices. Barely a year after the launch of the chain, some producers in the region applied pressure on the farmers who had joined the agri-chain and others who wanted to, as this new engagement sometimes impugned the commitments within the silage-equipment sharing cooperatives (CUMA). In addition, forage systems without silage were criticised and perceived by some producers as a 'step backwards' in a territory that had switched to silage and bale wrapping in the 1990s. The main apprehension was that milk production would drop significantly, and that farms would no longer be sufficiently profitable. This uncertainty was amplified by the fact that, in the early 2000s, there was a lack of reference structures and advisory bodies since this system was not common in the territory. The granting of aid to the region, as early as 2007, for the modernisation of livestock buildings, including for barn drying tools, was not enough to allay fears about unprofitable investments.

However, a proactive approach by the retailer and the processor to raise awareness of the practice of all-hay farming among other farmers helped ease tensions. Three features assisted in making the agri-chain attractive to farmers:

- exchange networks on the barn drying technique were created and contributed to strengthening relationships between producers, and in disseminating the agri-chain's advantages;
- the territorial context was favourable for a shift towards natural grasslands. Requests were made in the Auvergne PDOs to alter the specifications to reduce the use of silage, and tools to manage natural grasslands were proposed. In addition, discussions aimed at developing the region's grass resources were initiated. In parallel with these dynamics, advisory bodies worked together to offer producers new references on all-hay systems. They made a significant contribution by convincing some producers to diversify their incomes in line with the expansion of their structures and changes in their labour forces;
- more generally, the prospect of the abolition of milk quotas in 2015 had an effect on the willingness of producers to join the agri-chain, since they realised, as did territorial development actors, the benefits of the quality line as a means of providing a guaranteed outlet for their production.[5]

These developments overcame the last reservations of the producers, who feared that they would face difficulties in harvesting the hay and that they would suffer a drop in productivity due to fodder that was inferior in quality to that provided by silage.

Thus, in this favourable context, although these new practices were initially imposed on the farmers joining the agri-chain, they gradually spread throughout the Livradois-Forez region. For the actors in the territory, the practice of all-hay livestock farming seemed to be a distinctive feature of the territory and of an efficient animal husbandry system.[6] In the end, the search for solutions to the difficulties that were encountered helped establish dialogue networks that far exceeded the scope of the farmers involved in the system, thus reinforcing the activation and cohesion of a local professional group that could be considered a new territorial organisational resource.

[5] It should be noted here that from 2010 onwards, the dairy collected more or less sufficient quantities of milk to meet the retailer's requirement. Furthermore, the retailer experienced difficulties at that time (loss of market share) and changed its strategy (the development of quality agri-chains no longer seemed to be a priority). As a result, and despite the willingness of the producers, the dairy refused to accept new producers. The innovation thus produced here a kind of eviction, of exclusion of certain producers.

[6] Since the analysis was carried out in 2012, there was not enough data available to allow a quantitative accounting of the success of the innovation in terms of the number of adherents of the all-hay practice. The results presented here are based on the perception of the actors surveyed.

3 Conclusion

The examination of the links between hybridisation and innovation in our case study suggests that the two processes are closely linked. Hybridisation is based, first of all, on the relationships of actors with different scales of action, around values associated with territorialisation and the mobilisation of territorial resources, in a common economic interest, according to a mode of coordination which articulates formal and informal mechanisms. This hybridisation process is accompanied by organisational and technical innovations that not only affect the actors involved in the process, but also have an impact on the territory. The retailer occupies a central place in this dual process of hybridisation and innovation, taking on, as Burch and Lawrence (2005) have shown, the role of 'steering' the agri-chain, by participating in particular in the territorial qualification of products (Baritaux & Billion, 2016). This case study shows one of the ways in which such an actor intervenes by initiating the dynamics that engender a new coordination between actors, by mobilizing and activating specific territorial resources, and by encouraging and participating in the dissemination of the all-hay innovation at a territorial scale. The retailer thus plays a twin role in territorial development: by mobilizing and activating territorial resources, and also through the effects induced by the implementation of a technical innovation in the territory, which contributes to the creation of new material and immaterial territorial resources. Hybridisation thus appears to be a driving force for innovation within food systems and territories, and seems capable of changing the conventional model. Nevertheless, this innovation is carried out by a typical actor of the 'conventional' model. This raises the question of the innovative capacity of different models, 'conventional' versus 'alternative', to ensure a transition towards more sustainable food systems on a larger scale (Cleveland et al., 2014; Fournier & Touzard, 2014). This study thus refers to the debates on the processes of the 'conventionalisation' of the so-called 'alternative' model, which takes the form of the appropriation of the alternative model's values and objectives to support the agro-industrial model, on its effects on the power relationships between actors within food systems and, more generally, on their dynamics of territorial development.

References

Baker, S., & Mehmood, A. (2015). Social innovation and the governance of sustainable places. *Local Environment: The International Journal of Justice and Sustainability, 20*(3), 321–334.

Baritaux, V., & Billion, C. (2016). Les intermédiaires de la distribution dans la relocalisation des systèmes alimentaires: Perspectives de recherche. In *Énergie, Environnement et Mutations Sociales, 11th RIODD Congress, 6–8 July, Saint-Étienne, France.*

Baritaux, V., & Houdart, M. (2015). Relations fournisseurs-grande distribution dans les filières agro-alimentaires. Une analyse de la trajectoire d'une démarche de type «filière qualité». *Économie Rurale, 346*, 15–30.

Berges-Sennou, F., & Caprice, S. (2003). Les rapports producteurs-distributeurs: Fondements et implications de la puissance d'achat. *Économie Rurale, 227*(278), 192–205.

Beylier, R. P., Messeghem, K., & Fort, F. (2011). Les distributeurs à la conquête de la légitimité territoriale: Le cas de Carrefour. *Management et Avenir, 44*(4), 235–255.
Block, F. (1990). *Postindustrial possibilities: A critique of economic discourse* (p. 240). University of California Press.
Bloom, J. D., & Hinrichs, C. C. (2011). Informal and formal mechanisms of coordination in hybrid food value chains. *Journal of Agriculture, Food Systems, and Community Development, 1*(4), 143–156.
Bowen, S., & Mutersbaugh, T. (2014). Local or localized? Exploring the contributions of Franco-Mediterranean agrifood theory to alternative food research. *Agriculture and Human Values, 31*(2), 201–213.
Burch, D., & Lawrence, G. (2005). Supermarket own brands, supply chains and the transformation of the agri-food system. *International Journal of Sociology of Agriculture and Food, 13*(1), 1–18.
Cleveland, D. A., Müller, N. M., Tranovich, A. C., Mazaroli, D. N., & Hinson, K. (2014). Local food hubs for alternative food systems: A case study from Santa Barbara County, California. *Journal of Rural Studies, 35*, 26–36.
Corniaux, C., Baritaux, V., & Madelrieux, S. (2015). Les stratégies spatiales hybrides des laiteries entre (re)localization et globalisation. In M. Napoléone, C. Corniaux, & B. Leclerc (Eds.), *Voies lactees. Dynamique des bassins laitiers entre globalisation et territorialisation* (pp. 227–248). Cardère.
Daumas, J. -C. (2006). Consommation de masse et grande distribution: Une révolution permanente (1957–2005). *Vingtième Siècle. Revue D'histoire, 91*(3), 57–76.
Filser, M., Garets, V., & Paché, G. (2001). *La distribution: Organisation et stratégie* (p. 394). EMS.
Filser, M., & Paché, G. (2008). La dynamique des canaux de distribution. *Revue Française de Gestion, 182*(2), 109–133.
Ford, D., & Håkansson, H. (2006). The idea of interaction. *The IMP Journal, 1*(1), 4–20.
Forsell, S., & Lankoski, L. (2015). The sustainability promise of alternative food networks: An examination through "alternative" characteristics. *Agricultural and Human Values, 32*(3), 389–398.
Fournier, S., & Touzard, J. -M. (2014). La complexité des systèmes alimentaires: Un atout pour la sécurité alimentaire? *Vertigo, 14*(1).
Goodman, D. (2003). The quality "turn" and alternative food practices: Reflections and agenda. *Journal of Rural Studies, 19*(1), 1–7.
Goulet, F., & Vinck, D. (2012). L'innovation par retrait. Contribution à une sociologie du détachement. *Revue Française de Sociologie, 53*(2), 195–224.
Gumuchian, H., & Pecqueur, B. (2007). *La ressource territoriale* (p. 252). Economica.
Hinrichs, C. C. (2000). Embeddedness and local food systems: Notes on two types of direct agricultural market. *Journal of Rural Studies, 16*(3), 295–303.
Lamara, H. (2009). Les deux piliers de la construction territoriale: coordination des acteurs et ressources territoriales. *Développement durable et territoires.*
Le Velly, R. (2017). *Sociologie des systemes alimentaires alternatifs: Une promesse de difference* (200 p). Presse des Mines.
Marsden, T., Banks, J., & Bristow, G. (2000). Food supply chains approaches: Exploring their role in rural development. *Sociologia Ruralis, 40*(4), 424–438.
Maye, D., & Kirwan, J. (2010). Alternative food networks. *Sociopedia.isa.*.
Mazé, A. (2002). Retailers' branding strategies: Contract design, organisational change and learning. *Chain and Network Science, 2*(1), 33–45.
Mount, P. (2012). Growing local food: Scale and local food systems governance. *Agriculture and Human Values, 29*(1), 107–121.
Mendez, A. C. (2010). *Processus: Concepts et méthode pour l'analyse temporelle en sciences sociales* (260 p). Éditions Academia.
Milestad, R., Bartel-Kratochvilb, R., Leitnerc, H., & Axmann, P. (2010). Being close: The quality of social relationships in a local organic cereal and bread network in Lower Austria. *Journal of Rural Studies, 26*(3), 228–240.

Nogatchewsky, G., & Donada, C. (2005). Vingt ans de recherches empiriques en marketing sur la performance des relations client-fournisseur. *Recherche et Application en Marketing, 20*(4), 71–96.
Praly, C., Chazoule, C., Delfosse, C., & Mundler, P. (2014). Les circuits de proximité, cadre d'analyse de la relocalisation des circuits alimentaires. *Géocarrefour, 89*(1–2), 125–134.
Rastoin, J -L. (2008). Les multinationales dans le système alimentaire. *Projet, 6*(307), 61–69.
Renting, H., Marsden, T. K., & Banks, J. (2003). Understanding alternative food networks: Exploring the role of short food supply chains in rural development. *Environment and Planning A: Economy and Space, 35*(3), 393–411.
Sonnino, R., & Marsden, T. (2006). Beyond the divide: Rethinking relationships between alternative and conventional food networks in Europe. *Journal of Economic Geography, 6*(2), 181–199.
Watts, D. C. H., Ilbery, B., & Maye, D. (2005). Making reconnections in agro-food geography: Alternative systems of food provision. *Progress in Human Geography, 29*(1), 22–40.
Winter, M. (2003). Embeddedness, the new food economy and defensive localism. *Journal of Rural Studies, 19*(1), 23–32.

Chapter 5
Coexisting in Farm Machinery Cooperatives: Cooperation Between Heterogeneous Farmers

Véronique Lucas and Pierre Gasselin

Place-based coordinations are becoming strategically important to ensure the agroecological transition, for example to co-produce knowledge between peers or to design systems that optimise ecological processes. Indeed, collaborations between farms, in spite of all their diversity, are necessary to make agroecological innovation possible (Wezel et al., 2016; Lucas et al., 2019) since ecological and biogeochemical flows exceed the farm scale. But the segmentation and heterogeneity of 'agricultural worlds' can hinder innovation processes between farmers with different styles (Lémery, 2003; Hervieu & Purseigle, 2013). Nevertheless, an increasing number of public instruments designed to promote agroecology rely on collaborations among farmers, especially at the local level: Ecophyto, GIEE,[1] PAEC,[2] etc. The network of farm machinery cooperatives (CUMA[3]) is an illustrative example of this phenomenon, with nearly 12,000 CUMAs in existence, involving more than a third of French farms. Various studies have already noted these cooperatives' contribution to the co-design of sociotechnical solutions adapted to the needs of member farmers (Assens, 2002; Pierre, 2009). Agroecology is a expanding topic in the CUMAs, especially through the increased pooling of new resources (crops,

[1] Economic and Environmental Interest Group (French: *Groupement d'intérêt économique et environnemental, GIEE*).

[2] Agro-Environmental and Climate Project (French: *Projet agro-environnemental et climatique, PAEC*).

[3] French: *Coopératives d'utilisation de matériel agricole (CUMA).*

V. Lucas (✉)
UMR BAGAP, INRAE, Rennes, France
e-mail: veronique.lucas@inrae.fr

P. Gasselin
UMR Innovation, INRAE, Montpellier, France
e-mail: pierre.gasselin@inrae.fr

P. Gasselin et al. (eds.), *Coexistence and Confrontation of Agricultural and Food Models*,
https://doi.org/10.1007/978-94-024-2178-1_5

knowledge) between farmers. How do these collective innovation processes work despite the heterogeneity[4] of farmers and of local professional networks?

We propose a new understanding of the modes of localised cooperation in CUMAs, and in particular of their recompositions in the context of the growing emphasis on agroecology. We selected five CUMAs, whose equipment facilitates the development of two types of agroecological practices: conservation agriculture[5] and the cultivation of fodder legumes. Thirty individual interviews were conducted with farmers in order to analyse their individual and collective practices, as well as the conditions of their emergence and implementation.[6] After a literature review to identify the determinants of local professional cooperation between farmers, we analyse the recompositions taking place within the CUMAs surveyed and conclude with a discussion on the new research that is necessary.

1 Local Cooperation and Professional Networks

The current recompositions of local cooperation are reflected in a diversity of forms inherited from the transformations of French peasant societies. The ways in which these forms of cooperation take into account the heterogeneity of farmers, accentuated by the agroecological transition, are determined not only by social and technical factors, but also by cognitive and symbolic ones.

1.1 From the Old Mutual Aid in Villages to the Current CUMAs

In France, the passing of peasant societies has led to a recomposition of the modes of agricultural cooperation. Nicourt (2013) has analysed the replacement of traditional informal mutual aid in villages by formalised cooperation between modernist farmers. The CUMAs were thus formed, as were 'time' banks; they made it possible to record exchanges of material and labour between peers. These formalisation processes have led to an accelerated reduction in the size of collectives, which had already been induced by the reduction in labour requirements caused by agricultural mechanisation. This has favoured the technical and socio-economic homogenisation of these collectives in order to facilitate the balancing of exchanges, resulting in

[4] Instead of the notion of diversity, we prefer here that of heterogeneity, in the sense of an organised diversity. It is the heterogeneity of farmers (in terms of their resources, practices, results, statuses, values, membership of unions and cooperatives, and projects) in organised collective action and in networks. Heterogeneity invites us to look beyond the differences between farmers to explore the genesis, trajectories and functioning of collectives (Jollivet and Lepart, 1992).

[5] Conservation agriculture is aimed at restoring soil health through no-till practices, winter cover crops and crop diversification.

[6] For more details, see Lucas (2018), Lucas et al. (2019).

more selective pairings. Traditional mutual aid has gradually given way to smaller and more elitist collectives (Nicourt, 2013).

However, other studies, more nuanced, have discerned a tendency amongst farmers of creating a CUMA from a small nuclei of members with similar characteristics in order to limit coordination costs, but have observed that these groups can then evolve. These cooperatives thus find advantages in and can be strengthened by expanding to other farms with different configurations or systems, for example to improve the financial viability of collective investments (Cornée et al., 2020).

Nearly 12,000 local and self-organised CUMAs bring together an average of 25 farms each. Found in all regions of France, they encompass the diversity of social and technical forms of agriculture. While the CUMAs were initially aimed at promoting the modernisation of small and medium-sized farms, today they have more varied objectives, for example to process and market in short supply chains or to diversify production (Lucas, 2018). Different authors have observed that the quest for higher labour productivity on farms remains the most important objective, through access to high-capacity equipment (Harff & Lamarche, 1998; Jeanneaux et al., 2018). This may disadvantage or even marginalise small member farms that cannot invest in tractors powerful enough to pull this type of equipment (Mundler et al., 2010).

However, this heterogeneity can also be a result of the diversity of practices between one farm and the next even if both use the same equipment.

1.2 Continued Existence of Local Cooperation: Cognitive and Symbolic Issues

While alternatives to CUMA have emerged, such as agricultural contracting companies (ACC), these cooperatives continue to exist. Several studies have shown the continued reliance on mutual aid and on CUMAs, despite the presence of local ACCs (see, for example, Mundler & Laurent, 2003). Furthermore, in some of the mutual aid processes studied, many farmers do more work for their peers than they receive in return (Dedieu, 1993). These various analyses conclude that participation in mutual aid or in CUMA can go beyond the technical objective of improved management of labour and equipment to include issues of sociability and discussion between peers. Thus, local forms of cooperation create situations of coactivity, which farmers use for technical discussions with their peers, stimulating their thinking about their ways of working and improving their practices (Darré, 1996).

These discussions also have a symbolic dimension: requests for advice from one farmer to another are constrained by the effects of competition of symbolic status between them. Indeed, asking a peer about his or her way of working can mean granting credibility and importance, and can also imply an admission of a difficulty in knowing what to do (Chiffoleau, 2005). In a similar way, Deffontaines (2014) points out that by participating in mutual aid, farmers allow others to judge their ways of working. This can be disconcerting enough for some to choose not to participate, especially for those whose ways of working deviate from the local norm.

1.3 Impact on Local Networks of the Recent Emergence of Ecological Farming

Different ecological styles of agriculture have emerged in recent decades, contributing to the heterogeneity of production methods. Several authors have analysed the impact of the growth of organic farming on local professional networks. This development initially led to discussion groups involving specialised but distant peers, due to the local isolation of the first organic farmers (Le Guen & Ruault, 1994).

Organic practices differ according to places and producers, notably because of symbolic issues in local professional networks. Nicourt et al. (2009) highlight the different effects on the social interactions of organic sheep farmers according to their health management strategies, in one case selecting hardy animals to limit prophylactic interventions, in the other relying on a curative rationale, but one that is based on natural remedies. The curative strategy requires skills in monitoring and observing the herd to better anticipate problems and react quickly. These farmers thus maintain opportunities for dialogue with their conventional peers, and even acquire a role of 'experts' who are consulted by the latter in matters concerning animal health. On the other hand, livestock farmers who have opted for hardy selection tend to stand out in their local networks, which can increase their isolation.

In a similar way, some authors highlight the withdrawn position of some pioneers of ecological farming styles. These farmers may assert their singularity to the extent that they limit their opportunities to share experiences and technical dialogue with local peers. They may prefer to talk and meet only with similar pioneering peers within their specific domain, even if these latter are spatially distant, believing that this kind of meeting gives them more opportunities to progress (Brives & de Tourdonnet, 2010).

We therefore note that local professional agricultural networks reveal varied situations of coactivity between farmers. Thus, the continued existences of CUMAs, embedded within these local networks, is explained not only by the need to share resources and labour, but also by cognitive issues. Interactions between peers involving the sharing of resources obey implicit rules of social and symbolic exchange based on reciprocity. This is how the rationale of the gift, with its threefold obligation to give, receive and return, structures cooperation between farmers by conferring a social rank on each individual, and underpins the identity processes at play in the consideration of ways of working (Alter, 2010; Sabourin, 2012).

While cooperation seems easier between 'similar' peers, how do farmers manage their interactions with colleagues who are 'different'? To answer this question, we analysed the processes of pairing as well as the management of interdependence and heterogeneity within five CUMAs whose shared machinery allows the development of agroecological practices.

2 A Strong Reliance on Local Cooperation

The CUMAs surveyed were selected because their shared machinery facilitates two types of agroecological practices: conservation agriculture and the development of fodder legumes (see Table 1).

When producers implement conservation agriculture or develop the cultivation of fodder legumes, they mobilise their peers to find solutions to their questions. This gives rise to three types of cooperation, one of which is the CUMA. This latter is remobilised to collectively acquire the expensive equipment needed for these new practices, sometimes with the pooling of hay in dryers, or the hiring of employees to operate machinery, etc. In addition, new arrangements are made for sharing and exchanging other resources that are outside the CUMAs' statutory scope (exchange of

Table 1 Characteristics of the CUMAs studied

Geographical location	Farms surveyed in each CUMA	Main common activities in each group	Farm practices developed
Basque country	2 sheep dairy farms, 1 goat and sheep dairy farm (with direct sales)	Sharing of a collective hay dryer, training programme for members	Grassland legumes
Tarn	2 cow dairy farms with milking robot (1 organic), 4 grain farms (1 organic)	Sharing of no-till equipment, mutual aid, seed sharing	No-tillage and direct seeding, complex cover crops, crop diversification
Ain	4 cow dairy farms, 1 goat dairy farm (with direct sales), 1 grain farm	Sharing of a collective hay dryer, with a shared employee, mutual aid	Grassland legumes, crop diversification
Aube	2 sheep-meat farms (1 with direct sales), 1 cattle-meat farm (with direct sales), 3 grain farms	Sharing of no-till equipment, mutual aid through a time bank, seed sharing, cross-farm grazing of cover crops	No-tillage and direct seeding, complex cover crops, crop diversification
Touraine	2 goat dairy farms, 7 cow dairy farms (5 with milking robots), 1 cattle-meat farm	Sharing of hay-making equipment adapted to legumes, arrangements between grain farmers, collective experimentation programme	Grassland legumes and complex cover crops, crop diversification

farm seeds, etc.). The majority of farmers have long been involved in other collectives for sharing experiences and co-producing knowledge among peers in a formalised way with the support of facilitators.

2.1 *Learning About the Practices of Others Indirectly*

Two out of these three cooperation configurations allow farmers to learn about each other's practices without asking for advice directly, thus mitigating the symbolic cost of obtaining information.

In formalised peer-to-peer groups, the mediation of professional facilitators creates spaces for sharing in which each participant is both a provider as well as a receiver of information. Through comparisons of economic accounts and meetings in the field, the participants develop an in-depth acquaintance of their peers' production systems and technical and economic results. However, one of the farmers interviewed, whose farm was in financial distress during our fieldwork, withdrew from the local collective at that time. This shows that the revelation of one's professional results is easier when they are good or meet the norm expected by the group.

The coordination required to share equipment and other tangible resources such as seeds or fodder can also provide an opportunity for technical dialogue to learn about other producers' practices without appearing to do so. Joint operation on another farmer's farm, such as during the harvest of silage fodder, are also opportunities to learn in situ about the practices of one's peers.

2.2 *A Tacit Silence Within CUMAs*

In order to maintain the CUMA's primary function of machinery and labour sharing, topics that could bring differences to light (such as union membership) are avoided in a kind of tacitly agreed silence. The network's members often claim that in the CUMA, 'one leaves one's ideas in the locker room' so that the technical cooperation can take place. In order for farmers with different orientations to coexist within a CUMA, debates on the significance of new practices being adopted tend to be avoided and opinions remain unvoiced. Technical dialogue at CUMA meetings focuses on the machinery, i.e. the conditions for sharing and using it, as well as on the practical aspects of the farming systems.

3 Cooperation Despite the Heterogeneity of Technical Systems and Projects

Farmers naturally have a preference for working with peers who are similar to them or with farmers with whom they have good relationships (favourable joint experiences in the past, friendships, similar values, etc.), but they also choose to ally with colleagues with different systems and orientations, as shown by the ways in which they join CUMAs.

3.1 Farmers Prefer to Work with Similar Peers but Can Be Flexible

Various comments from farmers indicate a preference for creating small, homogeneous groups of similar farmers. For example, a farmer in the Ain region was pleased in being among peers within an economic interest group of eight farms and sharing the resources necessary for their herds' genetic improvement (embryo transplantation, marketing of embryos): 'I was the one who first pushed to create the group […] and I'm not ashamed to admit that I put some pressure. I told them I was joining, but on condition that some others didn't (laughs). […] We all think alike […]. Besides, we knew each other, we are all friends. […] We have nothing to fear from each other.' These opportunities for exchanges between peers with similar practices and backgrounds are welcomed when their level of trust is such that they are not afraid of the judgement of others.

But farmers can be flexible and make trade-offs between their instinctive desire to be among similar peers and the need to ally themselves with a wider range of colleagues to access strategic resources. This was the case in the Ain CUMA, created specifically to manage a collective fodder dryer. Initially planned by a group of eight farmers brought together by their common experiences as cooperative leaders, the large investment necessary led to this initial circle being expanded to include four other different farmers (one practising organic farming, a pluriactive farmer, a breeder of an exotic cattle breed, etc.).

3.2 'Pioneer' Farmers Cooperating with Different Peers

Several of the CUMAs include one farmer who can be described as a pioneer, i.e. one who has tried out new practices before the others. These pioneers play a 'mobilising' role and adopt strategies to convince their colleagues to invest collectively in the equipment required for the new practices. For example, in the Aube CUMA, the pioneer farmer, who was practising conservation agriculture, personally owned a direct seeding machine but wanted to invest in a better, more sophisticated and much more expensive one. He made his seeder available to others to try out direct seeding. This increased the interest of some farmers of the group in no-till agriculture, who relied on the experience of the pioneer farmer. Over time, this led to the collective acquisition of specific conservation agriculture equipment by the CUMA.

The involvement of these pioneer farmers in groups that also include less knowledgeable farmers is also explained by the fact that it is difficult for any one farmer to possess the entire range of skills now necessary in the farming profession. Indeed, farming is becoming increasingly complex; new skills are required to manage the diversification of marketing channels, the increased requirements of traceability, the adaptations required in the face of climate change, etc. However, these skills can be distributed and shared within a group. This is what a farmer in the Tarn CUMA noted, who stores his cereals using equipment shared with other colleagues and relies on their skills for marketing: 'There is also the storage group. I [...] talk less [with them], but I still like it anyway. [It's more] about managing the silo, marketing... I'm not good at grain marketing [...] so I rely on them.'

4 Managing Heterogeneity, Including Through Withdrawal

The most common strategy for managing intra-group heterogeneity is to increase coordination opportunities and to adjust rules so that the collective organisation can cater to different needs.

In the case of the Aube CUMA, harvesting hemp jointly in all the fields of different farmers requires a lot of labour and different equipment at the same time. This complicates the balancing of individual contributions, a difficulty that is exacerbated by the heterogeneity of the members' situations, especially in terms of availability: some are livestock farmers (and therefore have more routine work), others are cereal growers, and some sell directly on markets, while another is pluriactive. A time bank has been set up to account for the exchange of services, which makes it possible to record each person's individual contributions (time given and received, machinery lent or borrowed, etc.), and thus better balance the services.

These reconfigurations and this heterogeneity also generate tensions. The adoption by some farmers of new practices with specific needs leads to the reconfiguration of the original groups. For example, the two conservation agriculture farmers who created the farmer group in the Tarn CUMA did so following an initial failure. They had earlier been members of a neighbouring CUMA with conservation agriculture equipment. But after a few years of this CUMA's existence, two trends emerged: for some farmers, the CUMA was meant to facilitate no-tillage techniques, without however adopting direct seeding, which they considered risky, for the others (our two farmers), it was meant to help spread and generalise the practice of direct seeding. The diverging equipment needs of each of these two trends led to differences, and the subsequent exit by the two farmers wishing to continue with direct seeding. They thus joined another CUMA from which emerged the current group that was surveyed.

5 Tactics for Cooperating in Reciprocity

5.1 *Example of Interactions with Expert Colleagues*

In addition to the strategies farmers with heterogeneous technical orientations implement to organise themselves in groups, certain tactics are adopted aimed at ensuring an equal status in cooperation, especially bilateral cooperation.

Thus, a pioneer farmer can be considered an expert by his peers, who sometimes seek his advice and ask about his experiences. This sometimes explains the continued involvement of these pioneers in groups of peers with different practices.

Producers develop tactics to learn from the experiences of their expert colleagues at a low symbolic cost. In the Tarn CUMA, the farmer who is an expert in conservation agriculture receives much more labour from his colleagues during the joint operation of silage at his farm than he provides at theirs. Not only do the farmers not complain about the imbalance in exchanges between them and the expert, they maintain a complete silence about it. In this way, they learn from the expert farmer during joint operations on his farm, learning that is embedded within the dynamics of the exchange of labour, whose relative imbalance attenuates the symbolic effect of the recognition of his expertise.

5.2 *Between Organic and Conventional Farming*

Conventional farmers develop tactics to interact beneficially with organic producers.

In Tarn, the members relied heavily on the experience of the pioneer conservation agriculture farmer. He recently progressed to organic farming and has since been cooperating more with another organic farmer who has recently become a member of the CUMA. This development was viewed with apprehension by his conventional colleagues who feared they would no longer be able to rely on him. However, this fear has diminished as complementary interests have gradually become apparent. Indeed, the possibility of a ban on glyphosate, currently being debated in the political sphere, has transformed the experiences of these two organic farmers into potential sources of information should this ban be confirmed. Moreover, this farmer's transition to organic farming has eased the availability of the shared no-till direct seeder, which was previously in great demand during the sowing period. In fact, the two organic farmers sow later in the spring, which means they need the direct seeder later and which provides more flexibility in sharing this critical piece of equipment. This example shows that farmers can live with heterogeneity when they discover the potential for functional complementarities that it can offer.

While conventional farmers have a mostly positive view of organic farming, they rarely undertake technical discussions with geographically close organic farmers, except in the case of Tarn. Knowledge of organic practices is gained through other means: agricultural media, visits to organic farming research stations, development groups, etc. This is due to the symbolic cost of this type of dialogue, in which conventional farmers would risk an asymmetry of information and status by entering into discussions with an organic producer, to whom they fear they would have no useful experience to contribute in return, as illustrated by these words of a farmer from Touraine:

Interviewer: 'Have you had opportunities to enter into discussions with organic farmers [about grassland management]?' Farmer: 'No, and it's unfortunate because it's a technique they've mastered for a long time now, because I'd tell them [what I'm doing]. [And they'd reply]: "You're not teaching us anything new"…'.

Conventional farmers sometimes do take the initiative of entering into bilateral arrangements with organic producers. For example, a sheep farmer in the Aube region requested to have his herd graze on the fields of a nearby grain farmer who had recently converted to organic farming and wanted to include grasslands in his rotation. The sheep farmer advised the organic farmer on the species to choose for his grassland mix. In this arrangement, the former is undoubtedly in a situation of solicitation (asking whether his sheep can graze on his neighbour's fields), but also one of giving by providing advice, which brings a symmetry between him and his colleague.

6 Conclusion

The farmers surveyed showed a preference for cooperation between peers with similar systems. This simplifies coordination and the sharing of pooled resources, and facilitates the reciprocity of exchanges of services and equipment between farms. This type of cooperation also helps determine the 'right action' through fruitful dialogues between peers who share the same professional norms and who are in situations of mutual interdependence, and also notably manages the underlying symbolic issues of reputation. Nevertheless, despite this expressed preference, farmers belonging to a CUMA find it beneficial, especially in terms of functional complementarities and sources of knowledge, to share resources with farmers with dissimilar systems. However, specific means of coordination are required for this arrangement to be successful.

The move towards agroecology is introducing new sources of heterogeneity within the CUMAs, resulting in a gradual restructuring of cooperation processes. This sometimes comes at the cost of disagreements, or even splits within groups, when the new needs (long-term investments, pooling of material resources, etc.) do not suit everyone. Similarly, the social positions between farmers are reconfigured when some of them embrace an agroecological orientation. They have to rely on new tactics for cooperating in a reciprocal manner with their peers. For example, the symbolic cost of asking for advice from a peer who is more knowledgeable in agroecological practices can be mitigated by embedding this request in a reciprocal process that incorporates other services, such as labour exchanges. Tacit silence is also a useful means of managing divergences and maintaining the conditions for technical cooperation.

These results are complementary to those from existing studies on the conditions under which local agroecological innovation can take place amongst heterogeneous farmers, in particular by revealing the symbolic effects involved in peer interactions (Le Guen & Ruault, 1994; Sigwalt et al., 2012). In particular, they invite us to qualify observations that consider heterogeneity to be an inescapable hindrance to cooperation between farmers, while revealing the need for more research for a fuller understanding of the social mechanisms that farmers use to manage heterogeneity and mutual interdependence. While our case studies, based on successful cooperative experiences, ignore the situations of failure, they do show the importance of the role of 'mobilising' farmers, who are shown to have the time available and strategic skills to coordinate collective processes, as well as favourable social positions. Regional comparisons become necessary in order to better identify the social and geographical determinants of these collective dynamics.

Such research orientations would also make it possible to inform sociological debates between those who consider that individual and collective behaviour is now 'de-localized' thanks to modern means of mobility and communication (Giddens, 1990), and those who consider that the latter instead generate a superimposition of networks and places (Castells, 1997). The experiences of farmers in CUMAs are testament to their ability to build new identities with peers with whom they share common values and who are sometimes located at a geographical distance, while also cooperating with different peers in proximity. Furthermore, the mobilisation of ecological processes, which is site-specific, is also likely to generate new affinities between farmers who are a priori different through the continuous reproduction of the cycles of reciprocity that link them and to also produce new social values.

References

Alter, N. (2010). *Donner et prendre: La coopération en entreprise* (p. 238). La Découverte.

Assens, P. (2002). Les compétences professionnelles dans l'innovation: le cas du réseau des coopératives d'utilisation de matériel agricole (CUMA) (Doctoral thesis in economics, University of Social Sciences of Toulouse) 329 p.

Brives, H., & de Tourdonnet, S. (2010). Comment exporter des connaissances locales? Une expérience de recherche-intervention auprès d'un club engagé dans les techniques sans labour. In *International Symposium on "Innovation and Sustainable Development in Agriculture and Food" (ISDA)*, 28 June–1 July 2010, Montpellier, Montpellier SupAgro-Inra-Cirad.

Castells, M., (1997). *The power of identity. The information age: Economy, society, and culture* (Vol. II). Blackwell Publishers.

Chiffoleau, Y. (2005). Learning about innovation through networks: The development of environment-friendly viticulture. *Technovation, 25*(10), 1193–1204.

Cornée, S., Le Guernic, M., & Rousselière, D. (2020). Governing common-property assets : Theory and evidence from agriculture. *Journal of Business Ethics, 166*(4).

Darré, J.-P. (1996). *L'invention des pratiques dans l'agriculture. Vulgarisation et production locale de connaissance.* Karthala, 194 p.

Dedieu, B. (1993). Organisation du travail et fonctionnement d'exploitations d'élevage extensif du Massif central. *Études Et Recherches Sur Les Systèmes Agraires Et Le Développement, 27*, 303–322.

Deffontaines, N. (2014). La souffrance sociale chez les agriculteurs. Quelques jalons pour une compréhension du suicide. *Études Rurales, 193*, 13–24.

Giddens, A. (1990). *The consequences of modernity*. Polity.

Harff, Y., & Lamarche, H. (1998). Le travail en agriculture: Nouvelles demandes, nouveaux enjeux. *Économie Rurale, 244*(1), 3–11.

Hervieu, B., & Purseigle, F. (2013). *Sociologie des mondes agricoles* (p. 320). Armand Colin.

Jeanneaux, P., Capitaine, M., & Mauclair, A. (2018). PerfCuma: A framework to manage the sustainable development of small cooperatives. *International Journal of Agricultural Management, 7*(1), 1–12.

Jollivet, M., & Lepart, J. (1992). Hétérogénéité, diversité, complexité: Nuances et convergences. In M. Jollivet (Ed.), *Sciences de la nature, sciences de la société. Les passeurs de frontières* (pp. 373–380). CNRS Éditions.

Le Guen, R., & Ruault, C. (1994). La double appartenance professionnelle des agriculteurs biologiques. Réseaux de relations et évolution des qualifications. In J.-P. Darré (Ed.), *Pairs et experts en agriculture. Dialogues et production de connaissance pour l'action* (pp. 49–87). Éditions Érès.

Lémery, B. (2003). Les agriculteurs dans la fabrique d'une nouvelle agriculture. *Sociologie Du Travail, 45*(1), 9–25.

Lucas, V. (2018). L'agriculture en commun: gagner en autonomie grâce à la coopération de proximité. Expériences d'agriculteurs français en Cuma à l'ère de l'agroécologie (Doctoral thesis in sociology, University of Angers) 536 p.

Lucas, V., Gasselin, P., & van der Ploeg, J. D. (2019). Local inter-farm cooperation: A hidden potential for the agroecological transition in northern agricultures. *Agroecology and Sustainable Food Systems, 43*(2), 145–179.

Mundler, P., & Laurent, C. (2003). Flexibilité du travail en agriculture: Méthodes d'observation et évolution en cours. *Ruralia, 12–13*, 239–257.

Mundler, P., Guermonprez, B., Jauneau, J.-C., & Pluvinage, J. (2010). Les dimensions territoriales de la restructuration laitière. *Géographie, Économie, Société, 12*(2), 161–180.

Nicourt, C. (2013). *Être agriculteur aujourd'hui. L'individualisation du travail des agriculteurs* (287 p). éditions Quæ.

Nicourt, C., Benoit, M., Laignel, G., & Cabaret, J. (2009). Approches sanitaires comparées d'éleveurs ovins allaitants biologiques et conventionnels. *Innovations Agronomiques, 4*, 49–60.

Pierre, G. (2009). The biodiesel produced by farmers at a local scale using a traditional procedure: What kind of territorial construction for an agro-environmental project in social economy? *European Countryside, 1*(3), 141–152.

Sabourin, E. (2012). *Campesinos, mercados y políticas públicas: una lectura por la reciprocidad*, Lima, Pérou, Fundo Editorial Universidad Antonio Ruiz Montoya, Cirad, Serie Etica y Desarrollo, 518p.

Sigwalt, A., Pain, G., Pancher, A., & Vincent, A. (2012). Collective innovation boosts biodiversity in French vineyards. *Journal of Sustainable Agriculture, 36*(3), 337–352.

Wezel, A., Brives, H., Casagrande, M., Clement, C., Dufour, A., & Vandenbroucke, P. (2016). Agroecology territories: Places for sustainable agricultural and food systems and biodiversity conservation. *Agroecology and Sustainable Food Systems, 40*(2), 132–144.

Chapter 6
Emergence and Compartmentalisation of Advisory Subsystems for the Ecological Intensification of Agriculture in Burkina Faso

Aurélie Toillier, Saydou Bancé, and Guy Faure

In sub-Saharan Africa, the political and development aid spheres are looking for new models of agricultural production that are capable of feeding the population and addressing environmental challenges at the same time (HLPE, 2019). Civil society is also engaged in this search through the voices of producer organisations and NGOs (Coordination Sud, 2019). A new paradigm, adopted by several international research and policy organisations, has gained prominence in their discourses. It calls for the increased mobilisation of natural processes, reflected in the notion of ecological intensification (Caron et al., 2014; Tittonell, 2014). This new paradigm is backed by a large body of scientific literature on new agricultural practices, new ways of organising production and agri-chains, and new consumption habits that are all necessary to produce as much or even more, while reducing the use of synthetic inputs and being eco-efficient (Garnett et al., 2013). In contrast, there is far less research on the implications for the evolution of the agricultural advisory systems that will have to accompany these changes on farms and in territories. The trajectories of change of ecological intensification will necessarily differ depending on country, region or locality (Meynard, 2017; Lucas et al., 2018). An advisory system is understood as a social system that encompasses all the actors involved in the provision of advice and their interrelationships. Today, all actors in their broad diversity, including governments, the private sector and civil society, are considered stakeholders in the agricultural advisory system since these actors 'support and facilitate people engaged in

A. Toillier (✉) · G. Faure
Cirad, UMR Innovation, Montpellier, France
e-mail: aurelie.toillier@cirad.fr

G. Faure
e-mail: guy.faure@cirad.fr

S. Bancé
Programme d'appui à la décentralisation et à la participation citoyenne (Depac), Ouagadougou, Burkina Faso
e-mail: bancsaydou@yahoo.fr

P. Gasselin et al. (eds.), *Coexistence and Confrontation of Agricultural and Food Models*,
https://doi.org/10.1007/978-94-024-2178-1_6

agricultural production to solve problems and obtain information, skills, and technologies to improve their livelihoods and well-being' (Birner et al., 2009). Jayne et al. (2019) emphasise the need to develop adaptive local research and advisory systems, since such changes require incremental and collective learning based on local knowledge.

In practice, an advisory system is the result of agricultural development policy choices and complex social constructions (Faure & Compagnone, 2011). It is both a means of making farms evolve according to orientations defined by policies, markets or certain sectors of society (Davis, 2008), and a means of supporting the complex processes that take place within a broader innovation system involving different categories of actors (Hermans et al., 2015).

Our aim in this chapter is to explore the evolution of relationships between advisory actors with regards to incentives for ecological intensification of agriculture in Burkina Faso. The context is marked by a strong political period of promotion of agroecology, followed by commitments to sustainable intensification of agricultural production by a plurality of private and public actors (Côte et al., 2019). We are interested in particular in the possible emergence of advisory subsystems (Klerkx et al., 2017), i.e. of the multiple advisory systems which can coexist and are aimed at supporting the transformation needs of agriculture in different ways, from the farm to the value chain and the territory. In this perspective, we address the following questions: Who are the different advisory actors promoting ecological intensification in Burkina Faso today? What are their intervention methodologies? What roles do they play within the advisory system?

We first present the context of Burkina Faso, followed by the analytical framework we have developed, which combines structural analyses of networks of advisory actors and analyses of these actors' registers of action. We then present the three advisory subsystems we have identified, before concluding with the political and theoretical implications of these subsystems' existence.

1 Exploring Ways of Supporting Ecological Intensification

In Burkina Faso, as in many other African countries, the partial withdrawal of the State from the domains of agricultural advice and orientation has opened up a space for a multiplicity of actors (producer organisations, NGOs and associations, consultants, international agencies). They are using new methods to provide advisory services, and proposing and advocating alternative, more ecological models of agricultural production, at the fringe of the conventional intensification advocated by the Green Revolution (increased use of synthetic fertilisers, improved seeds and agricultural equipment). What results is a large number of actors and of interventions in support of farmers.

1.1 A Pluralistic and Poorly Coordinated Agricultural Advisory System

From the time of Burkina Faso's independence in 1960 to the early 1990s, agricultural extension followed a top-down, dirigiste approach, mainly focused on cash crops, primarily cotton, in which the producer was a 'supervised' actor who was asked to apply recommendations made to him. The State had a large network of supervisory agents for disseminating technological packages through 'training and visits' and relay farmers.

As in many African countries, the freeze on the recruitment of supervisory staff and the lack of funding for technical services, arising from the structural adjustment programme of the 1990s, weakened and contributed to the dismantling of the Burkina Faso extension and advisory system. Producer organisations, NGOs and other private sector actors reacted by building up their capacities to take over the functions that were earlier the State's prerogative and responsibility. They undertook initiatives and put the producer at the centre of their agricultural advisory mechanisms. Diversified advisory approaches were developed, such as field schools, farm management advice, model farms, as also more collaborative and open approaches such as discussion forums and peer-to-peer exchanges.

At the international level, during the African Union Summit in 2003 in Maputo, Mozambique, the adoption of the Comprehensive Africa Agriculture Development Programme (CAADP), the agricultural component of NEPAD,[1] marked a turning point. Indeed, it was decided to focus on agricultural advisory services, considered as a tool to achieve food security while better addressing farmer needs.

This encouraged Burkina Faso to set up, in 2010, the National Agricultural Extension and Advisory System (SNVACA[2]), whose guiding principle is the empowerment of the various actors (producer organisations, NGOs and associations, consultants, technical and financial partners) involved in the design and implementation of advisory support approaches that meet producer needs. Under SNVACA, producer organisations are seen as the pillars that should guide these approaches, with the State retaining the prerogative of regulating, orienting, steering and monitoring-evaluating extension and advisory services. However, given its limited resources, the State simply encourages the various actors in the agricultural sector to clarify their roles and responsibilities, leaving them relatively wide margins for taking initiatives.

1.2 A Diversity of Alternative Agricultural Models

Despite the Green Revolution, West African agriculture, and Burkina Faso's in particular, remains less productive than those in other regions of the world, with yields

[1] New Partnership for Africa's Development agency.

[2] French: *Système national de vulgarisation et d'appui-conseil agricole (SNVACA).*

increasing more slowly than elsewhere (Ouedraogo et al., 2016). On the margins of the conventional agricultural development model planned by politicians and implemented by the major economic actors in the rural world (public advisory services, agribusiness firms, producer organisations), alternative agricultural models based on ecological principles have developed over the past 30 years: organic farming (Toillier & de Lapeyre de Bellaire, 2017; de Bon et al., 2018), conservation agriculture (Dugué et al., 2015) and agroecological farming (Temple & Compaoré Sawadogo, 2018). These more ecological production models are anchored in different institutional processes, through markets, through the governance of resources and territories, or even through policies as was the case with agroecology during Thomas Sankara's presidency of the country in the 1980s. These models are not necessarily geared towards intensification, but cross-fertilisation between them in pursuit of sustainable intensification has been observed in various Burkinabe regions. Sustainable intensification is characterised by conventional intensification, combined with agroecological intensification strategies based on agricultural techniques borrowed from production models, such as the combination of cultivation and livestock husbandry, and the maintenance of trees in fields, as described by Vall et al. (2017) in mixed crop-livestock systems in western Burkina Faso.

These different dynamics of the parallel evolution of advisory systems and agricultural production systems have resulted in a great diversity of actors involved in supporting different models of the ecologisation of agriculture, mobilising various advisory support mechanisms that are not necessarily known and recorded by the State.

1.3 An Approach Based on the Networks of Actors Involved in Advisory Support and Their Registers of Action

To be able to characterise advisory systems supporting ecological intensification, we sought to identify the various types of actors who offer advisory services for agricultural models other than those of conventional intensification, their roles within the advisory system and their registers of action.

1.3.1 Registers of Action

The widening of the ambit of agricultural advisory services from guidance and supervision to accompaniment is reflected in a diversity of advisory approaches (Faure et al., 2018): decision-making support, problem solving, capacity building aimed at empowering farmers, or accompanying an individual or collective project. On the basis of the professional practices identified in the accompaniment sector (Paul, 2004), we propose to group these actions under three main registers: guidance, intervention, and incentive. Guidance refers to the co-construction of a project with and for the person or entity concerned. Intervention is initiated in response to a problem in order to solve it, usually with a solution found by people other than those affected. Incentive leaves the choice to those concerned of whether or not to apply the suggested changes.

1.3.2 Networks of Actors

In a context of pluralism and liberalisation, the provision of agricultural advisory services mobilises a range of actors who play different roles (Birner et al., 2009): funding of advisory services (Compagnone et al., 2015), governance of the system as a whole, identification of advisory support needs of final beneficiaries, design of innovative advisory approaches, creation of content suitable for illiterate populations, networking of advisory actors, intermediation between providers and clients (Klerkx et al., 2012), advisory service delivery in villages, training of agricultural advisors, etc. An analysis based on actor networks helps to understand how this collective action is organised by visualising the position of the different organisations within the network (Borgatti et al., 2009) as well as the nature of their roles (funding, governance, training, transfer of techniques and knowledge, co-production of solutions).

1.3.3 Data Sampling and Collection

Using a documentary search (websites, grey literature, brochures, activity reports), we built a sample of about 30 advisory service providers that seemed to play an important role (heads of networks, size of the structure and of intervention areas, reputation) in new models of agricultural production possibly linked to various forms of ecological intensification (sustainable intensification, agroecology, conservation agriculture or organic farming).

Interviews with operations managers allowed us to establish how these service providers justify their actions, design their offers, take the needs of producers into account, and interact with other organisations in the advisory system. The areas of intervention were also identified for each type of advisory service recorded. Specific interviews with beneficiaries of advisory services (producer organisations and farmers) made it possible to clarify the way in which the changes proposed by advisory actors are understood and interpreted.

2 Three Advisory Subsystems with Distinct Registers and Areas of Action

We have identified three types of advisory subsystems (ASSs) that differ in their registers of action: the first aims to solve the problems of sustainable intensification of conventional farming (ASS-CF); the second aims to encourage conversion to organic farming (ASS-OF); and the third aims to raise awareness of and provide training in agroecology (ASS-AE).

2.1 Registers of Action

The advisory services implemented within the different ASSs pertain to different registers of action: transferring techniques, solving problems in a participatory manner, educating to build up overall capacities.

The ASS-CF mainly mobilises approaches that enable technology transfers and/or problem solving (integrated soil fertility management, rational management of inputs), but does not really look at the issues causing the problems and generally does not undertake an evaluation at the end of the projects or recommend reorientation of actions. The ASS-OF and ASS-AE both rely on training and the use of model farms for teaching agroecological practices and integrated management of an overall farming system (management of spatial and temporal interrelations between cropping systems, livestock systems and fallow land, which cannot be taught via field schools). However, field schools are used extensively for teaching certain plot-level agroecological techniques (soil preparation, management of crop associations, fertiliser distribution). The ASS-AE tends to mobilise approaches based on exchanges of experience, peer-to-peer learning and action-research platforms. The intention here is to take advantage of the capacity of individuals and of local knowledge. The actors did not really identify contributions in terms of production of new useful knowledge, but it does not mean that this new knowledge is not produced.

2.2 Areas of Intervention

The geographical location of activities within each ASS is strongly correlated to different agroecological regions in the country and to the registers of action, yet this is rarely reflected in the actors' discourses.

The discourses on agroecology mainly concern the Sahelian context. Thus, activities of ASS-AE concern only the country's north-central region, where commercial agriculture is not very developed and access to production factors is limited, and the area around Ouagadougou, where most of the ASS-AE actors are based. This geographical localisation is also a legacy of the activities of Pierre Rabhi, who set up the first agroecological centre in the Sahel, at Gorom-Gorom, and thus laid the foundations of technical, social, cultural and economic references around agroecology for the Burkinabe context. It is on this basis that associations such as the Association for the Extension and Support of Agroecological Producers in the Sahel

(AVAPAS[3]) and the Association for Sustainable Resource Management (AGED[4]) have continued to promote agroecology. Their ambit of activities does not encompass the country's southern region, where they would no doubt have an important role to play. However, although there is no mention of agroecology in the cotton basin, other agricultural models such as organic farming or conservation agriculture, which promote the same practices (with the exception being the use of GMOs), are being tested. In this cotton zone, there is a lower overall presence of development aid associations. Pockets of development of organic farming mainly correspond to areas in which the production chains are well organised (cotton, fruit and vegetables) around large cities (Ouagadougou, Bobo-Dioulasso, Fada N'Gourma).

3 Interconnected Actor Networks

3.1 The Advisory System for Solving Problems of Sustainability in Conventional Agriculture

Support for the sustainable intensification of conventional agriculture follows the State's directions and vision, which are essentially to 'produce more, diversify, improve access to inputs and sell the products' (Government of Burkina Faso, 2011[5]). Environmental concerns are subordinate to these objectives. Achieving these objectives involves the application of research results and technical developments to find solutions to the problems of soil fertility and access to water that the majority of production systems in Burkina Faso face, solutions that have been validated by the State through its Ministry of Agriculture. Advisory services are built around the following aims: reducing the risks of pests and pesticides, adopting good agricultural practices, practising integrated pest management, encouraging the use of personal protective equipment, and producing transgenic cotton (*Bacillus thuringiensis*, or Bt), as also organic and fair-trade cotton meeting international standards to obtain better market value for Burkinabe cotton.

Organic farming therefore finds a place in this advisory system, as it is seen as a form of diversification and intensification, and provides access to international markets. Indeed, in cotton-based systems, organic farming allows 'cotton cultivation by those who do not have the capital to adopt the conventional system' (National Union of Cotton Producers of Burkina Faso, UNPCB[6]). It is mainly women who undertake organic farming of cotton, with a very low productivity since they were left the most degraded lands. But organic cotton cultivation enables the adoption of

[3] French: *Association pour la vulgarisation et l'appui aux producteurs agroécologistes du Sahel (AVAPAS).*

[4] French: *Association pour la gestion durable des ressources (AGED).*

[5] National Rural Sector Programme (PNSR) 2011–2015.

[6] French: *Union nationale des producteurs de coton du Burkina Faso (UNPCB).*

rotation systems based on sesame, soya and shea production, which become organic products from the organic cotton fields and for which there already exists a market. Organic cotton thus meets both the challenges of ecological intensification and the State's objectives (diversify, intensify, sell).

The actors guiding this advisory subsystem are the State, along with the processing and inputs industries, some producer organisations (such as UNPCB) and public research organisations, mainly the National Institute for Agricultural Research (INERA[7]) and the Institute for Research in Applied Sciences and Technologies (IRSAT[8]). Both the latter are public entities; private research entities do not exist in Burkina Faso. All these actors have been collaborating for many years (Fig. 1).

Producer organisations are technical partners in the provision of advisory services through their agricultural advisors. In this ASS, actors who promote AE or OF (such as AVAPAS or *Centre Écologique Albert Schweitzer*, CEAS) are the ones primarily involved, but only as trainers in more ecological practices. It is interesting to note that the entities that are promoting OF are expressing a growing interest in the results of agroecological experiments, but no formal links exist at this time. Playing a secondary role are a dozen development NGOs such as SOS Sahel, Ocades Caritas Burkina, *Office de développement des églises évangéliques* (ODE) and *Terre Verte*, which provide ad hoc support in the case of multi-donor programmes and which also intervene in the other advisory subsystems as agricultural advisors.

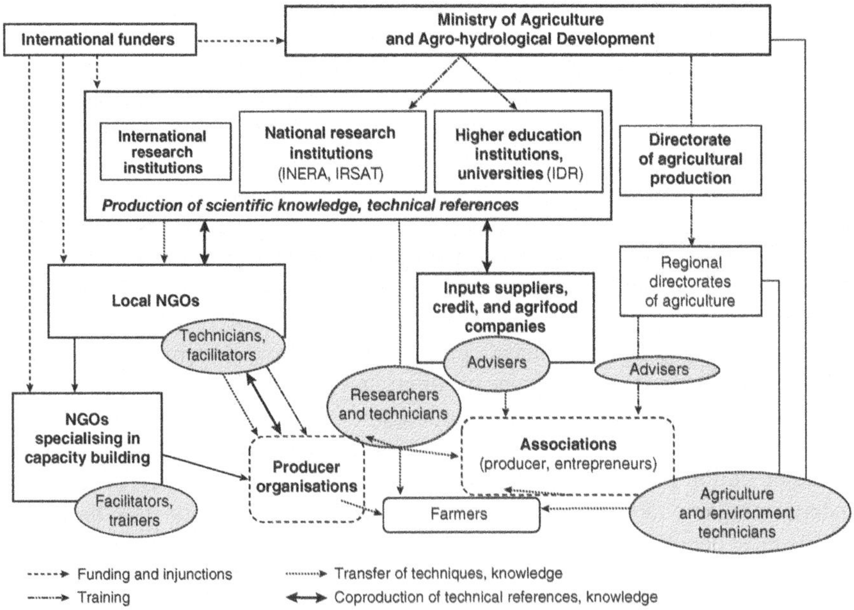

Fig. 1 Actors and interactions within ASS-CF. IDR: Institut de développement rural

[7] French: *Institut national de recherche agronomique (INERA)*.

[8] French: *Institut de recherche en sciences appliquées et technologies (IRSAT)*.

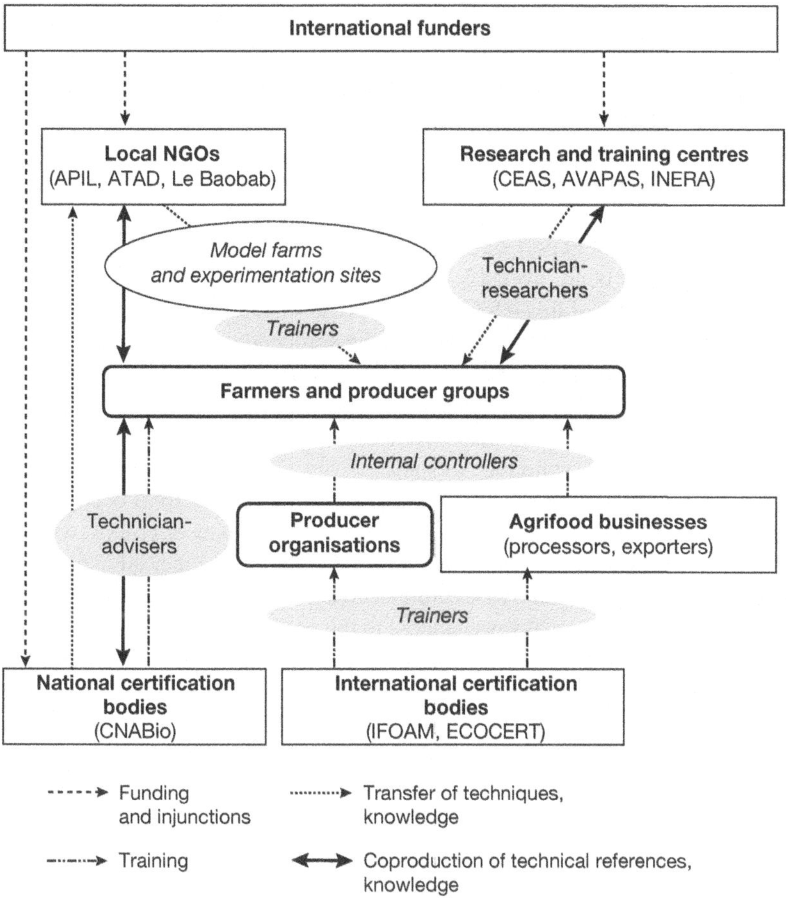

Fig. 2 Actors and interactions within ASS-OF. *APIL* Action pour la Promotion des Initiatives Locales; *ATAD* Alliance Technique d'Assistance au Développement; *CEAS* Centre Écologique Albert Schweitzer; *IFOAM* International Federation of Organic Agriculture Movements

3.2 The Advisory System for the Development of Organic Farming

This advisory subsystem (Fig. 2) aims on the one hand to develop organic farming for local markets, eschewing the export sector, in order to market 'healthy and quality products at a reasonable cost', according to CNABio,[9] address 'the uncontrolled use of chemicals that endanger consumer health', according to AVAPAS, and meet the needs of urban populations. On the other hand, it aims to respond to market incentives for organic products in countries of the Global North, as encouraged by international

[9] French: *Conseil National de l'Agriculture Biologique (CNABio).*

development organisations such as Helvetas or the International Development and Research Centre (IDRC), which mainly support producer organisations. The latter are then responsible for the entire process themselves (production training, collection, sales, internal controls).

The challenges for these actors are to develop a legislative and regulatory framework (monitoring, certification, specifications) through the adaptation of national specifications to the constraints and practices of local production, which will however still meet international export standards (Europe or sub-region), in order to ensure that certified products are available at a reasonable cost to the people of Burkina Faso. The main actors of this ASS, i.e. local NGOs and associations, pushed for the creation of CNABio in association with all the agroecology actors and by involving the Ministry of Agriculture. These actions are being supported by foreign donors (the *Agronomes et Vétérinaires Sans Frontières* association, the European Union, the *Action Solidarité Tiers-Monde* NGO). Knowledge and technical references are produced in conjunction with the ASS-AE.

Private companies involved in collection, processing and export occupy a minor place in this subsystem. They work mainly with ad hoc producer groups, which they train and certify collectively. They maintain few links with other advisory services (public or private) and national agricultural research entities. As a result, the technical support they provide is not always adapted to the production context. There is a lack of effective alternatives to chemical plant protection products. Producers are still not convinced of the effectiveness of biopesticides, especially for fruit and vegetable crops, which are prone to attack by a very high number of pests. Furthermore, the advisory actors of this subsystem are unable to respond to the constraints linked to the emergence and development of several GMO crops (Bt cotton, Bt cowpeas and Biofort sorghum) which limit the deployment of OF in these territories. They do not have the flexibility to offer advisory services geared towards consultation between production agri-chains, which would allow GMO and organic crops to coexist in the same territory.

3.3 The Advisory System for Awareness Raising of and Training in Agroecology

Although President Sankara introduced agroecology in a revolutionary way, it was the subsequent intervention of donors that led to the experimentation and development of more ecological production models. For a long time, however, these initiatives remained on the fringes of the dominant model of the Green Revolution. NGOs, associations and small producer organisations have nevertheless been able to create networks to accompany these changes. While these actors consider agroecology in all its three dimensions—technical, socio-economic and cultural, and socio-political—, their main activities consist of participatory design of new production systems based on agroecological principles, and the production and dissemination of technical references. The limited access to production factors (biomass, equipment) makes it necessary to define a set of practices that are similar to the already known 'good agricultural practices' (composting, water and soil conservation, rotations, agroforestry), with which indeed there exists consensus with the other ASSs. The emphasis is thus on empowerment and capacity building of farmers and on the fight against GMOs with the promotion of local seeds.

The actors managing this system (Fig. 3) are international and national NGOs often with ties to religious groups (*Terre et Humanisme*, Global Neighbours, Groundswell, Christian Aid), as well as international research organisations for the production of technical references and, to a lesser extent, local research entities. Some NGOs with ties to international research networks specialise in the production of technical references, such as ACT (African Conservation Tillage) for conservation agriculture. Consumer networks and public services are conspicuous by their absence, even though the objectives of the leading NGOs are to 'prepare a new model of society' according to AVAPAS. Close ties have been established with applied research entities through development programmes. Most often, the aim is to make technologies available that are adapted to agroecological practices, as is the case with CEAS, whose objective is to 'develop appropriate technologies for agroecology and environmental protection' (kassines, beehives, natural insecticides). Local producer organisations are essentially intermediaries in facilitating communication with and training of farmers, and support them in conducting experiments. There are no links with farmer unions or federations at the national level.

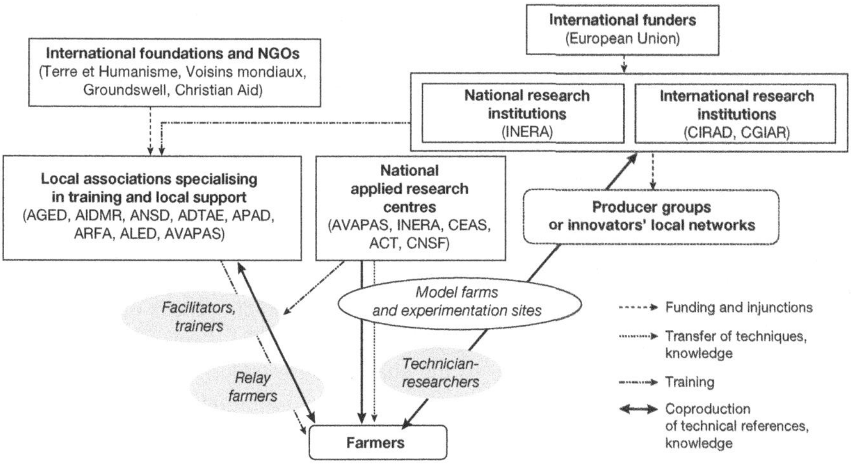

Fig. 3 Actors and interactions within ASS-AE. *CGIAR* Consultative Group on International Agricultural Research; *CIRAD* French agricultural research and international cooperation organisation; *AIDMR* Association Interzone pour le Développement en Milieu Rural; *ANSD* Association Nourrir Sans Détruire; *ADTAE* Association pour le Développement des Techniques Agroécologiques; *APAD* Association pour la Promotion d'une Agriculture Durable; *ARFA* Association pour la Recherche et la Formation en Agroécologie; *ALED* Association Les Enfants de Demain; *CNSF* Centre National de Semences Forestières

4 Political and Theoretical Implications

The different ways identified in which ASSs can support ecological intensification has both political and theoretical implications.

4.1 The Compartmentalisation of Advisory Subsystems: Obstacles to Ecological Intensification

Our results show that the ASSs are relatively compartmentalised due to the type of the actors they involve and by their geographical areas of operations within Burkina Faso. These compartmentalised configurations may be the root cause of hold-ups in innovation in the domain of ecological intensification.

The extension service providers of the ASSs are mainly from the public sector in the case of ASS-CF, and from the private and associative sectors in the case of ASS-AE and ASS-OF. The different actions of the ASSs are, moreover, subordinated to both ecological contexts and spatial logics of intervention that stem from historical, logistical, political or economic reasons, specific to the various networks of actors, and which are disconnected from the farmers' actual advisory and support needs. The regions concerned have already been the subject of numerous diagnoses which have ascertained the problems to be addressed and the actions to be taken (combating desertification, famine, soil degradation, adaptation to climate change) in the context of arguments that have been accepted for decades (lack of soil conservation techniques, overgrazing, overpopulation, lack of means of production).

The absence of spaces and time for re-examining and debating these old consensuses imparts a kind of rigidity to the different ASSs, whose actors are often caught up in the urgency of project implementations and in dealing with the lack of means. It is difficult for them to initiate and justify actions that are very different from those they have conducted up to now. For example, several projects to improve soil fertility have been undertaken in the past three decades, all promoting the same practices (animal manure and manure pits) even though there is no consensus on the origins of the problem and how to address it (Vall et al., 2017). The compartmentalisation of actor networks is not conducive to the sharing of knowledge and experiences, which is nevertheless an important factor in assigning a common sense to the actions undertaken and in supporting change in a given territory.

In this divided landscape, there are no links between unsolved problems, innovators, advisory, research and training systems, and political will. In this sense, both the rigidity of advisory systems and their propensity to embrace turnkey technical packages act as a brake on innovation in the search for original and territorialised forms of ecological intensification.

In the French context, Labarthe (2010) also shows how an advisory system can be subject to 'lock in' by its inclusion in institutionalised power relations, which prevent the construction of shared knowledge bases. As a result, certain agricultural production practices persist, even if they are not the most effective in preserving the environment.

It is therefore incumbent upon political actors to monitor the emergence and functioning of these subsystems and to become active participants in some of them, in accordance with the government's role as regulator of SNVACA, in order not only to mitigate the shortcomings of the subsystems, but also to take advantage of their complementarities. This observation about the State's essential role in such configurations has also been made in the context of the privatisation of advisory services in Europe (Klerkx et al., 2006).

4.2 Reconsidering the Boundaries of Advisory Systems

This analysis of actor networks shows that the governance, knowledge and financing structures of the different ASSs are rooted outside the territories in which they operate and even outside Burkina Faso's borders. The notion of subsystems, which, as defined by Klerkx et al. (2017), suggests a subnational level, should instead refer to extra-national advisory systems that operate in Burkina Faso.

Various authors have begun to highlight the importance of international linkages between regional and national advisory and innovation systems (Carlsson, 2006; Grillitsch & Trippl, 2013). They show, among other aspects, that a system's performance in the development and dissemination of innovations depends not only on the existence of coherent subsystems, but also on the possibility of structural coupling between them. This structural coupling takes place if specific actors, actor networks or institutions transverse or overlap various subsystems in a specific region or country, for example in a global NGO or a multinational corporation (Binz & Truffer, 2017).

These advances lead us to propose a deeper exploration of how the various paths to ecological intensification coexist even within the organisations that promote them. This will help us better understand the origin of the divides between ASSs. While we have emphasised geographical, technical and institutional divides, they can also be political or financial. Goulet (2019) shows how support for family farming in Argentina by the research and development system has become an alternative to the extension entities of public institutions promoting the Green Revolution.

5 Conclusion

There is little research being conducted on determining the specific configurations of an advisory system at the country scale, potentially involving the coexistence of advisory subsystems, each of which supports a different path to ecological intensification. The system of actors involved in the provision of advisory services for the ecological intensification of agriculture in Burkina Faso is complex and diversified. It is complex because the historical perspective of interventions and the political and economic issues play an important role in defining the objectives and modalities of action. Diversified because, in addition to the public system, there is a large number of national and international NGOs, producer organisations with widely varying capacities, and private companies, which are also expanding their activities rapidly as a result of the State's investment promotion programmes. Moreover, the term 'ecological intensification' does not have a common meaning and encompasses a diversity of agricultural development methods that differ based on geographical location and which are followed and advocated by different, relatively compartmentalised subsystems of actors. To support ecological intensification, we can, most importantly, position ourselves to help the various actors already involved in these subsystems in order not only to strengthen their capacities to guide, advise and support, but also to facilitate the production and exchange of knowledge between them. Forms of coordination at the national level involving political actors must also be pursued. It is necessary to overcome certain historical and geographical divides between the organisations involved, which, due to these organisations' limited room for manoeuvre to change production contexts, ultimately contribute to holding back innovation on farms.

References

Binz, C., & Truffer, B. (2017). Global innovation systems. A conceptual framework for innovation dynamics in transnational contexts. *Research Policy, 46*(7), 1284–1298.

Birner, R., Davis, K., Pender, J., Nkonya, E., Anandajayasekeram, P., Ekboir, J., Mbabu, A., Spielman, D., Horna, D., Benin, S., & Cohen, M. (2009). From best practice to best fit: A framework for designing and analyzing pluralistic agricultural advisory services worldwide. *The Journal of Agricultural Education and Extension, 15*(4), 341–355.

Bon H. (de), Temple, L., Malézieux, E., Bendjebbar, P., Fouilleux, E. & Silvie, P. (2018). L'agriculture biologique en Afrique: un levier d'innovations pour le développement agricole. *Perspective, 48*, 1–4.

Borgatti, S. P., Mehra, A., Brass, D. J., & Labianca, G. (2009). Network analysis in the social sciences. *Science, 323*(5916), 892–895.

Carlsson, B. (2006). Internationalization of innovation systems: A survey of the literature. *Research Policy, 35*(1), 56–67.

Caron, P., Biénabe, E., & Hainzelin, E. (2014). Making transition towards ecological intensification of agriculture a reality: The gaps in and the role of scientific knowledge. *Current Opinion in Environmental Sustainability, 8*, 44–52.

Compagnone, C., Goulet, F., & Labarthe, P. (2015). *Conseil privé en agriculture: acteurs, pratiques et marché* (252 p). Quæ/Educagri.

Coordination Sud. (2019). Quelles politiques publiques pour soutenir la transition agroécologique? *Les Notes de Sud, 19*, 4 p.

Côte, F.-X., Rapidel, B., Sourisseau, J.-M., Affholder, F., Caron, P., Deguine, J.-P., Faure G., Hainzelin E., Malézieux E., Poirier-Magona E., Roudier P., Scopel E., Tixier P., Toillier A., & Perret S. (2019). Agroecological transition of agriculture in the countries of the Global South: taking stock and perspectives. In F.-X. Côte, E. Poirier-Magona, S. Perret, P. Roudier, B. Rapidel, M.-C. Thirion (Eds.), *The agroecological transition of agricultural systems in the global south* (pp. 327–349). Quæ.

Davis, K. (2008). Extension in subsaharan Africa: Overview and assessment of past and current models and future prospects. *Journal of International Agricultural and Extension Education, 15*(3), 15–28.

Dugué, P., Djamen, N. P., Faure, G., & Le Gal, P. Y. (2015). Dynamiques d'adoption de l'agriculture de conservation dans les exploitations familiales: De la technique aux processus d'innovation. *Cahiers Agricultures, 24*(2), 60–68.

Faure, G., & Compagnone, C. (2011). Les transformations du conseil face à une nouvelle agriculture. *Cahiers Agricultures, 20*(5), 321–326.

Faure, G., Toillier, A., Havard, M., Rebuffel, P., Moumouni, I., Gasselin, P., & Tallon, H. (2018). Advice to farms to facilitate innovation: between supervision and support (chapter 11). In G. Faure, Y. Chiffoleau, F. Goulet, L. Temple & J.-M. Touzard (Eds.), *Innovation and development in agricultural and food systems*. Quæ.

Garnett, T., Appleby, M. C., Balmford, A., Bateman, I. J., Benton, T. G., Bloomer, P., Burlingame, B., Dawkins, M., Dolan, L., Fraser, D., Herrero, M., Hoffman, I., Smith, P., Thornton, P. K., Toulmin, C., Vermeulen, S. J., & Godfray, H. C. J. (2013). Sustainable intensification in agriculture: Premises and policies. *Science, 341*(6141), 33–34.

Goulet, F. (2019). *Faire science à part* (p. 264). Presses universitaires de Liège.

Grillitsch, M., & Trippl, M. (2013). Combining knowledge from different sources, channels and geographical scales. *European Planning Studies, 22*(11), 2305–2325.

Hermans, F., Klerkx, L., & Roep, D. (2015). Structural conditions for collaboration and learning in innovation networks: Using an innovation system performance lens to analyse agricultural knowledge systems. *The Journal of Agricultural Education and Extension, 21*(1), 35–54.

HLPE. (2019). *Agroecological and other innovative approaches for sustainable agriculture and food systems that enhance food security and nutrition*. Report by The High Level Panel of Experts on Food Security and Nutrition. HLPE report 14, FAO, Rome, 162 p.

Jayne, T. S., Snapp, S., Place, F., & Sitko, N. (2019). Sustainable agricultural intensification in an era of rural transformation in Africa. *Global Food Security, 20*, 105–113.

Klerkx, L., De Grip, K., & Leeuwis, C. (2006). Hands off but strings attached: The contradictions of policy-induced demand-driven agricultural extension. *Agriculture and Human Values, 23*(2), 189–204.

Klerkx, L., Schut, M., Leeuwis, C., & Kilelu, C. (2012). Advances in knowledge brokering in the agriculture sector: Towards innovation system facilitation. *IDS Bulletin, 43*(5), 53–60.

Klerkx, L., Stræte, P., Kvam, E., Ystad, G.-T., Butli, E., & Hårstad, R. M. (2017). Achieving best-fit configurations through advisory subsystems in AKIS: Case studies of advisory service provisioning for diverse types of farmers in Norway. *The Journal of Agricultural Education and Extension, 23*, 213–229.

Labarthe, P. (2010). Services immatériels et verrouillage technologique. Le cas du conseil technique aux agriculteurs. *Économies et Sociétés, 44*(2), 173–96.

Lucas, V., Gasselin, P., & van der Ploeg, J. D. (2018). Local inter-farm cooperation: A hidden potential for the agroecological transition in northern agricultures. *Agroecology and Sustainable Food Systems, 43*(2), 145–179.

Meynard, J.-M. (2017). L'agroécologie, un nouveau rapport aux savoirs et à l'innovation. *OCL, 24*(3), D303.

Ouedraogo, S., Vall, E., Bandagao, A.A., Blanchard, M., Ba, A., Dabire, D., & Saba, F. (2016). *Sustainable intensification of mixed farming systems in sub-humid savannah of Western Africa in relation to local value chains (maize, cattle, small ruminants, cotton...)*. PROIntensAFrica. In Depth Case study Final Report, Inra-Cirad, Bobo-Dioulasso, 57 p.

Paul, M. (2004). *L'accompagnement: Une posture professionnelle spécifique* (p. 352). Éditions L'Harmattan.

Temple, L., & Compaore Sawadogo, E.M.F.W. (Eds.), (2018). *Innovation Processes in agro-ecological transitions in developing countries* (187 p). Iste-Wiley.

Tittonell, P. (2014). Ecological intensification of agriculture: Sustainable by nature. *Current Opinion in Environmental Sustainability, 8*, 53–61.

Toillier, A., & de Lapeyre de Bellaire, L. (2017). Contribution of research to innovation within agri-chains. In E. Biénabe, A. Rival & D. Loeillet (Eds.), *Sustainable development and tropical agri-chains* (pp. 93–105). Springer.

Vall, E., Marre-Cast, L., & Kamgang, H. J. (2017). Chemins d'intensification et durabilité des exploitations de polyculture-élevage en Afrique subsaharienne: Contribution de l'association agriculture-élevage. *Cahiers Agricultures, 26*(2), e25006.

Part III
Adaptation

Adaptation: Necessity and Project of Coexistence. Introduction to Part III

Sylvie Lardon

The coexistence of and confrontations between agricultural and food models in territories compel the forms of territorial organisation to adapt to continuous processes or discontinuous changes, thus contributing to a transformation of the world (Caron et al., 2017). It is therefore important to understand what persists, what disappears and what emerges. Actors in the field find various solutions to ensure the continuation of their activities, despite the situations of uncertainty in which they find themselves. New forms of organisation emerge, linking actors, activities and spaces (Lardon, 2012) and responding to challenges of territorial development (Deffontaines et al., 2001).

What are these forms of organisation? How do they differentiate themselves in order to adapt to territorial dynamics? What is these forms' adaptive potential? How do they contribute to the sustainability of agricultural and food production systems? Under what conditions do adaptations constitute levers of territorial action? How are the scales linked, from the local to the global? How can we support these processes of territorial development? These are the questions that we will attempt to answer in this third part of the book.

We will first cover the diversity of territorial organisational forms observed. We will do so through a literature review that will lead us to formulate some research hypotheses on the adaptation modalities that are implemented. Next, a quick 'scan' of the case studies presented in this part and the previous part on innovation will make it possible to identify the key elements of the adaptation processes, analysed through the prism of the coexistence of agricultural and food models. Finally, we will draw some lessons on how actors organise themselves, implement their activities and link their spaces of action, contributing in this way to transformation of territories.

Transformation of Organisations to Adapt to Territorial Challenges

The transformations of recent decades have exacerbated perceptions of uncertainty in the face of increasing variability and unpredictability (Grossetti, 2004), whether

P. Gasselin et al. (eds.), *Coexistence and Confrontation of Agricultural and Food Models*,
https://doi.org/10.1007/978-94-024-2178-1

in terms of climatic conditions, market behaviour or the lack of stability of public policies. Our contemporary history is marked by numerous sudden, major and unpredictable crises, which put people, their activities and their social and environmental structures to the test (Chalas et al., 2009). These crises can concern food, health, finance, politics, civil and military nuclear use, climate, etc. Even though some of them are global, able to affect the entire planet, they can also have differentiated territorial expressions. Under these conditions, the capacity of farmers and territories to adapt is as much a necessity as a project (Gasselin et al., 2013).

It is a necessity because the finitude of exhaustible resources, especially biophysical ones, brings us back to the 'awareness of a finite world', in the sense of a closed space, controlled by humankind (Reghezza-Zitt, 2015). The main studies on sustainable development consider that the aim should be to combine environmental, socio-economic and territorial performances (in the sense of results of actions) or capacities (in the sense of means of action) of the system under consideration (the company, the city, the territory, etc.) in a moral obligation towards future generations (Godard, 2005; Villalba, 2017). And yet, this body of scientific work has struggled to incorporate the capacities of adaptation of the systems under study, resituated in their past dynamics and in the face of uncertainties.

Adaptation is also a project, because although it is certain that major demographic, climatic, energy and environmental changes will take place, they cannot be predicted with any precision. Changes in the future imply that the system must be able to adapt, whether it be to maintain its coherence, to reconfigure it, to learn, to absorb shocks or to create new opportunities. The ability of a system to maintain itself in an uncertain context, and thus to endure over time, necessarily implies a dimension of adaptability (Ancey et al., 2013). However, this dimension remains poorly understood in evaluations of the sustainability of farms, activity systems and territories, even though some authors have focused on it (Zaccai & Zuindeau, 2010; Vigne et al., 2017).

Therefore, adaptation is no longer just a question of 'coping with' hazards, but also, and at the same time, of 'acting on' our practices and our societies. The modalities of adaptation implemented are thus real choices between development models (Thérond et al., 2017). Given these conditions, any analysis of the adaptation of agricultural and food systems implies considering it from two angles. On the one hand, adaptation is a continuous process, but one in which the evolutionary trajectories must be studied and the actors' room for manoeuvre must be identified in order to influence the dynamics from a territorial development perspective (Deffontaines et al., 2001). On the other hand, it is a property, an adaptive capacity of forms of organisation, which must be inscribed in duration (time) and in extent (space) in order to understand the choices made and the modalities of action. This double tension, continuous–discontinuous, uncertainty-choice, is characteristic of this challenge of adaptation.

The first path to adaptation is the diversity of strategies of actors and territories. At the farm level, Darnhofer et al. (2010) highlight three types of strategy to develop a farm's capacity to adapt: experimentation and continuous analysis of results, flexibility in the organisation of activities, and diversification to spread risks. The ability to seize new opportunities to reconfigure a system is based on the revitalisation of diversity, the quest for flexibility, and the development of a learning capacity, mobilising past experiences to inform future decision-making (Dedieu & Ingrand, 2010). This evolution requires the capitalisation of existing know-how as well as of adaptation levers and strategies and calls for support approaches to clarify the prerequisites for change and the tools that can be mobilised (Rigolot et al., 2019).

For food systems, Lamine (2015) highlights the diversity of actors and institutions that reconnect agriculture, food and the environment in more resilient alternative trajectories. At the landscape level, Pinto-Correia and Godinho (2013) use the example of Montado in Portugal to show that by combining production, consumption and protection, land managers contribute to the multifunctionality of landscapes and the resilience of traditional farms by integrating newcomers and new forms of organisation (development of digital technology, remote urban markets, etc.). As for rural territories, Torre and Wallet (2016) envisage three types of situations to respond to territorial challenges: experimentation and exploration of forms of organisation based on local involvement and new technologies, differentiation based on the leveraging of local resources, and the development of integrated projects and of complementarity with other territories in pursuit of transversal cooperation.

The search for these diversities leads to a classic path of adaptation, that of differentiation. This is the case of nested markets that are developing away from the mass markets controlled by the multinationals. They reflect 'the concrete possibilities to counter distance with proximity, artifice with freshness, anonymity with identity and genuineness, standardisation with diversity, and inequality with fairness' (van der Ploeg et al., 2012). In this way, actors transform the conditions of competition by developing voluntary standards and contractual arrangements, for example in organic farming, short supply chains in countries of the Global North, and fair trade. Another route to adaptation is hybridisation. We consider it as a process of creating a new form of organisation by combining various elements inherited from different types of previous organisations. Hybridisation is therefore not only a factor of adaptation but also the result of a process of adaptation.

This state of knowledge on adaptation in situations of coexistence of agricultural and food models in territories reveals several favourable processes and strategies (diversification, differentiation and hybridisation). However, existing studies rarely raise the issue of scale. This leads us to formulate two hypotheses, which have not been widely investigated in the literature, and to ask related questions.

Hypothesis 1: *The coexistence of agricultural and food models confers capacities of adaptation on territories and the systems that constitute them because of their interactions.* These interactions (cooperation, competition, hybridisation, etc.) are only favourable for adaptation under certain conditions. So what are the interactions between agricultural and food models? How are they transformed? Under what conditions does the coexistence of agricultural and food models promote the sustainable and resilient transformation of territories?

Hypothesis 2: *Capacities of adaptation operate at different spatial and temporal scales.* There is a conjunction of issues at different scales (from the local to the global) and combinations of actors, activities and spaces at different levels (Lardon, 2012). What are the socio-spatial configurations of these adaptive processes? How are they articulated at local and global scales? Finally, in order to promote desired trajectories of adaptation, how can public policies and territorial actors support these processes?

Different Forms of Adaptation Observed in Territories

Agricultural and food activities are transforming, and new development models are emerging (Albaladejo, 2009), calling for renewed forms of adaptation in territories at different organisational levels. The case studies presented in the various parts of this book already provide some insights.

In Part II, which is devoted to innovation, Virginie Baritaux and Marie Houdart report on the evolutionary trajectories of agri-chains (Chap. 4). The case study concerns supermarket chain Carrefour's 'Quality Commitment' agri-chain, which involves some twenty farmers, a dairy and a supermaket retailer (Carrefour itself) in the Livradois-Forez Regional Natural Park in the Auvergne-Rhône-Alpes region in France (Baritaux & Houdart, 2015). The agri-chain produces two raw-milk PDO cheeses (Bleu d'Auvergne and Fourme d'Ambert) from 'all hay' elements, marketed in supermarkets under the Carrefour 'Quality Commitment' brand. For the livestock farmers, the hybridisation manifests in the coexistence of production practices that are more in line with an alternative agricultural production system (organic production, no more silage, only grass or hay as animal feed, barn drying) and distribution methods that are in line with the industrial model. At the level of the dairy, we find forms of hybridisation in the methods of deriving value from local products, with PDO products on the one hand, and standard products sold under its own brand or those of other retailers, on the other. There is therefore also a hybridisation of marketing channels. These forms of hybridisation correspond to the actors' adaptation strategies that meet economic objectives and which ensure the continuation of agrifood activity. For the milk producers, it is a means of obtaining a higher price for their milk and of putting their farms on a firmer economic footing. For the dairy, it is a way of diversifying its outlets and ensuring part of its sales, as well as part of its procurements through contractualisation to meet the specific demand of Carrefour. As for the supermarket chain, the hybridisation modalities correspond to a means of responding to competition and to evolving consumer demand.

In Part III, Chap. 7, Roberto Cittadini and Agnès Coiffard analyse the ProHuerta programme in Argentina, which combines domestic models, proximity models and naturalist models, on the basis of Fournier and Touzard's (2014) diversity of production and food models. They start from a context of stark contrasts, observed in

Argentina, between large companies and family farming. This context also encompasses the existence of other forms of organisation called 'territorialised family-run farm enterprises' (Chaxel et al., 2018). These producers can be key actors in advancing an alternative development model more oriented towards agroecological principles. The dynamics generated by the ProHuerta programme have triggered a series of public policies oriented towards family farming and agroecology. They have led to the adoption of a policy to combat the phenomena of social exclusion and loss of income by providing healthy and quality food to a poor section of the population. The ProHuerta programme has consolidated an agrochemical-free technological proposal far removed from the conventional practices recommended by INTA[1]. It has updated the skills of agricultural engineers in a process of re-learning, breaking away from the traditional technical vision tied to the Green Revolution. It has given rise to the figure of the 'promoter' who acts as a relay between the network of technicians and vegetable gardeners and who is a key element for the programme's functioning and success. The actors thus improve their capacity for action and the programme begins to invest in other spaces, such as local markets. The agroecological vision goes beyond the initial niche to become a more comprehensive proposal for a mode of production, a development model. This is a hybridisation that favours the emergence of a sector oriented towards agroecology, invested in by family-run farm enterprises that are successfully adapting and developing more resilient, profitable and environmentally friendly production models.

In Chap. 8, Rosalia Filippini analyses marketing strategies of peri-urban farmers on the Pisa plain, Italy. These strategies depend on the share of total production sold locally through alternative and local channels. Producers link traditional and alternative modes of production with traditional and alternative modes of marketing to meet new demands from consumers looking for different food products (Filippini, 2015). Peri-urban farmers adapt to the new possibilities offered by geographical proximity to urban areas by not only hybridising different forms of organisation of local commercial networks but also hybridising relationships with different local and non-local marketing actors. They develop relationships with local and non-local processing units, retailers and consumers, as well as with institutional actors who play a role in the recognition and legitimisation of the urban food system. The sustainability of the various initiatives depends on several elements but mainly on maintaining a balance between the urban and rural environment, on opening up to the outside world, and on the coordination between the different agri-chains.

In Chap. 9, Vanessa Iceri uses the concept of 'invented tradition' to understand the current transformations within a Brazilian farmer community (*Faxinal Emboque*), manifesting in new practices, new meanings, assimilations of novel practices or those of resistance, and links them to innovation processes. Her analysis highlights the capacity of local producers, members of civil society, to innovate by opening up to external markets in order to maintain their traditional know-how and to become firmly anchored to the territory (Iceri, 2019). Community members have taken the initiative

[1] National Agricultural Technology Institute (*Instituto nacional de tecnología agropecuaria, INTA*) in Argentina.

of seeking out new actors to promote access to local and external markets, derive greater value from local resources, and be able to access industry. For this, they benefit from the recognition of the Brazilian Ministry of the Environment to facilitate their search for external partners and funding. This situation demonstrates the farmers' capacity and willingness to change and to develop their activities; they are no longer satisfied with just producing food and protecting the forest. This form of organisation, which links scales ranging from the local to the global and which combines traditional and industrial models to better innovate in the territory, addresses various territorial development issues: maintaining farms and securing their futures, strengthening traditional practices to protect the forest, developing collective projects, disseminating know-how and knowledge (cooking, vegetable gardens, etc.) and recognising talent (Iceri and Lardon, 2018).

Finally, in Chap. 10, Christophe Albaladejo proposes a conceptual reading of the adaptation of development models. There is no longer a single model but instead a coexistence of development models characterised by their links to the State, the market, science and technology, as well as society. In order to consolidate, a model must be linked to the emergence of a form of agriculture as a 'territorial mediation', which characterises the manner in which agricultural activity is inserted into the local territory. A territorial pact is the matching of a territorial mediation to a development model (Albaladejo, 2020). In the Argentina of today, the situation is rather that of an unstable co-presence of incomplete territorial pacts, which give rise to a plurality of actions that no one coordinates. And yet, the coexistence of models requires the construction of a local public space and a profound change in the current strategies of the different models. Christophe Albaladejo examines the specific role that small and medium-sized towns could play in these adaptations by facilitating interactions between different models.

The Adaptive Capacities Conferred on Territories by the Coexistence of Agricultural and Food Models

These case studies provide the first answers to the initial questions on adaptation and confirm our hypotheses on the adaptation capacities conferred on territories by the coexistence of agricultural and food models, and on the conditions necessary for such an adaptation. There is a diversity of forms of coexistence that are organised along a gradient of complementarities, from forms of conflict and opposition, and of co-presence and cooperation, to those of hybridisation.

We thus recognise hybridisations between agricultural and food models which combine in a differentiated manner the modes of production, marketing or consumption, the power relations between different types of actors and the interactions with new actors. This transversal analysis of these few case studies shows that forms of hybridisation can confer capacities of adaptation at different levels of organisation under different conditions.

First of all, there is the availability of a certain number of local resources and means, such as forest resources for pig production in *Faxinal Emboque* (Chap. 9) or grassland resources for milk production in the Livradois-Forez Regional Natural Park (Chap. 4). But they can also be extra-territorial resources, originating from institutional and organisational dynamics, agricultural advisory systems and public policies, such as the development of vegetable gardens in Argentina (Chap. 7) or the food policy of the province of Pisa (Chap. 8).

Secondly, territorial governance is based on investments, partnerships and financing involving local actors and/or those from outside the territory. Territorial anchoring relies on this capacity of actors to link different scales, as Houdart et al. (2019) also show for different food initiatives in the Auvergne region in France. There is an integration of different scales, with hybridisation affecting both local and global processes, which confers on it an overall coherence and builds a common world, as Iceri and Lardon (2018) have analysed for *Faxinal Emboque* (Chap. 9).

Finally, hybridisation is achieved through the hybrid character of some of the actors themselves, who devise their own strategies to respond to the challenges that arise and engage in innovation processes. This is what the farmers, the local dairy and the supermarket chain are doing, each in its own way, in the Livradois-Forez Regional Nature Park (Chap. 4). This is also the case of the territorialised family-run farm enterprises, key structures in promoting an alternative development model in Argentina (Chap. 7).

Thus, the adaptation of territorial forms of organisation, through the complementarity of actors and activities, and through the linking of spatial and temporal scales, gives rise to new dynamics of territorial development, in which food production is a unifying factor (Lardon et al., 2017) and a lever for development (Loudiyi & Houdart, 2019). Coexistence can then be seen as revealing potentially interesting combinations for territorial development. Between the agroecology advocated by national discourse and that promoted by peasant agriculture, between the co-presence of models in the territory and the reappropriation by citizens of the links between agriculture and food, between the market economy and the social and solidarity economy, there are indeed new models to be invented and collective learning methods to be developed (Rey-Valette et al., 2008).

References

Albaladejo, C. (2009). Médiations territoriales locales et développement rural. Vers de nouvelles compétences d'accompagnement de l'activité agricole. *Les agricultures familiales dans les transformations territoriales en Argentine, au Brésil et en France* (p. 304). HDR (Accreditation to supervise research), Geography and Planning, Faculty of Social Sciences, Department of Geography.

Albaladejo, C. (2020). The impossible and necessary coexistence of agricultural development models in the Pampas: The case of Santa Fe province (Argentina). *Review of Agricultural, Food and Environmental Studies*, 1–28.

Ancey, V., Avelange, I., & Dedieu, B. (Eds.). (2013). *Agir en situation d'incertitude en agriculture: Regards pluridisciplinaires au Nord et au Sud*, PIE-Peter Lang.

Baritaux, V., & Houdart, M. (2015). Relations fournisseurs-grande distribution dans les filières agroalimentaires. Une analyse de la trajectoire d'une démarche de type «filière qualité». *Économie rurale, 346*, 15–30.

Caron, P., Valette, E., Wassenaar, T., Coppens, G., & Papazian, V. (Eds.) (2017). *Living territories to transform the world* (p. 278). Agriculture and Global Challenges Collection. éditions Quæ.

Chalas, Y., Gilbert, C., & Vinck, D. (2009). Comment les acteurs s'arrangent avec l'incertitude (p. 182). coll. Études de sciences, Archives contemporaines.

Chaxel, S., Cittadini, R., Gasselin, P., & Albaladejo, C. (2018). Family-run farm enterprises, territories and policies in Argentina. In P.-M. Bosc, J.-M. Sourisseau, P. Bonnal, P. Gasselin, E. Valette & J.-F. Bélières (Eds.), *Diversity of family farming around the world. Existence, transformations and possible futures of family farms* (pp. 163–176). Nature and Society Collection. éditions Quæ.

Darnhofer, I., Bellon, S., Dedieu, B., & Milestad, R. (2010). Adaptiveness to enhance the sustainability of farming systems. A review. *Agronomy for Sustainable Development, 30*, 545–555.

Dedieu, B., & Ingrand, S. (2010). Incertitude et adaptation: Cadres théoriques et application à l'analyse de la dynamique des systèmes d'élevage. *INRA Productions animales, 23*(1), 81–90.

Deffontaines, J.-P., Marcelpoil, E., & Moquay P. (2001). Le développement territorial: Une diversité d'interprétations. In S. Lardon, P. Maurel, & V. Piveteau (Eds.), *Représentations spatiales et développement territorial* (pp. 39–56). Éditions Hermès.

Filippini, R. (2015). Food production potential of periurban agriculture: Contribution of periurban farms to local food system. Thèse Scuola Superiore Sant-Anna (under the direction of E. Bonari, S. Lardon, & E. Marraccini) (p. 318), Speciality: Agricoltura, Ambiente, Territorio and AgroParisTech (ABIES), speciality: Agronomic and environmental sciences.

Fournier, S., & Touzard, J.-M. (2014). La complexité des systèmes alimentaires: Un atout pour la sécurité alimentaire? *VertigO—La revue électronique en sciences de l'environnement, 14*(1), made available online on 20 May 2014 (Retrieved 20 October 2021).

Gasselin, P., Cloquell, S., & Mosciaro, M. (Eds.) (2013). *Adaptaciones y transformaciones de las agriculturas pampeanas a inicios del siglo XXI*. Ediciones Ciccus.

Godard, O. (2005). Le développement-durable, une chimère, une mystification? *Mouvements, 4*, 14–23.

Grossetti, M. (2004). *Sociologie de l'imprévisible. Dynamiques de l'activité et des formes sociales* (p. 232). coll. Sociologie d'aujourd'hui, PUF.

Houdart, M., Baritaux, V., Iceri, V., Lardon, S., Le Bel, P.-M., & Loudiyi, S. (2019). The drivers of territorial anchorage of food. *Colloque ERSA*, Lyon (59th ed., pp. 29–30). August 2019.

Iceri, V. (2019). Actions collectives alimentaires en territoires ruraux: Un regard sur la diversité, une quête pour le développement territorial. Regard croisé entre Brésil et France. Doctoral thesis in geography, Université Clermont-Auvergne, Clermont-Ferrand, thesis defended on 28 November 2019.

Iceri, V., & Lardon, S. (2018). L'organisation sociospatiale, un commun pour le développement territorial. Le cas d'une communauté au Brésil. In L. Kebir, S. Nahrath, & F. Wallet (Eds.), *Biens communs et territoires Espaces et Sociétés, 4*(175), 87–104.

Lamine, C. (2015). Sustainability and resilience in agrifood systems: Reconnecting agriculture, food and the environment. *Sociologia Ruralis, 55*, 41–61.

Lardon, S. (Ed.). (2012). *Géoagronomie, paysage et projets de territoire. Sur les traces de Jean-Pierre Deffontaines* (p. 344). éditions Quæ.

Lardon, S., Houdart, M., Loudiyi, S., Filippini, R., & Marraccini, E. (2017). Food, integrating urban and agricultural dynamics in Pisa, Italy. In C. Perrin, C. Soulard, & E. Valette (Eds.), *Toward sustainable relations between agriculture and the city*, Urban Agriculture (pp. 15–31). Springer.

Loudiyi, S., & Houdart, M. (2019). L'alimentation comme levier de développement territorial? Les cas de la fête de la Pomme de Massiac et du projet al.imentaire territorial du Pays de Courpière, Auvergne, France. *Économie rurale, 1*(367), 29–44.

Pinto-Correia, T., & Godinho, S. (2013). Changing agriculture—Changing landscape: What is going on in the high value Montado landscape of Southern Portugal? In D. Ortiz-Miranda, A.-M. Moragues-Faus, & Arnalte-Alegre (Eds.), *Agriculture in Mediterranean Europe: Between old and new paradigms, research in rural sociology and development*, *19*, 75–90.

Rey-Valette, H., Lardon, S., & Chia, E. (2008). Editorial. Governance: Institutional and learning plans facilitating the appropriation of sustainable development. *International Journal of Sustainable Development*, *11*(2/3/4), 101–114.

Reghezza-Zitt, M. (2015). De l'avènement du monde à celui de la planète: Le basculement de la société du risque à la société de l'incertitude. HDR (Accreditation to supervise research), Geography. Unpublished volume, University Paris 1-Panthéon Sorbonne, Paris.

Rigolot, C., Martin, G., & Dedieu, B. (2019). Renforcer les capacités d'adaptation des systèmes d'élevage de ruminants: Cadres théoriques, leviers d'action et démarche d'accompagnement. *Productions animales*, *32*(1), 1–12.

Thérond, O., Duru, M., Roger-Estrade, J., & Richard, G. (2017). A new analytical framework of farming system and agriculture model diversities. A review. *Agronomy for Sustainable Development*, *37*(3), 1–21.

Torre, A., & Wallet, F. (Eds.) (2016). *Regional development in rural areas. Analytical tools and public policies*, Springer Briefs in Regional Science (p. 110). Springer.

van der Ploeg, J. D., Jingzhong, Y., & Schneider, S. (2012). Rural development through the construction of new, nested, markets: Comparative perspectives from China, Brazil and the European Union. *Journal of Peasant Studies*, *39*(1), 133–173.

Vigne, M., Vayssières, J., Wassenaar, T., Avadí, A., & Corson, M. S. (2017). Assessing the capacity of cropping systems to respond to challenges of sustainable territorial development. In P. Caron, E. Valette, T. Wassenaar, G. Coppens d'Eeckembrugge, & V. Papazian (Eds.), *Living territories to transform the world* (pp. 199–209). éditions Quæ.

Villalba, B. (Ed.) (2017). *Appropriations du développement durable: Émergences, diffusions, traductions*. Presses universitaires du Septentrion.

Zaccai, E., & Zuindeau, B. (2010). Équité territoriale et développement durable. In B. Zuindeau (Ed.), *Développement durable et territoire* (pp. 97–107). Presses universitaires du Septentrion.

Chapter 7
ProHuerta: From Subsistence Self-production to Throwing Down an Agroecological Challenge to Giants

Roberto Cittadini and Agnès Coiffard

In this chapter, we discuss the issue of the coexistence of agricultural and food models by analysing the role played by the ProHuerta programme[1] in the development of Argentinian agricultural, especially in that country's Pampean region (Fig. 1). This programme, which focuses on food security and sovereignty, was launched in 1990 and has led to the emergence of a vast network of vegetable gardens spread across the country that are cultivated according to agroecological principles. We will analyse the ProHuerta system and show how it has created and consolidated niches that are now calling the current dominant model of 'industrial agriculture' into question. We will see how ProHuerta, by actively promoting agroecology, has been a significant contributor to this change in orientation in practices.

We will first describe the context by presenting the main characteristics of the Argentinian agrarian structure, which is generally seen as a contrast between large companies (which now claim to be 'agribusinesses') and small farms (which lay claim to 'family farming'). We will show that this dichotomous vision needs to be nuanced in order to consider the existence of other agricultural variants between these two ideal types. We will then present the ProHuerta programme and the transformations it has shaped, which are having a profound impact in rural and urban territories in Argentina. We will continue by situating the role of ProHuerta within the various food models that coexist in Argentina. Finally, we will discuss the role played by the ProHuerta programme in coexistence's 'adaptation' dimension. We will analyse the types of adaptations taking place around this programme and understand how these

[1] http://prohuerta.inta.gob.ar/.

R. Cittadini (✉) · A. Coiffard
Instituto Nacional de Tecnología Agropecuaria (INTA), Universidad Nacional de Mar del Plata (UNMDP), Mar del Plata, Argentina
e-mail: racittadini@gmail.com

A. Coiffard
e-mail: agnescoiffard@hotmail.com

P. Gasselin et al. (eds.), *Coexistence and Confrontation of Agricultural and Food Models*,
https://doi.org/10.1007/978-94-024-2178-1_7

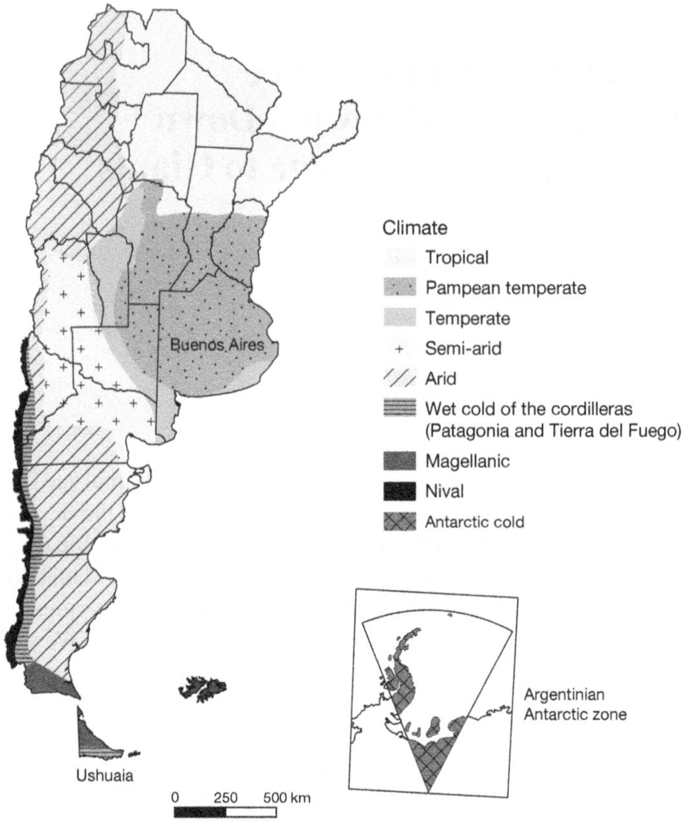

Fig. 1 Different climates in Argentina. *Source* INTA (1995)

adaptation processes spill over and contribute to the innovation process of the entire agrifood system, particularly to the process of transition to agroecology.

1 Argentina's Agrarian Structure

Argentina has a unique agrarian structure that has been shaped by the history of the occupation of its territory, and the changes engendered by the collaborations and confrontations between actors, on the one hand, and by public policies, on the other.

The main form of the current land use, especially in the Argentinian Pampas, follows from the historical allocations of land in large lots of more than 40,000 ha to various power elites: merchants, politicians, and military men who 'conquered' the lands of the native peoples. The rich soils here had a very high agronomic potential and had never been cultivated, especially those in the vast Argentinian Pampas. This process took place during the nineteenth century and reached its peak in the 1880s with the eviction and massacre of the local Indian populations (Gaignard, 1989; Sábato, 1988). The first type of agricultural development was the extensive grazing of sheep and cattle in large estancias.[2] The last decades of the nineteenth century and the first ones of the twentieth century saw a massive inflow of European migrants, many of whom established themselves in agricultural production as tenant farmers of latifundia proprietors. They had to adapt to the existing agrarian structures and would, in time, be the participants in the first agricultural revolution which, in the first decades of the twentieth century, turned Argentina into the 'granary of the world', with the country becoming a major exporter of beef and cereals. Mixed farming was soon introduced in the large estancias with the adoption of rotations between grasslands and field crops. These crops were cultivated by farmers who, at the end of a three-year lease, left the owners a more productive, sown pasture that allowed the intensification of the grassland production of meat (Scobie, 1968).

Coexistence, in this period, of large landowners, on the one hand, and farmers on the other, was punctuated by situations of collaboration and confrontation. A successful 'collaboration' could be said to have partially taken place, expressed through a market from which everyone benefited. The large landowner earned a rent from the land and benefited from the improvement of his pastures. The farmer benefited by the access to highly productive land that resulted in his family's financial growth, a situation that allowed him to hope of becoming a landowner himself one day. However, major conflicts intervened and gained momentum in 1912 in the course of a protest movement that came to be known as '*el grito de Alcorta*'.[3] A trade union organisation, the Federación Agraria, was consequently formed to represent farmers, with the goal of improving tenancy agreements (leases) and facilitating land ownership. However, it was not until the middle of the twentieth century, during the government of Juan Domingo Perón, between 1945 and 1955, that real progress was made. Indeed, this government announced a freeze on land rents and facilitated the purchase of tenanted land by introducing favourable credit facilities (Barsky & Gelman, 2001).

By the end of this process in the 1960s and 1970s, the Pampean agrarian structure was dominated by a large number of very dynamic and innovative producers who farmed between 200 and 2000 ha each (representing an average-sized farm in Argentina). This heterogeneous group largely consisted of former farmers who had managed to gain ownership of the land, and who often complemented their holdings with leased land. Some of them specialised in providing tilling services. Former latifundian owners who had undergone a process of property division through sales or inheritance were also part of this Pampean agricultural middle class, albeit in smaller numbers.

---

[2] *Estancia:* In Argentina, this term refers to a large agricultural estate dedicated mainly to cattle husbandry.

[3] In 1912 in Argentina, farmers mobilised and launched various protest movements to oppose the rules for setting prices and the conditions for renting agricultural land. This movement started in the village of Alcorta, hence the name 'the cry of Alcorta'.

Table 1 Agrarian structure in Argentina

	Distribution of farms (%)	Distribution of surface area (%)
Large producers (over 2000 ha)	15	71
Small and medium producers (up to 2000 ha)	85	29

Source Indec (2000) census

Between the 1960s and 1980s, these producers were the main actors in the second Pampean agricultural revolution, this time not due to the expansion of cultivation areas or the displacement of livestock husbandry to newer areas, but through the modernisation and intensification processes characteristic of the Green Revolution. These processes were supported by the National Institute of Agricultural Technology (INTA: *Instituto Nacional de Tecnologia Agropecuaria*) and by credit policies that facilitated mechanisation and the incorporation of new 'technological packages'[4] into existing systems.

Other transformations occurred in the Pampean agrarian structure in the 1990s, with the introduction of a new technological package that combined the cultivation of transgenic soya beans, direct seeding and massive use of glyphosate (GM soya beans are resistant to this herbicide). These technical innovations led to a high concentration of production, due largely to a simplified technical model that allowed vast cultivated areas to be managed. The cultivation of field crops gradually came to be dominated by agribusiness, which, in its quest for profitability, favoured the emergence of actors capable of farming 60,000 ha or more. These are the famous sowing pools which have received capital from hedge funds. This is how contract farming started in the country. The sowing pool manager rents land and hires companies that provide services. This model allows the continued growth of Pampean production, at the cost of a concentration of production, and even an increase in agricultural area at the expense of forested land that is stripped of its cover for cultivation, with soya bean being the preferred monoculture crop. This resulted in the disappearance of a large number of family farms, the degradation of productive resources and a pronounced negative environmental impact, especially due to glyphosate pollution of soil, water and air (Aparicio et al., 2017).

This process led many observers to characterise the current Pampean agrarian structure in a very dichotomous manner: on the one hand, large producers representing agribusiness and, on the other, family farming (Gras & Hernández, 2007; Guibert et al., 2011). This view of agriculture seems to be validated by censuses (Table 1).

Table 2 presents the main variables involved in illustrating this dichotomous view of Pampean agriculture.

However, our research in several districts of Buenos Aires province has allowed us to qualify this vision, especially with the observation of the persistence of a

[4] The term 'technological package' refers to a set of generally interdependent technical solutions (genetic improvement of seeds, chemical fertilisation, mechanisation, chemical control of weeds).

Table 2 Agribusiness and family farming

Variables	Agribusiness	Family farming
Labour relations	*Contratistas*, tertiarization	Family labour and possibly salaried workers
Farm size	Medium (200–2000 ha) Large (over 2000 ha)	Small (less than 200 ha) and medium (200–2000 ha)
Agronomy and technology	Industrial, intensive in external inputs, simple, specialised, generic packages	Agroecology or conventional, knowledge intensive, complex, diversified and location-dependent
Natural resources	Hybrids and GMOs	+ Conservation + Varieties + Biodiversity
Links with the territory	Deterritorialised	Strong link with the territory
Markets	Commodities, external	Diversified
Objectives	Profit	Various

large hub of family businesses linked to their territory and responsible for a major part of the production in these areas. We refer to them as 'territorialised family-run farm enterprises' (TFFEs) (Albaladejo & Cittadini, 2017; Chaxel et al., 2018). Although they are the heirs of the protagonist producers of the second agricultural revolution, they are barely visible today due to the predominance of agribusiness (Pengue, 2018) and small-scale family farming (Gisclard et al., 2015). These are landowning producers or agricultural contractors. The latter sometimes work on contract with sowing pools, and even directly with landowners. The latter type of contract involves the direct intervention of the producer and/or his family members at the landowner's farm.

These TFFEs have no doubt followed the influential trends of the technological model imposed by agribusiness, but it is also among them that we find producers who stand out with more sustainable practices (increased number of rotations, lower use of glyphosate) than those of the dominant model (Salembier et al., 2016), even if they seldom claim to be practising agroecological agriculture.

2 The ProHuerta Programme

ProHuerta is a programme that promotes vegetable gardens across Argentina. Initiated in 1990 in a challenging economic context, this programme is part of a strategy aimed at helping those who are excluded from the conventional agricultural production system in Argentina, especially due to very small surface areas, by compensating for the effects of structural adjustment policies. The main focus is the self-production of food by families in vulnerable situations in both the rural and urban sectors. A novel feature of ProHuerta is the complete non-use of chemicals for reasons of both limited resources and health protection.

ProHuerta's brochures clearly emphasise the same orientations described by Altieri (1983) as being at the heart of the agroecological approach:

- vegetation covers as an effective measure for soil and water conservation;
- the production and use of compost, the promotion of soil biological activity;
- crop rotations and associations to promote nutrient recycling, the regulation of pests and diseases through the intervention of natural enemies.

ProHuerta technicians like to say: 'We were practising agroecology without knowing it.' Indeed, the NGOs that have worked with ProHuerta (Cepar in Rosario, Cetaar in Marcos Paz, etc.) are among the most prominent actors to have helped introduce the concept of agroecology in Argentina.

The ProHuerta programme has seen the rapid and wide adoption—and even a complete appropriation—by all the actors involved (small market gardeners, promoters, technicians and their reference institutions). It has become a large-scale experiment in the development of socio-organisational and productive capacities, showcasing the values of solidarity and cooperation.

The programme is co-managed by the Ministry of Social Development within the framework of the National Food Security Plan, and by INTA, which views it as a key element in its extension strategy. It is organised around a national coordination mechanism, 24 provincial coordination units and a network of more than 700 professionals and technicians all over Argentina. At the same time, more than 15,000 volunteer promoters are helping in its implementation and form the core of this social network, working to support more than 600,000 vegetable gardens. These include family, community or institutional market gardens, 70% of which are located in urban or peri-urban areas and are geared towards food production. In addition to vegetable gardens, there is also the production of poultry and rabbits, and the cultivation of fruit trees. Some of the small market gardeners process their produce themselves. These small market gardens provide the backdrop for a variety of community actions that help consolidate the solidarity network.

ProHuerta has proven to be highly effective as a strategy for food security and sovereignty (Cittadini, 2010). This kind of food access is accompanied by the building of capacity of the people who work these small market gardens, resulting in improved self-esteem, a robust social capital, and the generation of productive skills within the agroecological paradigm. A growing number of small market gardeners are producing surpluses that are used for barter, for shared consumption or for sale. Experiments to create alternative markets (garden and home sales, food baskets, weekly markets, etc.) are growing in number, thus supporting the creation of short marketing chains in the social economy (Villagra et al., 2010).

Through its productive and socio-organisational dimensions, the programme is contributing to the development of an agroecological movement and agroecological practices in Argentina (Cittadini, 2012; Patrouilleau et al., 2017). Although it represents only a part of family farming, we found that the programme's dynamics, in combination with other initiatives within and outside INTA, have prompted a series of public policies oriented towards family farming and agroecology (Juárez et al., 2014; Gisclard et al., 2015).

3 Production and Food Models in Argentina

We will draw on the concept of food models used by Touzard (2015) to classify the three main production models that we have characterised in the agrifood systems in the Argentinian Pampas: family farming, of which ProHuerta is a part; TFFEs; and corporate farming (Table 3).

The domestic model is based on the self-consumption of food production, and is therefore the one that best characterises ProHuerta, although it may also be partially represented by other sectors of family farming.

The proximity model is characterised by the proximity between producer and consumer. It is mainly found in family farming and includes families that are connected to ProHuerta and that are invested in local markets.

The commodity and agro-industrial production model is characterised by mass production focused on generic markets, whether domestic markets or export markets. This is the preferred market for corporate farming and TFFEs. While some family farming sectors also participate in this food model, ProHuerta does not.

The naturalist model emphasises the relationship with, and the protection of, nature and the differentiation of products. We find organic farming and agroecology in this model. While it is still used by a minority in Argentina, this model is currently gaining in popularity. Farmers from across production categories have links to this model. It goes without saying that the production of ProHuerta is fully part of it. Indeed, ProHuerta's entire production is agroecological and its actors are ardent proponents of popularising agroecology.

Table 3 Production and food models

Contribution of production models to food models	Domestic	Proximity	Commodities/agro-industrial products	Naturalist
Family farming	+++ (including ProHuerta)	+++ (including ProHuerta)	++	+ (including ProHuerta)
TFFEs		+	+++	+
Corporate farming			+++++	+

TFFE territorialised family-run farming enterprise

4 Analyses of the ProHuerta Programme from an Adaptation Perspective

We can address several aspects of the notion of adaptation in terms of the establishment of the ProHuerta programme and of its continuation over time. This adaptation took the form of several innovation processes, which has allowed the programme to transcend its initial objectives and make a veritable contribution to the process of transition of the entirety of the Argentinian agrifood system to agroecology.

The programme itself was born of a policy of adaptation to the phenomena of social exclusion resulting from the economic structural adjustment programmes implemented in the 1990s. The families and groups that joined the programme did so in large part because it represented an adaptation strategy to compensate for the loss of income due to decreases in activity (loss of employment, reduction in working hours, increasing precariousness, etc.). The notion of adaptation is also present:

- in the decision to implement a technological proposal very different from the conventional practices of INTA, the institution that manages the programme. Agrochemicals are, in fact, not used in ProHuerta, partly because they are not easily accessible by poor families, and partly because of the risks involved in handling agrochemicals, especially in the families' living spaces;
- in the re-learning process that the agricultural engineers and technicians who are involved in the programme needed to develop, given that the only training that existed at that time was based on the principles of the Green Revolution (Bustos, 2017). It is interesting to note that, at the beginning, the majority of the agricultural engineers involved in ProHuerta were women, and thus it was they who had a strong influence in recasting the profession of the agricultural engineer, which was until then closely tied to a traditional technical vision inherited from the Green Revolution;
- in the fact that the population was mobilised on a large scale, with necessarily limited means. In this process, the role of the 'promoter' emerged, someone who acts as an intermediary or relay between the network of technicians and the small market gardeners, a role that rapidly became a key point in the programme's functioning and success. The promoter is usually a volunteer who succeeds in building up an identity as a socially respected and recognised actor in the community.

It is only natural that there will also be innovation in all these situations of adaptation that we have been able to identify. While the programme no doubt represented an adaptation to a period of crisis, the way that it was implemented is also innovative, as was its technical proposal, its successful involvement of community actors through the role of promoter, etc.

There is also a moment in the programme's life—it is not a specific moment, but rather an evolutionary process—when the dynamics of adaptation-innovation transcend the programme's initial objectives. We can say that the actors who are part of it build up their capacity for action and the programme begin to venture into other spaces. Thus, for example, around 2005, the programme went beyond its exclusive objectives of food for self-consumption. It was at this time that the process of the sale of surpluses was introduced into the programme, which led to the creation of local markets in which some of the small market gardeners began participating. Another transformation is linked to the programme's ideology (or 'vision'), which, far from remaining tied to the 'adaptation' approach, is progressively expanding to a vision of an active promotion of agroecology. This translates not only into individual technical practices, but also into a more global proposal for a mode of production, with the goal of becoming the dominant form of production for the entire agrifood system. A majority of ProHuerta actors are becoming agroecology activists: not only the promoters and small market gardeners in their communities, the different organisations or NGOs with which they are involved, but also INTA technicians, who promote agroecology in the territories as well as in their own institutions, which are still largely imbued with notions tied to industrial agriculture. Many of the ProHuerta technicians have moved on to other positions within INTA, including executive positions, and this has undoubtedly helped raise awareness about agroecology in some INTA teams.

Returning to Pampean agriculture, we can observe the emerging phenomenon in recent years of producers who practise a mixed farming system of field crops and livestock husbandry now turning to agroecological production. These farmers belong to the TFFE category. It is interesting to note that, even though these actors are far removed from ProHuerta from a socio-economic point of view, we can still observe the programme's influence on them. We can cite the example of a producer's son whose approach illustrates this influence. He was involved in alternative movements to promote agroecology during his university studies and also participated in the promotion of vegetable gardens as part of ProHuerta's activities. In 2013, wanting to return to the family farm, he convinced his father to convert a part of their 1000 ha farm to agroecology, which, in fairly rapid order, was expanded to include the entire holding. The son and the father have since become reference persons because of their successful transition to agroecology. Since the beginning of 2018, they have been participating in a group called Cambio Rural which consists of eight farmers implementing very diverse production systems. For example, two of them are developing vegetable production using environmentally friendly practices, which can be considered an extension of what was learnt in ProHuerta. What forges this group's identity is not the type of production, but their agroecological practices. Other similar groups have been formed in recent years, three of them in just the Sudeste region of Buenos Aires province. The common characteristic of these TFFE producers who have espoused agroecology is their satisfaction in developing a production model that conserves natural resources and is less risky and more profitable. In general, they are managing to maintain their productivity levels, or suffer a relatively small decline. Their profitability, however, is significantly higher due to the substantial savings resulting from the reduction in usage of inputs.

It is interesting to note that a large section of the technical managers who promote agroecology, regardless of the production systems, participated in the ProHuerta programme at the beginning of their professional careers. ProHuerta has thus played a major incubating role in raising awareness and in training technicians and young farmers, and has certainly influenced them in the transition to agroecology.

If we refer to transition theory (Geels, 2012), we can say that the 'niche' constituted by the ProHuerta programme has succeeded in challenging the dominant sociotechnical system. It has hybridised other production models and facilitated their adaptation towards agroecological models.

5 Coexistence and Adaptation

A priori, the analyses do not seem to provide much evidence that hybridisation has led to a greater overall adaptation of the agrifood system. Rather, we note a tension between models:

- on the one hand, agroecology (of which ProHuerta is a part) presents itself as an alternative to the dominant agricultural model in the Argentinian Pampas, even if it is still in the minority,
- on the other hand, agribusiness and its institutions extol their own path to productivity and exports, even while minimising the social and environmental costs of their model.

There is little room for discussions and experiences of hybrid production that would present coexistence as a virtuous process.

However, we have also shown that there is a core group of producers who, without being advocates of the idea of agroecology, are beginning to distance themselves from the simplified model promoted by the dynamics of agribusiness. And some of them have become agroecology activists, with a discourse that is similar to, and sometimes borrows from, the ProHuerta actors. This is a hybridisation that is favourable to the emergence of an agroecology-oriented family enterprise sector that is successfully adapting and developing more resilient, profitable, and environmentally friendly production models.

We can also find another type of coexistence of models within INTA, in the sense that the agents themselves are working towards a 'desired future'. The INTA mainstream still shares the paradigm whose main objective is the quest for productivity, achieved through the classic agronomic methods of the modernisation stage. However, the proponents of agroecology, new staff who joined INTA largely due to contracts under the ProHuerta programme, have succeeded in making their mark. Since 2005, five technical institutes for family farming (IPAF) have been created and mainly promote agroecological approaches. An 'Agroecology Network' was institutionalised within INTA in 2013. Its main activity was to develop a network of experiments spread over different experimental stations, and in different Argentinian territories. In early 2019, INTA and SupAgro in Montpellier, France, cooperated to develop the first MOOC (massive open online course) in agroecology in Spanish. It was remarkably successful, with more than 30,000 sign-ups in the first year, and more than 50,000 for the second edition, in early 2020. This course's coordinator, and one of the authors of this chapter, is a former coordinator of the ProHuerta programme. Within the framework of development practices (also an INTA responsibility in addition to that of research), agroecology occupies an important place, especially in the form of support of small-scale family farming, and thus also to ProHuerta.

6 Conclusion

On the basis of the ProHuerta programme, we have been able to provide an overview of the dynamics of the entire Argentinian agricultural sector, and more specifically, of the Pampean region. We have situated ProHuerta in relation to the types of producers who coexist in the Argentinian Pampas, as also in relation to current food models.

We have analysed the adaptation processes that have taken place within the framework of the ProHuerta programme. The programme itself has been such a successful adaptation that it has gone beyond the normal contours of a programme—which usually have short or medium term horizons—and has been continuously renewed for 30 years. Its capacity to adapt has also been supported by a strong capacity for innovation, which has enabled it to evolve and exceed its initial objectives of production for self-consumption, and thus develop several local markets.

Finally, what we want to highlight the most is the influence that ProHuerta has had in calling into question, and even helping to rethink and hybridise, the entire agricultural system, both on the ground and within INTA, Argentina's main agronomic research and development institution.

References

Albaladejo, C., & Cittadini, R. (2017). El productor silencioso: Destino del gran actor de la modernización de los años 1960–70 en la actual copresencia de agriculturas de la región pampeana Argentina. *Revista PAMPA, 16*, 9–34.

Altieri, M. (1983). *Bases agroecológicas para una agricultura sustentable* (280 p). Ediciones CETAL.

Aparicio, V., Gonzalo Mayoral, E., Costa, J.L. (2017). *Plaguicidas en el ambiente* (156 p). Ediciones INTA.

Barsky, O., & Gelman, J. (2001). *Historia del agro argentino. Desde la Conquista hasta fines del siglo XX* (460 p). Ediciones Grijalbo Mondadori.

Bustos, D. (2017). *Competencias que emergen de un campo de acción innovador. El caso de los técnicos de ProHuerta en la provincia de Tucumán, Argentina.* Tesis Maestría Plider, Universidad Nacional de Mar del Plata.

Cittadini, R. (2010). La seguridad y la soberanía alimentaria, un problema complejo y multidimensional. *Revista Voces en el Fenix*, 1, UBA, Buenos Aires, Argentina.

Cittadini, R. (2012). Limites et potentialités de l'agroécologie. Leçons d'une expérience à grande échelle: le programme ProHuerta en Argentine. In F. Goulet, D. Magda, N. Girard., V. Hernandez & S. Sarandon (Eds.), *Approches croisées de l'agroécologie en Argentine et en France* (pp. 117–133). Éditions L'Harmattan, France.

Chaxel, S., Cittadini, R., Gasselin, P., & Albaladejo, C. (2018). Family-run farm enterprises, territories and policies in Argentina. In P.-M. Bosc, J.-M. Sourisseau, P. Bonnal, P. Gasselin, E. Valette & J.-F. Bélières (Eds.), *Diversity of family farming around the world. Existence, transformations and possible futures of family farms* (pp. 163–176). Springer.
Gaignard, R. (1989). *La Pampa argentina, ocupación, poblamiento, explotación, de la conquista a la crisis mundial (1550–1930)* (512 p). Ediciones Solar.
Geels, F. W. (2012). A socio-technical analysis of low-carbon transitions: Introducing the multi-level perspective into transport studies. *Journal of Transport Geography, 24*, 471–482.
Gisclard, M., Allaire, G., & Cittadini, R. (2015). Proceso de institucionalización de la agricultura familiar y nuevo referencial para el desarrollo rural en Argentina. *Revista Mundo Agrario, 16*(31)..
Guibert, M., Sili, M., Arbeletche, P., & Piñeiro, D. (2011). Les nouvelles formes d'agriculture entrepreneuriales en Argentine et en Uruguay. *Économie Et Société, Série Systèmes Agroalimentaires, 33*, 1–20.
Gras, C., & Hernandez, V. (2007). L'agriculture argentine dans la globalisation: Connaissances et subjectivités. *Autrepart, 43*, 147–163.
Indec. (2000). *Recensement année 2000*. Buenos Aires.
INTA. (1995). Climas de Argentina. In *Atlas de suelos de la República Argentina*. Ediciones INTA.
Juarez, P., Gisclard, M., Goulet, F., Albaladejo, C., Cittadini, R., Elverdin, J., Patrouilleau, M., & Gonzalez, E. (2014). El caso de la agricultura familiar en la república argentina. In E. Sabourin, M. Samper & O. Sotomayor (Eds.) *Políticas públicas y agriculturas familiares en América Latina y el Caribe: balance, desafíos y perspectives* (pp. 51–73). Ediciones de la CEPAL.
Patrouilleau, M., Martínez, L., Cittadini, E., & Cittadini, R. (2017). La promoción de la agroecología desde las políticas públicas en Argentina. In E. Sabourin (Ed.), *Políticas públicas a favor de la Agroecología y la Agricultura Orgánica en América Latina y El Caribe* (pp. 20–43). Ediciones FAO.
Pengue, W. (2018). *Atlas del Agronegocio* (180 p). Ediciones Gepama.
Sábato, J.F. (1988). *La clase dominante en la Argentina moderna, formación y caractéristique* (280 p). CISEA y Grupo Editor Latinoamericano.
Salembier, C., Elverdin, J. H., & Meynard, J. (2016). Tracking on-farm innovations to unearth alternatives to the dominant soybean-based system in the Argentinian Pampa. *Agronomy for Sustainable Development, 36*, 1.
Scobie, J.R. (1968). *Revolución en las Pampas. Historia social del trigo argentino, 1860–1910* (245 p). Ediciones Solar.
Touzard, J.-M. (2015). Analyser la diversité et la coexistence des systèmes alimentaires face aux enjeux de sécurité alimentaire. *Séminaire Économistes du SAD*, Paris, January 13, 2015.
Villagra, C., Handan, V. & Cittadini R. (2010). Economía social y agricultura urbana: el caso de la feria verde de la ciudad de Mar del Plata. In R. Cittadini, L. Caballero, M. Moricz & F. Mainella (Eds.), *Economía Social y Agricultura Familiar: hacia la construcción de nuevos paradigmas de intervención* (pp. 245–276). Ediciones INTA.

Chapter 8
Hybridisation of Food Chains in Peri-urban Production Systems: The Example of Pisa in Italy

Rosalia Filippini

1 A Farming Model that Has to Adapt

Peri-urban farms can react to urban pressure in different ways. Farmers can give up farming altogether and sell the land, profiting from high land prices, maintain their existing farming and marketing practices within the context of a reduced territory, or partially or completely modify these practices in order to take advantage of increasing urban demand. None of these solutions is ideal and their applicability and relevance depend on the opportunities and constraints of the farmers' situations and farming activities, the socio-political context in which they are embedded, and the physical and geographical conditions of the land. At the same time, since they are part of the dynamics of this urban and peri-urban context, the farmers' actions and marketing choices can have an impact on other economic activities and populations.

Our overall objective in this chapter is to illustrate the use of hybrid marketing systems by peri-urban farmers in order to adapt to urban pressure by taking advantage of their proximity to urban markets. We examine the complexity of this adaptation and explore its impacts. We base our analysis on research, undertaken in the case of Pisa (Italy), on assessing the capacity of farmers to integrate into the local food system.[1]

[1] We refer here to two research projects: the Daume project (Sustainability of Urban Agriculture in the Mediterranean, http://www1.montpellier.inra.fr/daume/), whose objective was to analyse the conditions of sustainability of peri-urban farming systems in the Mediterranean region (Soulard et al., 2017); and Ph.D. student Rosalia Filippini's (2015) research project, aimed at understanding the integration of peri-urban farming systems into the local food system as a means of territorial development.

R. Filippini (✉)
Department of Economics and Management, University of Parma, Parma, Italy
e-mail: rosalia.filippini@unipr.it

UMR Territoires, INRAE, Clermont-Ferrand, France

P. Gasselin et al. (eds.), *Coexistence and Confrontation of Agricultural and Food Models*,
https://doi.org/10.1007/978-94-024-2178-1_8

We first present the conceptual framework we mobilise before characterising our study area and explaining our methodology. We then present our results on the hybridisation of food chains, with a focus on the motivations of peri-urban farmers and on the territorial scales involved. We end with a discussion and conclusion.

1.1 Urban Pressure and Agriculture

The term 'urban sprawl' refers to rapid, unregulated, low-density development that extends urban space out from an urban centre (Snyder & Bird, 1998). Urban sprawl is linked to peri-urbanisation, a process in which rural and urban areas intermix (EEA, 2006). In Italy, this phenomenon is very significant (EEA, 2006): almost all rural land in the country is characterised by 'diffuse' urbanisation (Ispra, 2015). The peri-urbanisation phenomenon has multiple consequences (EEA, 2006) and its impacts on agriculture have been so significant that many analyses refer to urbanisation as a phenomenon that is essentially 'against agriculture' (Tolron, 2001; Pascucci, 2007). A Peri-urban Farming System (PFS) is defined here as a farming system functioning in close proximity to an urban area. PFSs appear as a topic in which different issues of sustainability are concentrated in a dynamic process: these agronomic, environmental, economic and socio-political issues emerge when thinking about how to better integrate agriculture in an urbanised area. At the same time, PFSs are seen as a key element for addressing and resolving many of these issues. Indeed, studies on PFSs have highlighted the social and environmental functions of peri-urban farming activities that benefit urban dwellers and urban environments (Zasada, 2011), as well as their contribution to issues of food security (Filippini et al., 2019). The FAO (2010), for example, argues for the development of more local food systems.

1.2 Local Food Systems and Peri-urban Farming Systems

A Local Food System (LFS) is defined here as a 'food system in which foods are produced, processed and retailed in a defined geographic area' (Kneafsey et al., 2013). Alternative food chains (AFCs) are seen as a way to localise the food system. As a reaction to the dominant agro-industrial model—sometimes referred to as Conventional Food Chains (CFCs)—, AFCs have the primary aim of restoring direct connections to consumers themselves (Marsden et al., 2000). The expected benefits of such processes rely on the fact that consumers are given the opportunity to seek out and choose food products, while farmers get appropriate economic recognition by consumers and thus generate higher profits. Renting et al. (2003) define three types of AFCs, depending on the degree of proximity:

- 'Face-to-face interaction', where 'consumers purchase products directly from the producer or processor, and authenticity and trust are mediated through interpersonal interaction';
- 'Proximate AFCs', based on the distance between relationships in time and space such as shops and smallholder associations where actors share cultural values;
- 'Extended AFCs', where 'products are sold outside the production region to consumers who might not have personal experience with the locality'.

In some of these AFC initiatives, such as fair trade or labelled foods, it is not the geographical distance that matters, but the fact that consumers 'make connections with the place/space of production and, potentially, with the values of the people involved and the production methods employed' (Marsden et al., 2000, p. 425).

Since alternative food chains bring together urban consumers and rural farmers, they are seen as a factor of adaptation of farms to the new challenges posed by peri-urbanisation (Lamine & Perrot, 2008). Urban pressure as well as the crisis of traditional production systems based on conventional marketing channels are encouraging peri-urban farmers to change their farming practices and marketing strategies. Furthermore, several food scandals have led consumers to demand more traceability and options in their food choices. AFCs are thus seen as an opportunity to help agriculture adapt to the problems of peri-urban areas, while fostering sustainability in territorial development and contributing to community food security.

Ilbery and Maye (2006) employ the concept of 'hybridisation of food chains' to discuss the boundaries between 'conventional' and 'alternative' food systems. For example, after analysing the marketing practices of livestock farmers in the Scottish-English border region, they conclude that 'a straightforward polarity between […] mainstream food systems and […] locally dedicated food systems is unlikely'. According to them, local farmers combine local or alternative systems with conventional systems because farmers 'have not really moved from one system to another' (Ilbery & Maye, 2006). Other studies go further: Forney and Häberli (2015) note that not only do farmers combine CFCs and AFCs, they generally have transformative power, such as large agribusinesses implementing a process of 'conventionalisation of organic production' or adopting policies of social inclusion of local farmers. On the other hand, in local AFCs, conventional marketing rationales can be found, partly because farmers are not always able to establish prices when selling through local distributors (Bloom & Hinrichs, 2011). At the same time, while analysing the actual impact of AFCs, other studies discuss their ability to promote sustainable development (Tregear, 2011). Nevertheless, the expected benefits of AFCs, such as social inclusion, more sustainable farming practices and increased economic viability for farmers, should not be taken for granted (Bloom & Hinrichs, 2011). Sonnino and Marsden (2006), for example, suggest establishing a 'new process of re-localisation of economic activities and practices', which would also help highlight the variability of AFCs as noted by Venn et al. (2006). This should ultimately help the research community better identify the place of these initiatives in the processes of sustainable development (Izumi et al., 2010).

2 The Pisa Case Study and Surveys of Peri-urban Farmers

The case study pertains to the urban area of Pisa in Tuscany, Italy, consisting of six municipalities—Pisa, Cascina, Calci, Vecchiano, Vicopisano and San Guiliano Terme (Fig. 1)—commonly called 'Area Pisana'. This area is geographically encir-

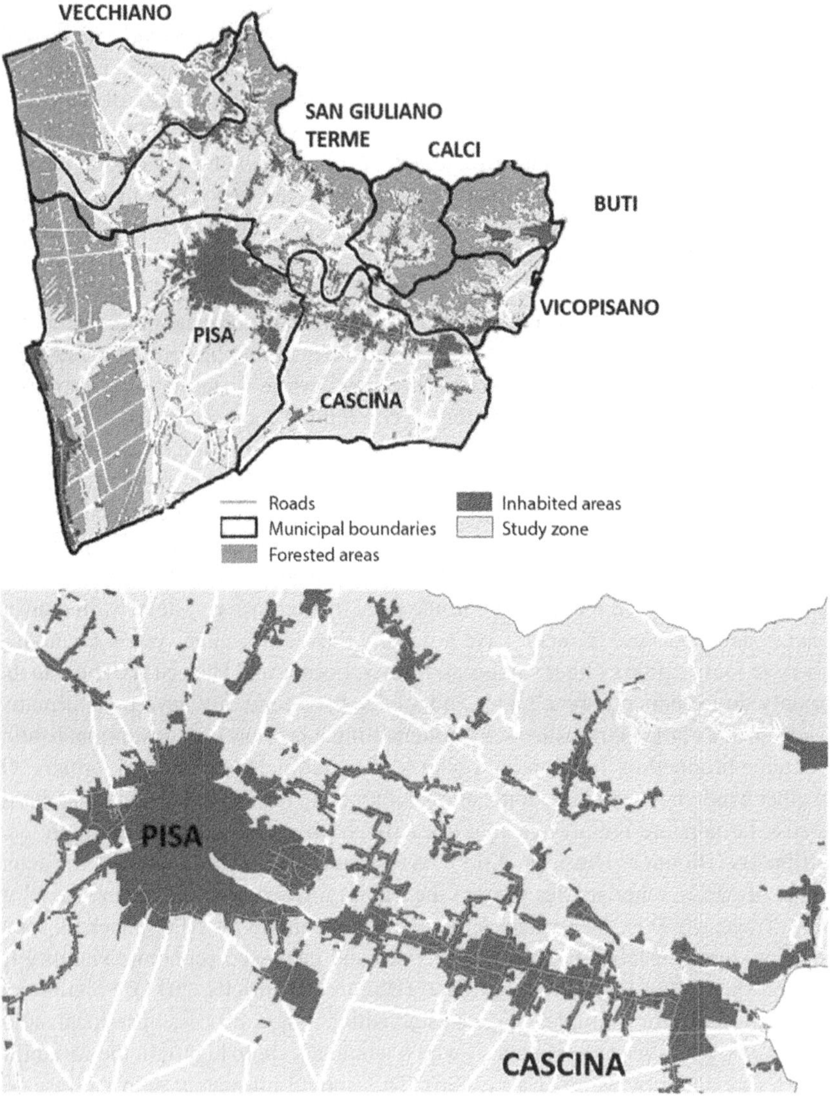

Fig. 1 The study area: the urban area of Pisa in Tuscany, Italy. *Source* Filippini (2015)

cled by the coast to the west, Monte Pisano to the north, the province of Livorno and the hills to the south, and by the Valdarno plain to the east.

The area is particularly apt for the analysis. First, the urban area is illustrative of the urban sprawl in medium-sized cities: in the last national census (Istat, 2011), the number of city inhabitants decreased by 4%, while the populations of nearby urban centres increased by an average of 8%.

Second, the heterogeneity of farming systems that characterises the area's agriculture allows us to analyse the different possible implications of the development of PFS and LAS for different farming systems and food chains. The area's agriculture is following the Italian and European trends of the development of Mediterranean farming systems. From 1981 to 2000, the total Usable Agricultural Area (UAA) came down in both peri-urban and non-peri-urban areas almost equally (6.2% vs. 5.9%). At the same time, the number of peri-urban farms decreased by 30% (on an average), compared to about 20% in rural areas. In both cases, this decrease in the number of farms has been particularly marked over the last decade, but especially so in the peri-urban area. Between 2000 and 2010, Pisa's peri-urban area lost 60% of its farms (Istat, 2011).

This analysis is based on data from interviews with 56 farmers between 2013 and 2014. To properly represent the PFS, farmers were selected according to their main farming system, its size and its distance from the nearest urban centre. The semi-structured interviews were designed to elicit information on farming practices (rotation cycles, inputs, livestock production), the farm's structure (buildings, manpower, machinery), the farm's social composition and origin (family support, education, ages), the marketing of agricultural production (buyers, prices, quantity of production) and the farm's relationship with the urban area, with a focus on constraints and opportunities arising from urbanisation.

The results presented here are of the analysis of the strategies adopted by farmers in order to participate in Alternative and Local Food Chains (ALFCs), which are defined as AFCs in which the final sale is to local consumers. In other words, these farmers have knowledge and control of the food chain, and marketing is aimed at local urban consumers. These food chains can thus contribute to local food security. To analyse these strategies, the sample was first divided according to the farmers' participation in alternative and/or conventional food chains (Filippini et al., 2016a).

Three groups were identified: farmers selling all production in AFCs, farmers selling all production in CFCs, and farmers selling to both types of food chains, i.e. 'mixed' food chains (MFCs) (Filippini et al., 2016a). Considerable difference was observed between the number of farms participating exclusively in AFCs (10%) with those selling exclusively in CFCs (47%). An interesting result is the large percentage of farmers (43%) belonging to the MFC group, i.e. who combine AFCs and CFCs. These results instigated an in-depth analysis of mixed food chains in order to understand their characteristics and the factors that lead farmers to this combined choice. Thus, Filippini et al. (2016b) specifically analysed the different strategies farmers adopt to gain access to local markets. This study only considered farmers who take part partially or fully in the LFS via AFCs, thus participating in alternative and local food chains (ALFCs). For this reason, out of the initial number of 56 farmers studied by Filippini et al. (2016a), only the 26 farmers surveyed who produce food for local urban consumers were finally selected for the analysis (Tables 1 and 2).

Table 1 Main characteristics of the farm sample

Farm	Farming system	UAA (ha)	LSU	Total % of production delivered to ALFCs	Farm type
F01	Vegetables	14	–	100	Family farm
F02	Livestock	65	87	100	Family farm
F03	Olive oil	3	–	100	Family farm
F04	Livestock	250	116	2	Family farm
F05	Olive oil	2	–	100	Family farm
F06	Livestock	140	29	90	Family farm
F07	Livestock	280	213	70	Family farm
F08	Olive oil	6.5	–	50	Family farm
F09	Cereals	145	–	60	Family farm
F10	Olive oil	1.6	–	80	University experimental farm
F11	Olive oil	10	–	60	Family farm
F12	Cereals	80	–	50	Family farm
F13	Vegetables	6	–	50	Family farm
F14	Olive oil	11	–	65	Family farm
F15	Olive oil	5	–	2	Family farm
F16	Vegetables	7	–	50	Family farm
F17	Livestock	126	52	80	Family farm
F18	Olive oil	11	–	90	Family farm
F19	Livestock	284	63	95	Family farm
F20	Livestock	31	11	100	Family farm
F21	Livestock	29	275	80	Family farm
F22	Livestock	30	41	90	Family farm
F23	Livestock	110	150	20	Family farm
F24	Cereals	595	–	2	Cooperative
F25	Vegetables	11	–	50	Family farm
F26	Vegetables	22	–	5	Family farm

UAA usable agricultural area; *LSU* livestock unit; *ALFC* alternative and local food chain

Table 2 Main characteristics of the sampled farming systems

Farming system	Number of farms	% of the farm sample	Average UAA (ha) and (standard deviation)	% of production delivered to ALFCs
Cereals	3	12	134 (103)	27
Livestock	10	38	6 (4)	73
Olive oil	8	31	12 (6)	68
Vegetables	5	19	273 (280)	51

The number of farms, the corresponding percentage, the average area in hectares, the percentage of production delivered to alternative and local food chains (ALFCs) were taken into account. UAA: Usable agricultural area.

3 Hybridisation Between Food Chains: Motivations of Peri-urban Farmers and Territorial Scales

3.1 Hybridisation Between ALFCs and CFCs

The farmer surveys show that most of those involved in ALFCs hybridise alternative and local marketing with conventional forms of marketing. Figure 2 shows the number of farmers who adopt each strategy and the percentage of each farmer's production sold in ALFCs.

3.1.1 Passive-Strategy Group

Four farmers, depicted at the very bottom in Fig. 2, comprise the first group. They sell only a small part of their production through ALFCs. Our analysis of the interviews with these farmers showed that the main reason for selling to ALFCs is the proximity of buyers (e.g. neighbours and friends) and the opportunity offered by professional or personal relationships. Farmer F04, who follows this strategy, sells almost all (98%) of his sheep's milk through conventional marketing to a regional milk factory, but the rest is sold to local consumers and to the nearest milk factory. 'I sell milk to the local cheese factory whenever they run short of sheep's milk' (F04).

3.1.2 Opportunistic-Strategy Group

The 17 farmers of the second group try to maximise the benefits of both marketing strategies. Livestock is the main production (41% of the group), followed by olive oil (29%), vegetables (18%) and cereals (12%). Under this strategy, there is significant variability both in the quantity delivered to local markets (from 15 to 69%) and in way marketing is organised. These farmers practise this dual marketing strategy for several reasons: to maximise profits, to use pre-existing CFCs with which the family

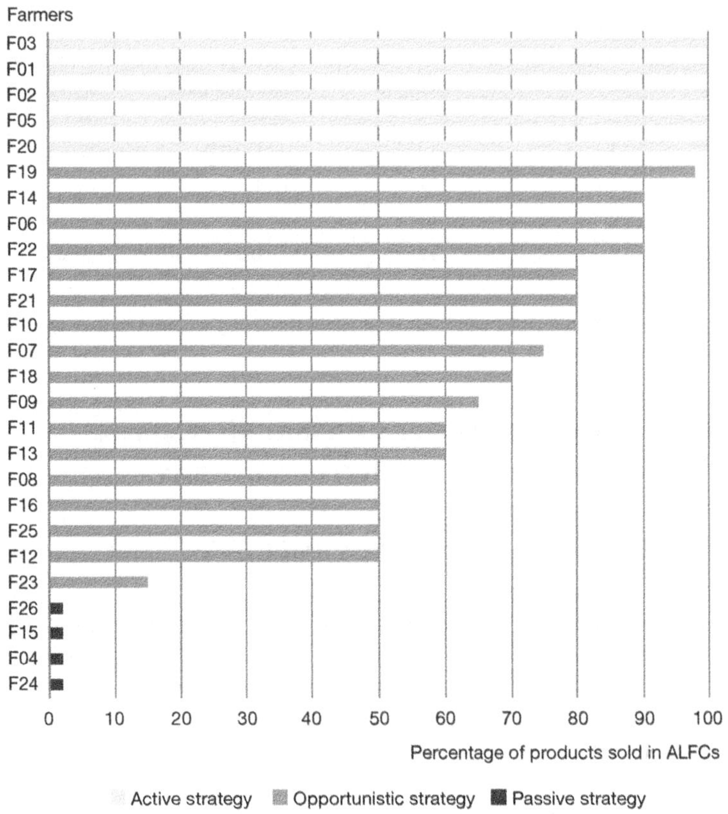

Fig. 2 Percentage of production sold in ALFCs for each farmer. Based on Filippini et al. (2016b)

has previously developed relationships, and to take advantage of new networks. Some of them supply the same products to AFCs and CFCs, with the same product going to different food chains; others supply different products to different chains. For example, farmers F16 and F25 (vegetables producers) sell locally, both on-farm and at local farmers' markets. They also invest in new structures or in the diversity of products offered to customers. Despite these efforts and investment in ALFCs, both farmers admit the need to maintain a relationship with the wholesale market (seen as part of a CFC) in order to ensure that their entire production is sold and to spread future business risk.

In this way, the farmers express doubts about the long-term sustainability of ALFCs. They also point out that even the CFCs are backed by social relationships and trust: in order to be able to sell at the right price and at the right time, the farmer needs to establish personal and long-term relationships. Even in wholesale markets, farmers 'have to gain the buyer's trust' (F17).

Several farmers (F19, F22, F23) sell through CFCs and ALFCs, but direct different products to these different channels. This is the case, for example, of farmers who need to cultivate fodder and specific crops to maintain their crop rotations. These productions are not easily sold on the local market. In most of these cases, 100% of the main production (meat, milk and cheese) is sold through ALFCs, while cereals are sold through cooperatives that collect the product and sell it on national and international markets (CFCs).

3.1.3 Active-Strategy Group

The five farmers of the third group, shown at the very top in Fig. 2, sell 100% of their production through ALFCs. As with the farmers in the opportunistic-strategy group, increased control over the product's destination and better traceability of quality motivate farmers to participate in ALFCs. Personal satisfaction is usually the reason the farmers provide for belonging in this category, which is also linked to a certain social recognition of agricultural activity. Most of these farmers note the advantages of shorter food chains: simplified procedures and negotiations, and the possibility of adding value to products. Some of them do not want to deal with supermarkets, which require relatively constant production levels and therefore lead to the problem of sale of surplus production. Through ALFCs, it is also often possible to obtain higher prices and faster payments.

3.2 Hybridisation of Food Chains and Territorial Scales

Figure 3 shows the socio-spatial configuration of an olive and fruit farm from the active-strategy group. This farmer markets all his products through ALFCs by selling directly at the farm (70%), to nearby restaurants (15%), and at farmers' markets (15%) organised by the municipalities. The farmer thus combines different ALFCs and makes a considerable effort to ensure that all his production is consumed locally. Several farmers in the active-strategy and the opportunistic-strategy groups sell through different types of local outlets, each involving different commercial actors: direct sales other than those on the farm (44% of the sample), shops (19%), restaurants (14%), farmers' markets (12%), Solidarity Purchasing Groups (9%) and schools (2%). All of these commercial outlets have different product requirements, especially in terms of quality and quantity, but no coordination to this end exists at the territorial level (Filippini, 2015).

As can be seen in Fig. 3, fruits are not processed locally. This practice is common in the area and also applies to the processing of meat, milk and cereals. The reason is that, first, there are few processors operating in the area. Second, local processors do not always have the facilities or the expertise that the farmer needs. Thus, it seems that an 'alternative' marketing logic is possible for local sales, but not for local processors, who tends to adhere to more standard and industrial processing methods typical of CFCs. This is especially true for direct on-farm sales. This example thus shows that:

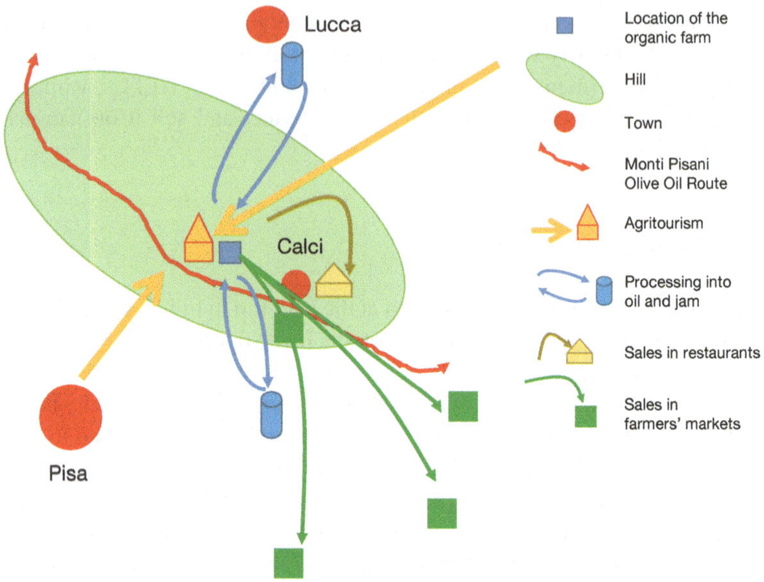

Fig. 3 Hybridisation of an ALFC between different spaces of marketing and product processing. In this example, olive oil is processed locally, but fruit is not. *Source* Lardon et al. (2017)

- even when farmers sell all production locally, the sales are often split between different markets, which leads to a hybridisation of markets;
- even if the sale is alternative and local, the processing is not necessarily local, creating a potential for hybridisation of territorial scales in the analysis of the food supply from a single farm.

Figure 4 shows an example of a farm of a farmer-retailer. The farmer belongs to the opportunistic-strategy group and combines CFCs and ALFCs. All of the farmer's sunflower production and half of his wheat production is sold through the local cooperative which uses CFCs. The rest of the wheat is sold via ALFCs by processing the wheat into flour at a local factory and in his own bakery. This bakery transforms the flour into bread, which the farmer then markets in various stores in the city and peri-urban areas. He has also set up a door-to-door sales network in his own village and also sells in another municipality as part of a project to supply school canteens. Finally, he has opened a retail shop, in association with another bakery in Pisa, where consumers can also find other products from neighbouring farms. Thus, in this case, not all the production is sold through ALFCs. For the production that is sold through ALFCs, the processing is done locally. Of the flour produced locally, some of the production is sold elsewhere in the province. Thus, he combines ALFCs with the non-local AFCs. In this case too, there is a hybridisation of territorial scales. Finally, it should be noted that the diversity of his activities in these local networks are supported by his revenue from his sales through CFCs.

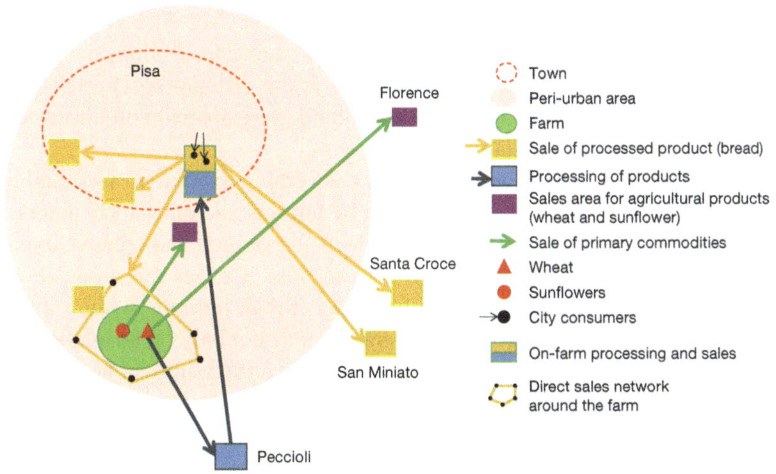

Fig. 4 Example of a farm: hybridisation of ALFCs and CFCs, and hybridisation of different ALFCs. *Source* Lardon et al. (2017)

4 Discussion and Conclusion

The purpose of this study was to show the hybridisation of peri-urban farmers' marketing strategies. This hybridisation is a strategy of adaptation used by farmers who are subjected to the effects of urban pressure. Participation in AFLCs is a response to the economic and territorial constraints of the farming world, as well as to the consumers' growing interest in locally produced food. Farmers are adapting to the new demand and new territorial conditions. The farmer's personal social ties, perception of risk and production system can influence the extent and nature of this hybridisation.

This study has shown that marketing hybridisations that combine AFCs and CFCs can be of different kinds. To begin with, farmers may combine ALFCs and CFCs for the same and/or for different productions. Furthermore, not only do farmers combine ALFCs and CFCs, but there can also be interdependence between them, for example in the economic support that one form of marketing may offer to another, or in the possibility offered by CFCs to sell what cannot be sold through ALFCs. With the agricultural production system being in a state of flux, the hybridisation of CFCs and ALFCs can offer an opportunity to divide risks.

Even if we consider only those farmers who sell through ALFCs, our analysis suggests that there is a hybridisation of marketing strategies that farmers use to meet the demand of local consumers and traders for products and product quality. This can raise questions about the ability of farmers to ensure a constant supply to all commercial actors in the market.

There is also a hybridisation of the territorial scales at which farmers act along the local food chain. Even if sales and production are local and consistent with the rationale of ALFCs, processing is not always local. Not all actors in the food chain are able to apply alternative rationales, and 'being local' does not guarantee a more sustainable development of the food system (Bloom & Hinrichs, 2011). This reflection also leads us to take the social and geographical conditions and the territorial impacts of such initiatives into account.

Finally, during the three years of this study, several individual projects and initiatives were launched, modified or abandoned by farmers and other local actors. This process is of interest, not only because it reflects the general and creative dynamics of adaptation to new opportunities, but also because it raises questions about the conditions necessary for the sustainability of peri-urban farming systems and local food systems. The sustainability of a system depends not only on adaptations to the hybrid characteristics of supply and demand, but also on stability, economic viability and resilience (López-Ridaura et al., 2005). This requires understanding how these initiatives can be profitable for farmers and other actors in the system and to what extent local food systems can stabilise and sustain agriculture in peri-urban areas that are under pressure from urban expansion.

References

Bloom, J. D., & Hinrichs, C. C. (2011). Informal and formal mechanisms of coordination in hybrid food value chains. *Journal of Agriculture, Food Systems, and Community Development, 1*(4), 143–156.

EEA. (2006). *Urban Sprawl in Europe: The Ignored Challenge*, European Environment Agency, Office for Official Publications of the European Communities.

FAO. (2010). *Food, Agriculture and Cities: Challenges of Food and Nutrition Security, Agriculture and Ecosystem Management in an Urbanizing Word.*

Filippini, R. (2015). Food production potential of periurban agriculture: contribution of periurban-farms to local food systems. Agricultural sciences, AgroParisTech, Scuola superiore Sant'Anna distudi universitari e di perfezionamenton, Pisa, Italy.

Filippini, R., Mazzocchi, C., & Corsi, S. (2019). The contribution of urban food policies toward food security in developing and developed countries: A network analysis approach. *Sustainable Cities and Society, 47*, 101506.

Filippini, R., Marraccini, E., Lardon, S., & Bonari, E. (2016a). Is the choice of a farm's commercial market an indicator of agricultural intensity? Conventional and short food supply chains in periurban farming systems. *Italian Journal of Agronomy, 11*(1), 1–5.

Filippini, R., Marraccini, E., Houdart, M., Bonari, E., & Lardon, S. (2016b). Food production for the city: Hybridization of farmers' strategies between alternative and conventional food chains. *Agroecology and Sustainable Food Systems, 40*(10), 1058–1084.

Forney, J., & Häberli, I. (2015). Introducing "seeds of change" into the food system? Localisation strategies in the Swiss dairy industry: Introducing seeds of change into the food system? *Sociologia Ruralis, 56*(2), 135–156.

Ilbery, B., & Maye, D. (2006). Retailing local food in the Scottish-English borders: A supply chain perspective. *Geoforum, 37*, 352–367.

Ispra. (2015). *Il consumo di suolo in Italia*, Istituto Superiore per la protezione e la Ricerca Ambientale.

Istat. (2011). *15 Censimento Generale della Popolazione e delle Abitazioni*, Istituto nazionale di statistique, Dati warehouse.

Izumi, B. T., Wynne, W. D., & Hamm, M. W. (2010). Market diversification and social benefits: Motivations of farmers participating in farm to school programs. *Journal of Rural Studies, 26*, 374–382.

Kneafsey, M., Eyden-Wood, T., Bos, E., Sutton, G., Santini, F., Paloma, S.G., Venn, L., Schmutz, U., Balázs, B., & Trenchard, L. (2013). *Short Food Supply Chains and Local Food Systems in the EU: A State of Play of Their Socio-Economic Characteristics*. European Commission, Joint Research Centre, Institute for Prospective Technological Studies, Seville, Spain, p. 128.

Lamine C., & Perrot N. (2008). *Les AMAP: Un nouveau pacte entre producteurs et consommateurs?* Éditions Yves Michel, Gap.

Lardon, S., Houdart, M., Loudiyi, S., Filippini, R., & Marraccini, E. (2017). Food, integrating urban and agricultural dynamics in Pisa, Italy. In C. Perrin, C. Soulard, & E. Valette (Eds.), *Toward sustainable relations between agriculture and the city* (pp. 15–31). Springer.

López-Ridaura, S., Keulen, H. V., Ittersum, M. K., & van, Leffelaar P.A. (2005). Multiscale methodological framework to derive criteria and indicators for sustainability evaluation of peasant natural resource management systems. *Environment, Development and Sustainability, 7*(1), 51–69.

Marsden, T., Banks, J., & Bristow, G. (2000). Food supply chain approaches: Exploring their role in rural development. *Sociologia Ruralis, 40*, 424–438.

Pascucci, S. (2007). Agricoltura periurbana e strategie di sviluppo rurale. Working paper 2/2007. Università degli Studi di Napoli Federico II, Italy.

Renting, H., Marsden, T. K., & Banks, J. (2003). Understanding alternative food networks: Exploring the role of short food supply chains in rural development. *Environment and Planning a: Economy and Space, 35*(3), 393–411.

Snyder, K., & Bird, L. (1998). *Paying the Costs of Sprawl: Using Fair-Share Costing to Control Sprawl*, US Dep. Energys Cent. Excell. Sustain. Dev.

Sonnino, R., & Marsden, T. (2006). Beyond the divide: Rethinking relationships between alternative and conventional food networks in Europe. *Journal of Economic Geography, 6*(2), 181–199.

Soulard, C.-T., Valette, E., Perrin, C., Abrantes, P. C., Anthopoulou, T., Benjaballah, O., Bouchemal, S., Dugué, P., Amrani, M. E., & Lardon, S. (2017). Peri-urban agro-ecosystems in the Mediterranean: Diversity, dynamics, and drivers. *Regional Environmental Change, 18*(3), 651–662.

Tolron, J.-J. (2001). L'agriculture périurbaine: paradigme et paradoxes d'une péri-agriculture. Illustration en région méditerranéenne. *Ingénieries eau-agriculture-territoires, 28*, 65–74.

Tregear, A. (2011). Progressing knowledge in alternative and local food networks: Critical reflections and a research agenda. *Journal Rural Studies, 27*(4), 419–430.

Venn, L., Kneafsey, M., Holloway, L., Cox, R., Dowler, E., & Tuomainen, H. (2006). Researching European "alternative" food networks: Some methodological considerations. *Area, 38*(3), 248–258.

Zasada, I. (2011). Multifunctional peri-urban agriculture. A review of societal demands and the provision of goods and services by farming. *Land Use Policy, 28*(4), 639–648.

Chapter 9
Marketing Tradition: Leveraging the Know-How and Identity of the Brazilian *Faxinal Emboque* Community

Vanessa Iceri

As Edgar Morin said, the first thing that strikes us when we observe a starry sky at night is the seeming disorder of the scattered stars. A second glance reveals a cosmic order, with every star in its place, on every day that we look up. A third look discloses a disorder once again, exposing an expanding universe that is constantly in movement, with stars being born, exploding and dying. We see that the universe is organizing itself even as it is disintegrating. With this third look Edgar Morin demonstrates the dual capacity of our mind to conceive order and disorder at the same time. For science, the desire to comprehend this complexity depends on taking into account the links that are broken by the divides between cognitive categories, types of knowledge and disciplines (Morin, 1990, p. 164).

As with the stars, transformations in agriculture can be seen, at first glance, as a disorder (Allaire & Daviron, 2017). The second glance, that of science, allows the classification and naming of these transformations (as models or forms), sometimes to the point of compartmentalising these transformations into broad categories such as conventional agriculture and alternative agriculture (Lockie & Halpin, 2005; Le Velly, 2017). The third look would then consist of going beyond the dualism promoted by this distinction between the two major agricultural models, and more generally of looking beyond simplifications and preconceived ideas (Renting et al., 2012; Wilson, 2012).

Even though radical transformations in the system of production and organisation have been widely described and analysed since the start of the modernisation of agriculture (Mendras, 1960; Bodiguel, 1975; Chonchol, 1986; Rattin, 2008), tradition in agricultural practices has not disappeared (Hervieu & Purseigle, 2008; Bouche et al., 2010; Vizeu et al., 2015). In fact, new strategies have emerged, some of which rely on tradition (Bérard & Marchenay, 2007; Cruz, 2012), to help actors survive in

V. Iceri (✉)
UMR Territoires, CNPq (Brazil), AgroParisTech, Clermont-Ferrand, France
e-mail: vankimie@hotmail.com

P. Gasselin et al. (eds.), *Coexistence and Confrontation of Agricultural and Food Models*,
https://doi.org/10.1007/978-94-024-2178-1_9

this context and find a place in the market (Chazoule & Lambart, 2011; Diestchy, 2015).

The traditional aspect of a community, once it becomes a source of innovation, can be used as an asset for its development. We suggest using the concept of adaptation to understand how this happens in a community project of traditional farmers in Brazil (the *Terra Faxinalense* project[1]). How does the transformation and adaptation of agriculture and socio-spatial organisations take place? We have chosen an approach based on socio-spatial organisation, which is understood not only as the relationships that actors maintain with their surroundings for their activities, but also as the way they coordinate with each other for territory-anchored collective actions (Lardon, 2015).

In order to report on this adaptation based on maintaining tradition, we will first introduce the analytical framework. We will then describe the Brazilian national agricultural context and the transformations of agriculture in an attempt to understand where the analysed case study fits and why these producers have created a collective project. This will be followed by a description of the place of tradition in the community project in order to discuss the forms of adaptation that are revealed. We will end with a conclusion.

1 The Invented Tradition: A Concept to Understand Adaptation?

We will address the topic of tradition on the basis of the concept developed by Hobsbawm and Ranger (1983) in *The Invention of Tradition*, a book in which these authors analyse the relationship between the large-scale production of new traditions and the acceleration of modernising social, economic and political relationships in the context of Europe's industrial revolution.

Although this concept, developed in the 1980s, was formulated in a European context, Babadzan (1999) suggests its reuse on the basis of more recent situations in Europe and elsewhere. He shows the similarity between the process of invention of tradition and other phenomena of symbolic and ideological production, marked by a tendency to suppress agrarian societies and the emergence of new forms of social and economic organisation. These latter are a response to modernity, driven by the need for political legitimisation of traditional societies (Babadzan, 1999).

In this perspective, we find two types of tradition: the first is an authentic tradition, in which the continuity with the past is real, not fictitious, i.e. when it does not pursue objectives that are no longer traditional (e.g. the legitimisation of the modern political order). The second form is the invented tradition, characterised as 'a set of practices, normally governed by overtly or tacitly accepted rules and of a ritual or symbolic

[1] *Faxinalense* (singular) or *faxinalenses* (plural) are ethnonyms to describe what pertains to traditional communities of the '*Faxinal*' category in southern Brazil, mainly in the state of Paraná.

nature, which seek to inculcate certain values and norms of behaviour by repetition' (Hobsbawm, 1995, p. 174).

The invented tradition aims to express or ensure a group's cohesion and identity. It also serves to structure social relations in a context of rapid societal transformations, which are weakening or destroying the social models for which the old traditions were developed (Babadzan, 1999). In this framework, social groups, weakened by the transformations of their environment, will adapt in order to survive and evolve more easily.

Adaptation, according to Simonet (2009), takes into account the surrounding's influence on man. Man creates a response to mitigate the constraints he is up against. Adaptation is the result of an event that has already occurred. Two important questions thus arise in relation to collective action. The first refers to the reasons for the adaptation. These may be a desire to reduce the vulnerability of social systems in the face of crises (Burton et al., 1993) or to help avoid the feeling of social displacement or exclusion (Rouillon, 1996). The second question concerns how this change takes shape, how it materialises or translates into the reality we observe. This allows us to distinguish four forms of adaptation: changes in behaviour; changes in functions; assimilation of a new practice; and resistance to transformations with gradual change. 'Faced with change (social, economic, technological), there exist organisations that are resistant to adaptation, but which must transform to avoid disappearance by accepting a gradual and permanent change, rather than an endured and brutal change' (Simonet, 2009, p. 397).

2 The Coexistence of Agricultural Models in Brazil

2.1 Transformations of the National Context

When we talk about agrarian transformations in Brazil, a process that is not homogeneous,[2] we need to distinguish three major movements: modernisation, industrialisation, and the creation of agro-industrial complexes (Kageyama, 1990). These three movements at a national scale have influenced the history of agrarian transformations at the country level, especially in Paraná State, located in the country's south.

Modernisation, marked by technical changes from the 1960s onwards, has resulted in a reduced production of artisanal and hand-produced goods, in the specialisation of labour and in agrarian concentration. At the same time, the agricultural production system has moved closer to an industrial model, with the introduction of fertilisers and the supply of raw materials to other sectors. Finally, the agro-industrial complexes of the 1970s have brought agriculture within the ambit of financial processes and under the influence of international capital and the State (Sepulcri, 2005).

[2] Modernisation varies from farmer to farmer. It can be partial or full within a farm. Only certain actors have access to the different phases of agrarian transformations.

These transformations have contributed not only to increased production, improved productivity and reduced prices, but also to land concentration. This has resulted in an impoverishment of populations because of income concentration, rural exodus, reduction of areas devoted to food crop cultivation, and increased pollution and water consumption (Sepulcri, 2005).

Some farmers have remained outside this process of agricultural transformation, thus finding themselves in a 'marginalised' non-homogeneous category.[3] Some of them are attempting to leverage their specificities in order to strengthen their ways of life and production, which are no longer in sync with the new paradigms of a society transformed by modernisation, industrialisation and globalisation.

Despite this trend in Brazil, particularly in Paraná, which has led to a significant migration of people from rural to urban areas and has precipitated the transition to industrial agriculture, there is an area of large concentration of family farmers located in the south-central part of Paraná State, a region also known as 'Traditional Paraná'. This area is not only home to the majority of the traditional *faxinal* communities, but also has a larger rural population compared to the rest of the state. This hasn't however prevented the advent of temporary monocultures, mainly tobacco, corn and soya bean.

2.2 General Characteristics of the Traditional Agricultural Model of the **Faxinalense**

The status of 'traditional community' in Brazil has evolved from a social distinction to a legal and administrative one since the 1990s (Kohler, 2009). This evolution has allowed certain populations to live and produce in environmentally protected areas. The decree on the status of the *faxinalenses* was approved as recently as in 2007–2008.[4]

A *faxinal* community is defined by its traditional peasant system, marked by the collective use of land for raising livestock and for environmental conservation. This system is based on three components: collective, extensive, and community animal husbandry; agricultural production (mixed-crop food production for self-consumption and production for the market); and sustainable use of forest resources (mate tea, araucaria fruit and other forest species).

The very existence of the *faxinalense* communities will be threatened if their way of life, intrinsically linked to agricultural production, disappears or if it loses its traditional character. Indeed, several *faxinalense* communities in Paraná have

[3] The marginality of farmers can be characterised by practices, values, postures, and actions that are different from those of the conventional model, while remaining heterogeneous: family farming, *campesin*ato, Landless Workers' Movement, agroecology, traditional communities (*quilombolas*, *faxinalenses*, *ribeirinhos*, etc.).

[4] The 'Lei Estadual 15.673/2007' (at the level of Paraná) and 'Lei Municipal 1.780/2008' (at the level of the São Mateus do Sul municipality).

disintegrated as a result of the sale of their lands, or after adopting crops that are not allowed in a traditional community zone (conventional soya, tobacco). Strengthening their social organisation is one way to meet the challenges of emerging from obscurity, participating in political decisions, and carving out a niche in the market. This highlights the importance of renewing traditions and passing them on to younger generations, so that they find a coherence in their way of perceiving reality and act in accordance with this perception.

2.3 The Socio-spatial Organisation of This Agricultural Model

The socio-spatial organisation of a *faxinal* (Fig. 1) is based on an 'inner land', in which 'livestock-rearing land' is clearly demarcated by fences, which are characteristic landscape elements that denote the entrance to the community. This fenced-in area encompasses the communal pastures, the forest area where the livestock roams about freely, the houses, the food-crop gardens and the production infrastructure. Surrounding the fenced area is the 'outer land', which fills an important economic function: the production of grain and other market crops. The community's outer land is divided into plots for individual use, in contrast with the inner land, which is for collective use.

The inhabitants of the community formulate their rules together for managing such a socio-spatial organisation collectively. These rules are formalised through a

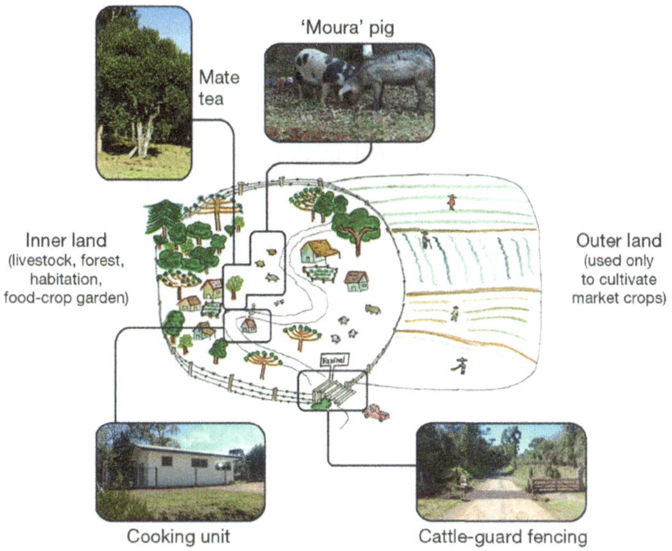

Fig. 1 Socio-spatial organisation of the *Faxinal Emboque* with elements that serve as an identity for the community. Based on Iceri & Lardon, 2018, © Érès

document called 'Community Agreement'. It lists the general regulations concerning the organisation of actors for decision-making (annual general assembly to decide on resource management plans, community work sessions for maintaining fences). It also includes the rules on taxation (creation of a local commission, notification of irregularities), rules for animal husbandry and agriculture (permission to produce mate tea, unless it obstructs the access of pigs to the area), rules for transportation and collaboration (reciprocity of land use).

3 The *Faxinal Emboque* Case Study

The São Mateus do Sul municipality, home to the *Faxinal Emboque* community, is located in the south-central region of Paraná (Fig. 2).[5] This community, comprising of 68 families, is spread over an area of 480 ha. Most of these families are of Polish descent and have been living here since the nineteenth century. The main economic activities are pig farming, gathering of forest produce (mate tea and araucaria fruit) and cultivation of commercial crops on the outer land (mainly maize, beans and tobacco, but also some rice, wheat, soya, potatoes and cassava). The production system is characterised by family labour, use of draught animals and low incomes.

Tese communities resort primarily to political activism as a way of dealing with the challenges of the confrontation between agricultural models. They form networks to demand increased rights and recognition from various public institutions, as well as more participatory agricultural policies. But this type of networking does not take into account the specific needs and day-to-day difficulties of each community. This is the reason why *Faxinal Emboque* decided to act in a different manner. In addition to its involvement in political activism, the community also planned and implemented a development project to address the local problems it faced: the *Terra Faxinalense* project.[6]

This project was initiated in 2013 and involved a total of 170 people, including 56 *Faxinal* member families and 27 neighbouring small family farmers. It was conceived following a controversy around an event suffered by the *Faxinal's* farmers, which is still being debated. In 2011, the pigs being raised in the community were culled by the municipality's health department following an anonymous allegation of the presence of diseases in the animals, something that was apparently never confirmed. This episode was the trigger for the mobilisation of the community to undertake the project.

[5] This area was studied as part of my on-going doctoral thesis. The data collected were obtained from semi-structured interviews with the farmers of the *Faxinal Emboque* community, during a three-month stay in Brazil (June to August 2016). Other data were obtained from observations and from a workshop with a group of the community's women producers. On my return to the field in November 2018, I organised a workshop in the community to share my results with its members.

[6] Video presentation in Portuguese of the Emboque community's project: https://www.youtube.com/watch?v=uQyrq_9QzPU (retrieved 30 October 2021).

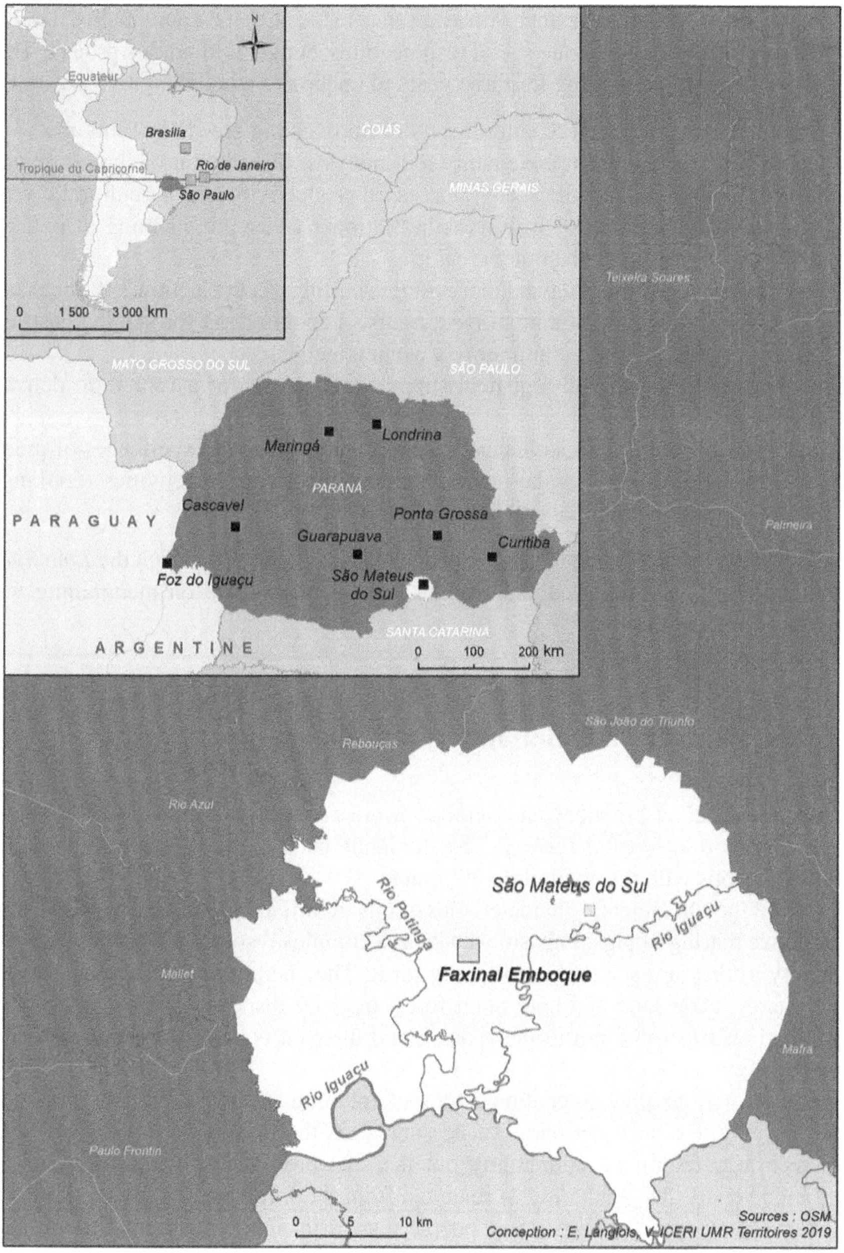

Fig. 2 Location of the *Faxinal Emboque* community. Execution: Langlois and Iceri

A major national oil company, with a presence in the farmers' municipality, agreed to use funds from its corporate social responsibility budget to fund this project. The project thus received funding for three years to undertake work along five axes:

- livestock: reintroduction of naturalised or improved pig breeds by Embrapa[7];
- agriculture: technical improvement (soil analysis, distribution of lime, organic fertiliser and green manure seeds), increased productivity and agricultural diversification (introduction of fruit trees in the inner areas, and planting of potato, wheat, rye and soya in the outer areas);
- processing: acquisition of machinery for rice hulling, seed selection, soya bean oil extraction, production of animal feed, as also a cooking unit for making pastries and bakery items by the community's women members;
- marketing: technical and logistical support (acquisition of a vehicle to deliver food items);
- exchange of knowledge: collective training and sharing of experiences of practices, techniques and regulations concerning the project's activities (cooking, health, agriculture, animal and human health, etc.).

These five themes of the collective project reflect the way in which the *Emboque* community planned on strengthening the *faxinal* system and on maintaining the families that comprise it.

4 The Place of Tradition in *Faxinal Emboque*

A large number of practices and symbols were recognised and asserted as being traditional within *Faxinal Emboque* by the inhabitants in the course of interviews and workshops with the community's farmers.

One of the fundamental characteristics of this traditional production system is the free-range rearing of pigs. Pigs are viewed as a complex resource for the community territory as they are a symbol of identity for it. They help transform organic waste into energy, clear land and help plant forest trees by dispersing seeds, since they also feed on fruits that fall to the ground, and their fat is used as medicine by the community.

The farmers exhibit other dimensions of tradition in their subsistence mixed-crop farming. Exchange or barter of peasant seeds that cannot be found elsewhere are seen as a traditional community practice that promotes self-reliance, sharing and biodiversity. The gathering of mate tea from the forest and its consumption in '*roda de mate*' (get-togethers where people share drinks) are also seen as traditional practices in the *faxinal*. The area for the movement of pigs and its fences are managed and maintained collectively. These practices point to the producers' organisational tradition.

---

[7] Embrapa: *Empresa Brasileira de Pesquisa Agropucària* (Brazilian Agricultural Research Corporation).

The fence, the cultivation of mate tea in the forest, the free-range pigs, and the collective cooking unit, highlighted by the *faxinalenses*, can be regarded as socio-spatial elements that are representative of the community (Fig. 1). Some of these elements directly reflect a *faxinal's* intrinsic traditional aspects (fences, pigs, mate tea), while others serve as support for newer activities (collective kitchen and new products derived from pigs) which are, however, based on traditional production. Tradition and innovation are complementary here.

As far as food production is concerned, the community associates its tradition with recipes or a few artisanal products that are marketed, some of which are an extension of their European (especially Polish) cultural heritage—Polish soup and ham, *pierogi* (Polish ravioli)—while others are more linked to Brazilian culture, such as peasant *cabocla* (pork crisps, sausage) and *queijo de porco* (or pig's head cheese, a type of sausage).

All these variations illustrate how tradition is practised and how it is intrinsic to the way of life and daily existence of the *faxinal* farmers. Thus, it is the practices and symbols as a whole that form the basis and coherence of their tradition.

4.1 Swimming Against the Tide? The Marketing of Tradition by **Faxinal Emboque**

Because of its project, *Faxinal Emboque* is beginning to view its work, production and the marketing of its products in a new light. It is a matter of a restructuring to adapt its system to the market, without losing its production system's 'traditional' character. This adaptation involves a reorganisation of production and compliance with health and business standards imposed by the market. For example, the women's group has been able to supply its products to the school nutrition programme as well as to weekly markets. This group is undergoing training, for example in vegan cooking, to access new markets in order to meet emerging market demands.

Another example of adaptation is the production of dry-cured ham, a specialist product that could find acceptance by the Brazilian market, especially in large urban centres such as Curitiba, the region's capital (located 150 km away). In fact, *Faxinal Emboque* has been contacted by a local entrepreneur interested in the qualities of its livestock. A final example is the creation of an ice cream made with gaviroba,[8] a regional fruit, in partnership with an entrepreneur in the community. At present, gaviroba is little used for human consumption; it mainly serves as animal feed. This ice cream illustrates an innovation that finds root in the *faxinal* tradition.

The five axes of the *Terra Faxinalense* project and the innovation processes underway demonstrate the will to use traditional community products in order to conquer new markets. As a farmer of the community affirmed: 'We want to sell a story behind a product…. For me, innovation simply means deriving value from what already exists' (M.W., *Faxinal Emboque*).

[8] *Campomanesia lineatifolia*, or gaviroba, a fruit that is well-known in the region.

This shift between traditional practices and innovation illustrates changes in ways of thinking and acting by producers in order to deal with the transformations of context confronting them.

5 An Adaptation Based on Four Forms

Using the analytical framework described above, we characterise and discuss the four forms of adaptation of the *Faxinal Emboque* project.

5.1 New Practices

The event that triggered the collective project in the *Faxinal Emboque* (the culling of the supposedly diseased pigs) led the *Emboque* community (collective action) to adopt a different strategy from those of other communities in the region that resort only to political action (militancy). It has developed various collective actions to aid new production practices, new techniques (new pig breeds) and new activities, such as the collective kitchen and the seed processing unit (organisational adaptation).

5.2 A New Meaning Given to Livestock Farming and the Faxinalense *Way of Life*

Over and above the productive advantages of free-range livestock rearing, it also helps in the conservation of collective lands and forest areas. A new meaning is thus attributed to it.

The marketing of *faxinal* products corresponds to the economic desire to increase community income. However, the new outlets also function as a means to disseminate this agricultural model and the families' way of life. In this manner, the community has found a way to make itself known more widely, to raise societal awareness and to identify support for its cause.

5.3 Ways of Assimilating the New

Learning and experimentation have been essential to these processes of adaptation and innovation. New recipes, especially vegan ones, have been introduced as a result of the training provided and emerging food demands. Experimentation has resulted in the production of dry-cured ham and gaviroba ice cream, which are novel products in the region.

5.4 Practices of Resistance

This adaptation is in line with Hobsbawm's (1995) notion of invented tradition. In fact, the 'traditional community' status only appears after these social groups find themselves weakened and feel the need to reaffirm themselves as such. The qualitative term 'traditional' implies an adaptation as resistance, where the process of change is undoubtedly present, even as tradition is reaffirmed.

6 Conclusion: Adaptation as a Way of Conceiving Tradition and Innovation Together

The clear divide between the modern and the traditional no longer applies in the process of the modernisation of Brazilian agriculture. In fact, we identify porosities in these frontiers that manifest in a hybrid form of socio-spatial organisation. Hybridisation is understood here as the 'capacity to renegotiate some of the traditional characteristics of a mode of functioning' (Le Velly & Dubuisson-Quellier, 2008, p. 7).

Using the example of *Faxinal Emboque*, we have examined the complementarities between tradition and innovation in the transformation of practices and products, and in the adaptation of the socio-spatial organisation. In a desire to maintain the *faxinalense* tradition, the community's farmers adapt their action strategy (creation of a project), their behaviour (new techniques, new practices), their products (new products) and the functions of certain practices (usefulness of forest protection for livestock farming, usefulness of social awareness for gaining access to commercial outlets).

The study of *Faxinal Emboque* shows that tradition developed and enhanced by the community and the market favours innovation (social, knowledge-related, technical, commercial, new products), conceived from the intersection of socio-spatial scales. It is a matter of a process of adaptation that links ideas, behaviours and functions that are a priori antagonistic.

References

Allaire, G., & Daviron, B. (2017). *Transformations agricoles et agroalimentaires. Entre écologie et capitalisme*, coll. Synthèses, éditions Quæ, Versailles, 432 p.

Babadzan, A. (1999). L'invention des traditions et le nationalisme. *Journal de la société des océanistes, 109*(2), 13–35.

Bérard, L., & Marchenay, M. (2007). *Produits de Terroir. Comprendre et Agir*, CNRS/Alimentec, 59 p.

Bodiguel, M. (1975). *Les paysans face au progrès* (p. 177). Presses de Sciences Po.

Bouche, R., Bordeaux, C., & Aragni, C. (2010). Ancrage territorial de savoir-faire collectifs: les fromages corses. In: J. Muchnik & C. de Sainte-Marie (Eds.), *Le temps des Syal: Techniques, vivres et territoires* (pp. 79–99) coll. Update Sciences & Technologies, éditions Quæ, Versailles.

Burton, I., Kates, R. W., & White, G. F. (1993). *The environment as hazard* (p. 290). Guilford Press.

Chazoule, C.,& Lambert, R. (2011). Ancrage territorial et formes de valorisation des productions localisées au Québec. *Économie rurale. Agricultures, alimentations, territoires, 322*, 11–23.

Chonchol, J. (1986). *Paysans à venir. Les sociétés rurales du tiers monde* (298 p). La Découverte.

Cruz, F. T. (2012). Produtores, consumidores e valorização de produtos tradicionais : um estudo sobre qualidade de alimentos a partir do caso do queijo serrano dos Campos de Cima da Serra – RS. Doctoral thesis, Speciality: rural development, Porto Alegre, 292 p.

Diestchy, M. (2015). Tensions et compromis dans les valeurs spatiales du slow. *Carnets De Géographes, 8*(1), 2021 (retrieved on 30 October).

Hervieu, B.,& Purseigle, F. (2008). Troubled pastures, troubled pictures. French agriculture and contemporary rural sociology. *Rural Sociology, 73*(4), 660–683.

Hobsbawm, E. (1995). Inventer des traditions. *Enquête, 2*, 171–189.

Hobsbawm, E. J., & Ranger, T. (1983). *The invention of tradition* (p. 320). Past and Present Publications.

Iceri, V., & Lardon, S. (2018). L'organisation socio-spatiale, un commun pour le développement territorial. Le cas d'une communauté *faxinal* au Brésil. *Espaces et Sociétés, 4*(175), 87–104.

Kageyama, A. (1990). O novo padrão agrícola brasileiro: do complexo rural aos complexos agroindustriais. In: (G. Delgado et al. (Eds.), *Agricultura e Políticas Públicas* (Vol. 127, pp. 113–223). IPEA.

Kohler, F. (2009). Commentaire sur « De la "communauté" aux "populations traditionnelles": aspects de la modernité amazonienne » (Roberto Araújo). In: *Des catégories et de leurs usages dans la construction sociale d'un groupe de référence : « race », « ethnie » et « communauté » aux Amériques*, ou Colloque *Nuevo Mundo Mundos Nuevos*, 7 July 2009.

Lardon, S. (2015). L'agriculture comme potentiel de développement des territoires péri-urbains. Analyse par les configurations socio-spatiales. *Journal of Urban Research*, special issue 6.

Le Velly, R. (2017). *Sociologie des systèmes alimentaires alternatifs, une promesse de difference* (200 p). Presses des Mines.

Le Velly, R., & Dubuisson-Quellier, S. (2008). Les circuits courts entre alternative et hybridation. In: G. Maréchal (Ed.), *Les circuits courts alimentaires. Bien manger dans les territoires* (pp. 105–112). coll. Références, Educagri.

Lockie, S., & Halpin, D. (2005). The 'Conventionalisation' thesis reconsidered: Structural and ideological transformation of Australian organic agriculture. *Sociologia Ruralis, 45*(4), 284–307.

Mendras, H. (1960). Exode Rural Et Industrialisation. *Diogène, 30*, 116–130.

Morin, E. (1990). *Introduction à la pensée complexe* (p. 160). Éditions du Seuil.

Rattin, S. (2008). Évolution des structures. L'exploitation française est devenue une entreprise. *Rapport du Service central des enquêtes et études statistiques (SCEES)* (p. 24). et Ministère de l'Agriculture et de la Pêche (MAP).

Renting, H., Schermer, M., & Rossi, A. (2012). Building food democracy: Exploring civic food networks and newly emerging forms of food citizenship. *International Journal of Sociology of Agriculture and Food, 19*(3), 289–307.

Rouillon, F. (1996). Expressions psychiatriques des nouvelles inadaptations. In: P.-F. Chanoit & J. de Verbizier (Eds.), *Les nouvelles inadaptations* (pp. 25–35). Éditions Érès.

Sepulcri, O. (2005). Estratégias e trajetórias institucionais da empresa de assistência técnica e extensão rural do Paraná (Emater-PR). Doctoral thesis, Universidade Federal do Paraná, 161 p.

Simonet, G. (2009). Le concept d'adaptation : polysémie interdisciplinaire et implication pour les changements climatiques. *Natures Sciences Sociétés, 17*(4), 191–199.

Vizeu, F., Seifert, R. E., & Hocayen-da-Silva, A. J. (2015). Non-capitalist organizations in Latin America: Lessons from the Brazilian Faxinal Grassroot community. *Cadernos EBAPE, 13*(2), 2021.

Wilson, A. D. (2012). Beyond alternative: exploring the potential for autonomous food spaces. *Antipode, 45*(3), 719–37.

Chapter 10
History and Coexistence of Agricultural Development Models. The Cases of Argentina, France and Brazil

Christophe Albaladejo

In order to consider the adaptation of agriculture, one has to introduce the question of time into the analysis of agricultural activity, in other words, into the analysis of one of the most significant activities of humanity that concerns nature and forms of occupation of geographical space. We therefore need a theory of social change, and of the modes of relating to territories and to nature. Change, and therefore time, can be conceptualised in three major ways: systemic time, which highlights functioning (Delattre, 1985), but struggles to take evolution into account; adaptation time, which makes it possible to address the evolutionary process of systems, notably through Piaget's (1975) concept of 'majoring equilibration', but which does not explain revolutions; and historical time, which addresses the long term through a division of time into periods (called 'historical blocs' in Gramsci, 2012) that explains profound upheavals. Thus, functioning, adaptation and history are three different ways of understanding change and therefore time.

Why then emphasise adaptation? One possible reason could be that we humans have become aware of the earth-bound nature of the human condition, as Arendt (1958) explained, i.e. of the limits of our world. She quotes the naive exclamation of a journalist at the time of the launch of Sputnik into orbit around our planet in 1957, believing that we were finally escaping from our imprisonment to the earth.[1] Arendt points out that, on the contrary, since that event, humanity has continued to notice that its inhabited, and above all inhabitable, space is so small that it is increasingly filled by objects that are the result of our own actions, or these actions' involuntary consequences. Sputnik, albeit temporarily and in a very modest way, even managed to intrude into what we see in the sky: the eternal celestial bodies. The functioning

[1] A 'step toward escape from men's imprisonment to the earth' (Arendt, 1958, p. 21).

C. Albaladejo (✉)
ACT Scientific Division, AgriTerris network, INRAE & UNLP FCAyF-Conicet, Buenos Aires, Argentina
e-mail: albalade@me.com

P. Gasselin et al. (eds.), *Coexistence and Confrontation of Agricultural and Food Models*,
https://doi.org/10.1007/978-94-024-2178-1_10

of the world in which we live, and in particular its dysfunctions, can thus no longer be resolved by escaping or through expansion. Yet we resist exploring the unknowns of history, which would convince us of the necessity of adapting within our world's limits by exposing the sedimentation of our own actions and objects, most notably because we can no longer 'wipe the slate clean' of our technological past.

A comparative study between three large agricultural and rural countries has revealed significant differences in the collective awareness of this earthly condition, but which in all three cases forces us to adopt a vision that is different from that of 'classical' modernity. This voluntary abandonment of modernity, and in particular of its essential concept of universality, does not, however, mean that we know clearly towards which goal we are headed. The concepts of post-modernity (Lyotard, 1979), or liquid modernity (Bauman, 2003), or, more simply, late modernity (Dubet, 2002), which are concepts designating a situation in which relativity and diversity replace the notion of universality, are not clearly defined—which is indeed one of their essential characteristics. And yet, the socio-cultural condition of coexistence is based on these principles of diversity and relativity, precisely because there is only one model, and it necessarily refers to a 'post-modernity'. French agriculture is, by and large, concerned with the environment. However, this reference to the environment is more ambiguous in a large modern country such as Brazil, whose flag bears the positivist slogan of Auguste Comte (*Ordem e Progresso*[2]), and in Argentina, where 44 million inhabitants in a country with a surface area comparable to that of India can sometimes forget the limits of the earth-bound nature of the human condition.

We will therefore briefly present a theory that makes it possible to account for the coexistence of agricultural models, and then analyse the modern origins of the current situation. We will then discuss the adaptation processes at work in this new context and, finally, we will examine the future of this situation of coexistence and ask whether it is temporary.

1 The Paradox of the Impossible and Necessary Coexistence of Development Models: The Theory of Territorial Pacts

The agricultural modernisation of the 1960s and 1970s created a unique situation of transformations across the world, but it was also accompanied from its very beginning by criticisms, alternatives, resistances, remanences, resiliencies, etc., which were so explicitly in opposition to, and therefore also in reference to, the dominant model that they became inseparable from it. However, in the 1990s, in the midst of the global push for globalisation, we discovered original experiences of production and ways of life in the countryside in France and Argentina that had no need to refer to a dominant model to define themselves (Albaladejo, 2005a, 2005b). These emerging forms of agriculture, which proved to be highly coherent internally, also possessed, in a modest

[2] 'Order and progress'.

but stable way, the four components that made the 1960s and 1970s agricultural model a 'development model'. Indeed, these forms of agriculture encouraged a stable link to a national or local (municipal) State sector, a durable and stable presence in a market, a specific relationship with a science and technology sector, and a capacity to link itself to emerging concerns of society in general. For this reason, these innovations seemed to us to be more than just resistances, and we called them 'discreet innovations', coherent and connected to the State, markets, science and society, but discreet nonetheless. Indeed, in the 1990s, especially in Argentina, the State apparatus and public policies turned their backs on them, and society as a whole ignored them.

Changes in government, and subsequently in a more or less important State sector over the following two decades, made these innovations less and less discreet, and even helped to promote some of them as 'development models' in their own right. For example, after the 2001 crisis in Argentina, family farming became, in the same way as in Brazil (Albaladejo, 2003), an agricultural model identified not only in public policy, society, and the agricultural sector itself, but also in science. In France, there has been a gradual commitment by the government to change production models, as attested by the *Grenelle de l'Environnement* in 2007, the launch of an 'Agroecological Project for France' in 2012,[3] the 2014 law on the Future of Agriculture, Food and Forestry, etc.

This visibility accorded by public policy in the 2000s to certain discreet innovations has enabled us to refine our notion of an 'agricultural development model' through inductive observation of changes in three different countries (Fig. 1). A model is the result of a fourfold convergence of changes that, taken together, help consolidate the emergence of a form of agriculture that goes hand in hand with a type of territorial insertion of agriculture (which we represent by the concept of 'territorial mediation'), most often having previously passed through a stage of discreet innovation (Albaladejo, 2017). These are changes first in society, or more precisely in the 'social agenda' (linkages with urban demands, new identities, etc.); then in markets and the State, and more generally in the 'public agenda'; and finally in science and technology (emergence of knowledge, currents of thought, or even 'specific' sciences for certain forms of agriculture such as agroecology). There is therefore indeed a co-presence of development models within the same country, and several relationships exist between these models which interact and exchange with each other horizontally at the international level during a period of globalisation. The main legacy of the classical modern period of agricultural modernisation and the nation-state is the invention of 'development' as a relationship between public or private agencies and farmers or rural inhabitants. It is the continuity of this development relationship that leads us in particular to the idea of a development model. But the major difference that coexistence introduces is the multiplicity of models, a multiplicity that goes hand in hand with their indefiniteness, their inexorable incompleteness, and the disappearance of hegemony (not dominance, which can be brutal) of one model over the others.

[3] https://agriculture.gouv.fr/le-projet-agro-ecologique-en-france, retrieved 9 October 2021.

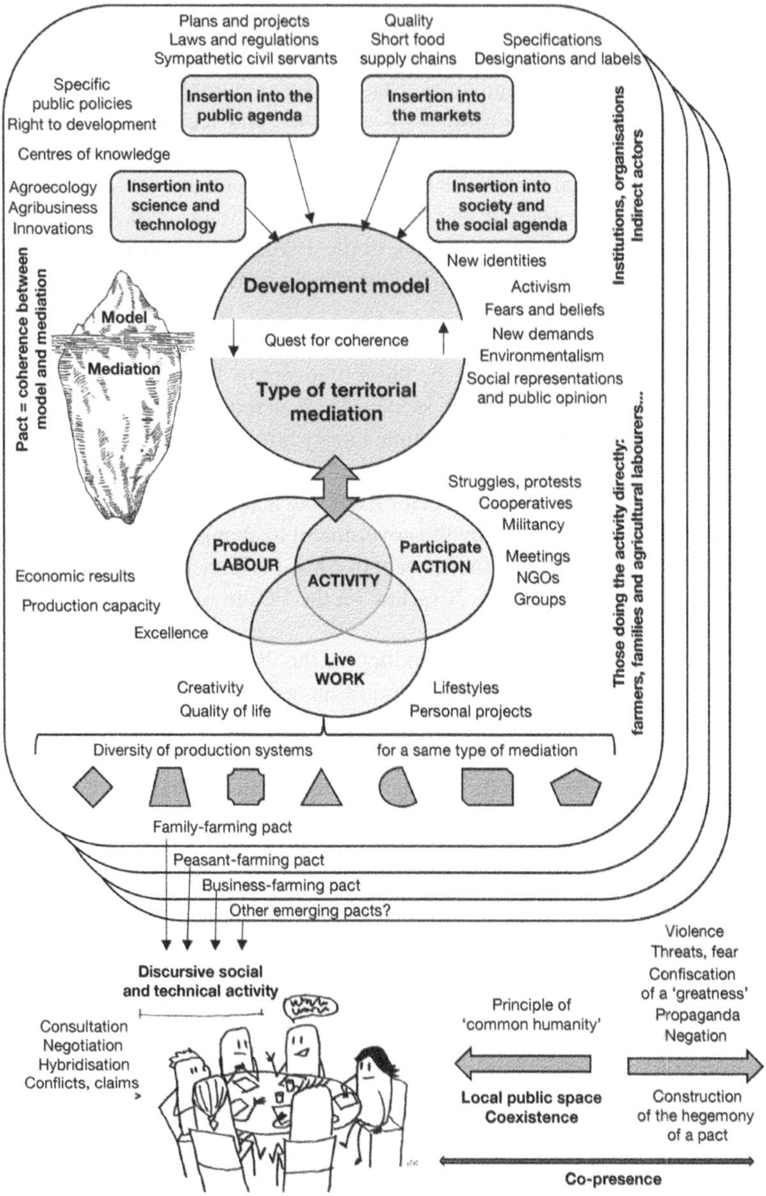

Fig. 1 The coexistence of agricultural development models, a 'late modernity'. Based on Albaladejo, 2017

We use the image of an iceberg in Fig. 1 to show that the development model is what we see, but that it is not the most stable or the most important part of this phenomenon of co-presence. In contrast, territorial mediation is stable and profound, anchored in the daily modes of production and living of those who actually undertake the agricultural activity. Mediation characterises the mode of insertion of agricultural activity in the local territory, and it is always an individual's ad hoc invention, even if it results from an alignment with what a development model proposes.[4] In order to account for the fact that agricultural activity is not just an activity of labour or of production, we have modelled it with Arendt's (1958) theory of human activity, which considers two other dimensions: work (individual creation), which gives personal meaning to things and productions and which is most often, in the case of agriculture, associated with the invention of a way of life; and action, which is characterised by participation in political and associative life in the *polis* (Albaladejo, 2005b). The modes of arrangement between these three dimensions of agricultural activity—labour, work and action—make it possible to define types of territorial mediation that follow the same logic. When a type of territorial mediation corresponds to a development model, we believe that agricultural activity defines what Santos (2000) calls a 'territorial pact'. A territorial pact is 'indispensable for civil society [in our case the social base of the mediation in question] to acquire a legal form [an institutionalisation in our case, in other words, a development model], in order to be able to intervene legally in the politico-legal process [i.e., in our case, in the development relationship and process]' (Santos, 2000, pp. 104–105[5]).

There is one notion that is missing to complete our theorisation of coexistence: that of the local public space (in Fig. 1 it is symbolically represented by actors around a table[6]). This is the space for action in the sense of Arendt (1958), the space for speech that allows us to reason and negotiate the place of each of the territorial pacts. To succeed in constructing this public space at the local level, i.e. at the level of the multiple forms that co-presence takes, means moving from co-presence to coexistence. This discursive treatment is only possible within the limits of situations of 'common humanity', as Boltanski and Thévenot (1991) note, in other words, of situations of respect for the speech of others. But we know, especially in Argentina and Brazil, that co-presence goes beyond this common humanity, and also most often includes situations of violence, intimidation, propaganda, etc. It is most common for co-presence to translate into a juxtaposition resulting in the negation of other pacts. These common situations of co-presence are therefore far from being forms of coexistence. Indeed, coexistence can even be said to be an atypical form of co-presence.

---

[4] Such is the case with organic farming, where the farmers who are the least militant are usually content to align themselves with the model, without inventing their own territorial mediation, compatible with the model but original.

[5] Translation by the author.

[6] Of course, it is just an image, supposed to represent the multiple forms of spaces and interactions in everyday life, taking place at different times and in very diverse places that are not necessarily, or even usually, around an actual table.

A few years ago, we deliberately and provocatively stated that this local discursive activity of coexistence was a 'necessary utopia' (Albaladejo, 2003): it is theoretically impossible, yet it is a socio-political goal that is impossible to renounce. This is why coexistence is a challenge for actors in a democracy, even in countries where democracy has been replaced by brutality, because each model of agriculture must learn to justify itself with appropriate discourses. This is what the considerable semantic and discursive work done by researchers and, in particular, by agronomists, attempts to explain. However, we also felt during this research (*op. cit.*, 2003) that the creation at the time of two Ministries of Agriculture in Brazil represented a considerable opportunity to consolidate new forms of agriculture (family farming in particular) without the obligation to ensure coherence (territorial, commercial, etc.) with other agricultural models. However, this isolation seemed to us to be necessarily temporary, because this form of adaptation through fragmentation cannot forever desist from the quest for major coherences. Recent events, which seem to prove us right (the abolition of the two ministries in Brazil), are, however, part of another process: that of an authoritarian attempt to rebuild a hegemony based on a single model imposed by force.

2 Is the End of the Hegemony of the Territorial Pact of 'Modern' Agriculture Equivalent to Its Disappearance?

Armed with the concepts that have helped us comprehend the current situation of coexistence, we will turn to the past to attempt to understand better where this situation originated. It is clear that the classic modern agricultural pact of the 1960s and 1970s did not possess the necessary adaptive capacities to endure or evolve, but why? What remains of it in the current situation in which no other pact has been able to impose its hegemony? The answer requires a longer-term perspective. In particular, it is clear that the sectorisation of agricultural activity, in other words its relative autonomisation in the national economy and society, which is one of the pillars of the modern pact, was achieved for agricultural activity that had already been 'ruralised', i.e. removed from the cities by the emergence of an urban phenomenon that itself developed from some of the largest cities. The ruralisation of agricultural activity was in turn itself one of the pillars of what we have called 'agrarian pacts', or, in contrast to the succeeding pact, the 'traditional pacts'. In fact, in the pacts that preceded the agrarian pacts, whether in the colonial pacts of Brazil and Argentina or in the bourgeois pacts and, before them, the feudal pacts in France, the actors of agricultural activity were not contained within specific centres (agricultural towns, agri-towns), nor within a clearly distinct culture and space (the rural). The 'rural' (and therefore also the 'urban') and the 'agricultural' (understood as 'sectorised') are therefore constructions and legacies of previous pacts, whose reinterpretations can be understood as adaptations of the succeeding pacts to the 'rugosities' (according to the concept of Santos, 2000) of their past.

Since the 1960s, agricultural modernisation has strengthened the role of the nation-state and has organised agricultural activity into a national 'sector' regulated by public policies. This was the golden age of 'national' disciplines such as agronomy, zootechnics and quantitative genetics. Agriculture was oriented towards the national market or external markets, and it ceased to be organised in production basins intended to supply nearby towns. The model of modernised agriculture based on science and the State constructed a hegemony. According to Gramsci (2012), hegemony is more than domination, because it goes beyond coercion and succeeds in imposing itself through the spontaneous consensus of all, including and especially that of the dominated. Classical modernity, built on the principle of the universalisation of interests, firstly of agriculture's social base, and secondly of society with its farmers, was fully ready for the construction of a hegemony. One can even go further to say: there is no classical modern model without hegemony.

During the same period of the 'ruralisation of agriculture', cities, especially the larger ones, gained in autonomy, and a phenomenon of metropolisation has occurred in parallel with globalisation. Since the late-1980s, the emphasis on globalisation, metropolisation and the expansion of urban culture have weakened the nation-state in the face of a civil society and businesses in networked metropolises. This is the golden age of internationalised disciplines such as biology, molecular genetics, etc. At the same time, the trust between citizens and their modern agriculture, and even between citizens and science, has been dented by environmental problems and health crises, and by increasing levels of education and information of citizens. After three or even four decades of adapting to its own crises (through the addition of an adjective: community, endogenous, sustainable, local, territorial, etc.), the modern model is losing its hegemony in the face of these disturbances, and development models originating from other forms of agriculture are consolidating (peri-urban agricultural models that have their own public policies, family, organic, financial, business farming, etc.) These forms of agriculture no longer fit into the concept of the 'sector', either because they are (re)territorialising and localising themselves (proximity farming, family farming), or because they depend on non-agricultural actors and capital (financial farming).

What has become of the farmers at the social base of the modern model, those who are still called 'professional farmers' in France and 'conventional producers' in Argentina? They are still very much present in the territory and represent a very significant number of farmers in rural localities and, indeed, very often they form a majority. Many elements of a model's four dimensions, most often inherited from the preceding model, remain available and can be mobilised, and sometimes undergo significant adaptations. What these farmers no longer have, even though they were the 'aria' of the previous period (of Gramsci's 'historical bloc'), is a discourse of their own, which is why we have dubbed them the 'silent farmers' of coexistence. They express themselves, even very prominently, but with the vocabularies and rhetoric of other emerging models. And we know the importance of one's own semantic activity in order to exist and survive as a model within the historical bloc of coexistence.

Furthermore, we have determined that their discourse does not correspond to that of business farming models, nor to that of family farming models. To take the example of technology, as heirs of earlier modernisation, new models are burdened by the sedimentation of numerous objects, representations and discourses, whereas agroecology and agribusiness start with a clean slate. Agribusiness does not yet have a technical past, or at least its logic is not beholden to a 'context'; the environment is supposedly virgin. Agroecology is based on a technical past that is sufficiently remote, or mythical, to allow it to be completely 'patrimonialised' and thus to be invented and mastered. The technical past of silent farmers cannot be patrimonialised, because it is literally 'incorporated', in their bodies, but also in ways of thinking and in the territory. This past cannot be set aside or manipulated. This is why this model of conventional agriculture, which is neither that of agroecology nor that of agribusiness, must undergo a profound process of adaptation, and not of invention as other models do. And the research community is not currently of any help.

3 Adapting to a New Context

How do farms adapt in this new context of the historical bloc of coexistence? Jean Piaget's (1975) theory of equilibration, which we have presented in detail in an earlier study (Albaladejo, 2013), seems to us particularly well-suited to represent these adaptations, given the importance of the so-called 'transition' processes. A system's activity of adaptation is always driven by a disturbance, whether external (markets, climate, policies, opportunity, etc.) or internal (illness, accident, project, etc.). Internal disturbances include the 'directed constructions' (Piaget, 1975), i.e. the system's internal programme of transformation, which is most often conceived 'on the fly'. These constructions can be compared to what is called the 'agroecological transition', or even to transitions towards agribusiness. The term transition is problematic, however, for two reasons. On the one hand, it is not in harmony with the 'postmodern' culture of this historical bloc of coexistence, since it presupposes that one knows where one is going—which seems to be much more in line with the culture of the earlier bloc of the classical modern pact. On the other hand, the term 'transition' obscures the importance of the intermediate stages, which are perhaps the only interesting ones, if not the only possible ones. Coexistence itself cannot be approached as a transition from a hegemony to a counter-hegemony, and this book rightfully assumes it to be a historical bloc in its own right.

Systems function with a continuous absorption of small disturbances, through a double process of accommodation/assimilation. Assimilation is the incorporation of external elements into the forms expected for them by the system. It leads to a system that continues to function completely 'normally', without any hitch. For its part, accommodation, in the face of the inevitable small disturbances in system-environment relations, is the successful incorporation of these disturbances into the system's normal functioning (and thus enabling assimilation) by making a small change in the system that does not transform it fundamentally. If this normal accommodation/assimilation process is unsuccessful when confronted by a disturbance, the system moves from functioning to equilibration, which can be of three kinds.

'Alpha equilibration' consists of denying the disturbance, and waiting for it to stop when it has consumed the system's reserves. This is costly in the long run, but it is, for example, the most common reaction to climate change. This attitude of denial of disturbance is also, incredibly enough, the attitude of the agribusiness model in Argentina to environmental problems. We encounter statements such as, 'Yes, glyphosate is classified as carcinogenic by WHO, but so is coffee.'

'Beta equilibration' consists of modifying the structure to respond to the disturbance, without changing the rest of the structure. This is the case of family farmers who lease part of their fields to companies to grow soya, and thus receive part of another model's rent (indeed, the difference in rent is a very strong disturbance). This is also the case in Argentina, in the Pampas, with the agroecology movement, which is calling for the exclusion of phytosanitary products in buffer zones of 100–500 m around towns, to avoid (but for how long?) the issue of coexistence with agribusiness. This is also the case for agribusiness, which, in the face of concerns about glyphosate, is adding a system of expertise and controls at the level of technical operations ('good agricultural practices') without actually modifying the farm's functioning.

'Gamma equilibration' is the only 'heightening equilibration' (*equilibration majorante*) according to Piaget (1975), because it modifies the structure and therefore produces a change in the system. The relationships between the elements are modified to permanently incorporate the disturbance into the system's functioning (consisting of assimilations and accommodations). This is what one would expect from models involved in 'heightening confrontations' in a public space. And yet, we will come up empty if we look for examples because we are clearly witnessing, far from anything resembling this process, a radicalisation of positions and a hardening of discourse over the last decade, particularly in the logic of a form of militancy that requires this kind of confrontation, partly to reinforce certain forms of leadership internally.

4 Conclusion and Discussion: Coexistence is not the 'End of History'

Unless we embark on speculations about the 'end of history', such as those of an eminent American professor after the fall of the Berlin Wall (Fukuyama, 1992) but which have not come to pass, we must believe that there will necessarily be another historical bloc after that of coexistence—but which one? To ask this question about the 'after' is to shift the problem of adaptation from the farm level to that of the model, and, if there is no single model, to that of the socio-political 'regime' that is the base of the current historical bloc. It is difficult to assess the adaptability of a regime that is neither a system (like the 'sector') nor even a pact, but an unstable coexistence of incomplete pacts interacting in undefined public spaces. Will one of the current models then succeed in imposing its hegemony? Our observation of the radicalisation of militant positions shows that this is what some observers believe, and that the idea of the emergence of a new pact negotiated through majoring equilibration or the confrontation and hybridisation of current pacts may not necessarily be the one that prevails.

This 'prospective' effort will, in our opinion, be useful in helping us better understand our current era, in particular the nature of coexistence. All thinkers who more or less accept the idea of a postmodernity, or liquid modernity—including Hardt and Negri (2004) who openly call themselves 'postmodernists'—should be able to admit that coexistence constitutes a historical bloc in itself, and not a mere 'transitional' phase from one bloc to another that would equate it to a period of crisis during which 'the old is dying and the new cannot be born' (Gramsci, 2012: p. 8). When they do so, the notion of hegemony serves as a backdrop, but paradoxically it is, at the same time, rejected outside this period. Hegemony is the construction of a grand narrative, necessarily unique, based on the utopia of universality. It serves to build spontaneous consent, even and especially among the dominated. It is therefore more than domination. It refers to the notion of 'civil society', which Gramsci defines not in opposition to the State, as is most often the case in the literature, but as complementing the State, notably through ideological and symbolic apparatuses (media, schools and training centres, churches, etc.). While the State provides the physical and symbolic coercion necessary for domination, civil society constructs the hegemony. But this is what it does not do in this period of coexistence. We could then suggest the idea of a fragmentation of 'civil society' linked to agriculture and food, with each 'fragment' corresponding to a model. Each of these fragments will try to impose a grand narrative (necessarily partial because it is not universal): agribusiness for business farming, agroecology for family farming, etc. We find here two of the dimensions of our notion of 'model' (Fig. 1): the State and political society (public agenda and government agenda), and civil society (social agenda), with only the market and science missing. Dominant actors will then try to impose an order (a development model) each in its own fragment, before perhaps imposing its hegemony on society as a whole. This is a possible vision of the processes of adaptation and evolution, which would more or less respect a reading of Gramsci, even if this step-by-step vision is not the most convincing. On the other hand, it shows the essential role, for this author of the national scale, of the nation-state. Hegemony is either national or it is not, and the hegemonic class needs the (national) State to defend its interests. Neo-Gramscian authors refer instead to a transnational capitalist class and a global imperial State (World Trade Organization, International Monetary Fund, World Bank, etc.) as well as the law of the market (Laperrière & Bachand, 2014). However, some authors also note that the hegemonic class is not denationalised, and that divergences in their interests can appear and close the national level in on itself. This is what political developments of recent years show, especially in Brazil, where each country may come to define its (unique) agrifood model according to the interests of its dominant class, unfortunately more through State coercion than through a grand narrative that may well remain unconvincing.

We should remember that coexistence is a product of globalisation, and may therefore be susceptible to the fate of other related concepts. For a better understanding of its nature, we can turn to authors with a theory of globalisation, such as Hardt and Negri (2004), and refer to their notions of empire and multitude in a globalised world and a networked society. The notion of empire is opposed to that of imperialism, just as the notion of multitude is opposed to those of mass, people or class. Empire has no well-defined centre, its borders are moveable, its action is deterritorialised. Multitude is the daughter of social networks and the individualisation of society. It is far from being an artificial unit like the 'people' or a simple productive force like the 'mass' or the 'class'; it is instead a 'multiplicity of singularities' (Hardt & Negri, 2004, p. 127). Like the 'private' dimension of territorial mediations, it is based on the creativity and the contingent personal work of individuals. How then to move from singularity to collective action? From mediation to development model? Thus, the notions of 'cooperation' and 'hybridisation' are essential for these authors to understand the aggregation of singularities and therefore the capacity of the multitude to act in the face of empire. Hybridisation is the capacity to mix with others, to learn from their creativities, to see others not as enemies or dangers, but as resources, even allies. The power of the multitude is that of a plurality of actions that no one coordinates and that has no leader. Like the creative but dispersed singularities of discreet innovations, the multitude would not be capable of action, especially of generating models, without a unity that somehow manages to bring these elements together and without an intermediary space for their meeting. The *auto-convocados* movements (and in general the *nuevos movimientos sociales*) in Argentina and Brazil, the *gilets jaunes* in France, and collective action in general, lead us to think that the coexistence of agricultural and food models still has some way to go. The question is to know what forms it will take in the face of existing disturbances that will become more pronounced (environmental and global in particular) and new ones that are emerging (coercion, control, hardening of positions). It may well be that for the first time in a long time there will be strong divergences in experiences depending on the country, without, however, any collective emergence from this regime of action.

References

Albaladejo, C. (2003). Changement social et développement rural: La notion de « pacte territorial » à l'épreuve en Amazonie. In J. Picard (Ed.), *Le Brésil de Lula* (pp. 227–253). Éditions Karthala.

Albaladejo, C. (2005a). Les « innovations discrètes » : Vers un pacte territorial citoyen pour les espaces ruraux français ? *Hégoa, 25*, 87–100.

Albaladejo, C. (2005b). Une Argentine « discrète »... Repérage de nouvelles territorialités en région pampéenne à partir de parcours d'entrepreneurs issus de l'agriculture familiale. Le cas du district de Saavedra (Pigüé). *Norois, 4*(197), 7–22.

Albaladejo, C. (2013). Las capacidades de adaptación de las explotaciones agropecuarias analizadas a través de la teoría de la equilibración de Jean Piaget. In: P. Gasselin, S. Cloquell, & M. Mosciario (Eds.), *Adaptación y transformaciones de las agriculturas pampeanas al inicio del siglo XXI* (pp. 315–348). Ediciones Ciccus.

Albaladejo, C. (2017). Coexistencia en el territorio de diferentes modelos de desarrollo agropecuario: la teoría de los pactos territoriales aplicada al caso argentine. In: D. Nieto, P. Palacios, P. Carricart, C. Albaladejo, & A.L. de Carvalho Fiúza (Eds.), *Transformaciones territoriales y la actividad agropecuaria* (pp. 27–52). Ediciones UNLP.

Arendt, H. (1958). *The human condition* (p. 370). University of Chicago Press.

Bauman, Z. (2003). *Modernidad líquida* (232 p). Fondo de Cultura Económica.

Boltanski, L., & Thévenot, L. (1991). *De la justification. Les économies de la grandeur* (483 p), Éditions Gallimard.

Delattre, P. (1985). *Système, structure, fonction, évolution* (p. 185). Éditions Maloine.

Dubet, F. (2002). *Le déclin de l'institution* (p. 421). Éditions du Seuil.

Fukuyama, F. (1992). *The end of history and the last man* (p. 418). Free Press.

Gramsci, A. (2012). *Guerre de mouvement et guerre de position. Textes choisis et présentés par Razmig Keucheyan* (344 p). La Fabrique Éditions.

Hardt, M, & Negri, A. (2004). *Multitude. War and democracy in the age of empire.* The Penguin Press.

Laperrière, M. -N., & Bachand, R. (2014). L'hégémonie dans la société internationale: un regard néo-gramscien. *Revue québécoise de droit international*, special issue, 1–13.

Lyotard, J.-F. (1979). *La condition post-moderne* (p. 140). Les Éditions de Minuit.

Piaget, J. (1975). *L'équilibration des structures cognitives* (p. 185). PUF.

Santos, M. (2000). *O espaço do cidadão* (142 p). Nobel.

Part IV
Transition

Considering Transitions Through the Coexistence and Confrontation of Agricultural and Food Models: Scales, Actors and Territorial Trajectories. Introduction to Part IV

Salma Loudiyi and Claire Cerdan

Studies on the transformation of agricultural and food models and the processes of transition towards sustainability are mainly based on the framing of sociotechnical regimes. These studies have relied on these regimes to describe, analyse and support the transition trajectories, the actors involved and the innovations induced. For the most part, there is little clarity on the issues of the coexistence and confrontation of agricultural and food models engendered by these trajectories, or they are little recognised as such by these analytical frameworks. The territorial conditions during sustainability transitions, which depend on the situations of coexistence of models in these territories, are also little addressed by the scientific literature. The chapters in this part aim to contribute to the exploration of the links between transition and coexistence of territorial agricultural and food models.

In this introduction, we first review the analytical frameworks used to understand the dynamics of transition in sociotechnical systems and the way in which some research originating from transitions studies is gradually integrating the spatial dimensions of these dynamics by paying particular attention to local contexts and by placing the issue of territories at the centre of analyses. This quick review shows how the issue of coexistence is implicit in these studies and reaffirms the need to take the interplay of scales, actors and trajectories of local development into account. It also allows us to formulate two working hypotheses: one on the links between transitions and modalities of coexistence of territorial agricultural and food models; the other on the issues of governance of coexistence at the territorial level. We then present the four chapters that make up this part, which inform the formulated hypotheses and open up new questions for longer-term research.

P. Gasselin et al. (eds.), *Coexistence and Confrontation of Agricultural and Food Models*,
https://doi.org/10.1007/978-94-024-2178-1

Analytical Frameworks for Understanding Processes of Transition Towards Sustainability in Agricultural and Food Systems

Over the last two decades, various studies on transition processes have been undertaken in order to try to understand the dynamics of socio-economic and environmental changes in new ways (Lawhon & Murphy, 2012). These studies recognise that climate change, biodiversity loss and resource scarcity, and now the health crisis, are major societal challenges (Kölher et al., 2019). To face these challenges, a growing number of analytical frameworks on sustainability transitions have emerged over the last decade to help understand how radical changes can be implemented, even while the societal functions provided by these systems are maintained (Grin et al., 2010).

A Predominance of Theoretical Frameworks Oriented Towards the Analysis of Sociotechnical Systems: Regimes and Niches

Transition is defined as a process of transformation in which a complex system moves from one state of dynamic equilibrium to another. This concept assumes the presence of a desired goal, the transition to sustainability in our case. It also assumes that a progressive path is possible: 'Transition tends not to be revolutionary in its occurrence' (Hinrichs, 2014). Finally, this concept refers to our capacity to act on the temporal trajectories. In general, these studies not only describe the processes and trajectories of sustainability transitions but examine above all the ways in which they can be implemented (Hölscher et al., 2018). The 'how to do it' question has led several authors to suggest that the transformation process is the result of the simultaneous occurrence of multiple convergent changes at different levels and in different sectors of society (technology, the economy, institutions and norms, culture, etc.).

There are several theoretical and analytical frameworks that can be used to address these transition processes. One of the most prominent is the analysis of the transition of sociotechnical systems using Multi-Level Perspective (MLP) (Geels, 2002; Geels & Schot, 2007; Smith et al., 2010). In this perspective, transitions are considered the result of interactions between three levels: (1) the sociotechnical landscape, which encompasses the environment in which society is embedded, (2) a stable sociotechnical regime, composed of rules, practices and interdependent actors who orient or constrain the actions of operators, and (3) niches, which are spaces in which more radical innovations are constructed. The transition from one sociotechnical regime to another is the result of pressures exerted by the landscape on the regime or of the progressive integration of radical innovations (new rules, new practices) into the regime. In this approach, niches (innovations) are understood as incubation spaces (Geels, 2002), places where learning processes take place and where new economic networks are constructed. They are intended to host the construction and consolidation of alternative systems (Meynard et al., 2013). In Geels and Schot's (2007) graphic representation of the transition of sociotechnical regimes, niches tend to gradually integrate the dominant regime by evolving its different dimensions (norms, actors, knowledge, etc.). This representation underscores the transformative or non-transformative character of these innovations vis-à-vis a dominant model.

Other complementary and equally important approaches can be used to address particular dimensions of these transitions. The technological innovation systems (TIS) approach explains how new technologies flourish using different functions such as knowledge development, market formation or legitimisation processes (Negro & Hekkert, 2008; Markard et al., 2015). Strategic niche management (SNM) approaches are widely used to analyse the emergence of innovations and the creation of 'protected' spaces (Geels & Raven, 2006; Schot & Geels, 2008). Finally, the so-called transition management (TM) approaches show how certain actors, in particular public policy actors, can shape transition processes using a set of activities, whether strategic, technical, operational or reflexive (Rotmans et al., 2001; Loorbach, 2010).

All these theoretical and analytical frameworks are based on the analysis of transition processes of different sociotechnical systems (electricity, mobility, buildings, etc.). Over the last decade, analyses of sociotechnical systems associated with agriculture and food have increasingly focused on the transformation of agricultural and food production, processing and marketing systems, and the reconfiguration of interactions and power relations between actors of these food systems (Hinrichs, 2014). Among these studies, some contributions highlight the importance of approaching the transition of food systems through a plurality of objects and complementary themes: global transition (Spaargaren et al., 2013; Hinrichs, 2014), agroecological transition (Lamine, 2012; Ingram, 2015; Levidow, 2015; Bui et al., 2016), and sustainable consumption transition (van Gameren et al., 2015; Vittersø & Tangeland, 2015). In the majority of cases, these studies mobilise MLP's theoretical frameworks. But while they envisage the existence of two well-stabilised regimes (generally the conventional and the alternative) that coexist in the same place, they do not delve into the diversity of situations, nor their specificities and variations with regard to geographical conditions or the modalities of their territorial embeddedness. Even though MLP is based on the presence of a single dominant regime, these studies help understand the coexistence of several sociotechnical regimes in the same context (Dumont et al., 2020). Several studies address the multiplicity of possible and existing trajectories of sustainability transitions. For example, El Billali et al. (2018) show that different transition pathways can be proposed or implemented for achieving food and nutrition security. They identify 'efficiency-oriented pathways' (or sustainable intensification), 'demand-restraint pathways' (or sustainable diets) or 'food systems transformation' (or agrifood transition) leading to an in-depth transformation of the entire food system. According to these authors, these different pathways reflect different visions of what is desirable and achievable in terms of practices, visions that are based on fundamentally different, even opposing, models, ideologies and values. Considering that 'food system transitions thus do not have one easy, obvious, or uncontested pathway but will be characterised by a diversity of options, approaches, places, voices, and historical contexts' (El Bilali et al., 2018, p. 13), these studies underscore the challenges of the coexistence of different approaches, their specificities, their plurality according to the contexts in which they are placed, and, indeed, the challenges of governing the coexistence of these different models and trajectories (Bui, 2015; Bui et al., 2016).

The Emergence of New Analytical Perspectives: The Territorial Conditions During Transition Processes

Despite the important advances in MLP-based research, a few authors (Lawhon & Murphy, 2012; Murphy, 2015) have shown some of its limitations. These limitations include the focus accorded to technological artefacts in these studies or to certain actor categories that shape transitions (leaders, innovators, scientists, government agents, to the detriment of consumers or workers, for example); an approach, seen as 'naive', to the spatial dimensions of transitions towards sustainability (i.e. different scales and spatialities); and the avoidance of analysis of the power games between actors.

The geographical dimension has indeed long been ambiguous and even misunderstood in MLP analyses. The three MLP levels (niche, regime and landscape) are often implicitly aligned with specific territorial boundaries (Raven et al., 2012; Truffer et al., 2015): regimes tend to be presented as national characteristics; sociotechnical landscape dynamics equated with those of international scales; and niches are often equated with sub-national or even local scales. Thinking of national contexts as key elements in which regimes and niches are located, while important, does not capture the territorial differentiations and complex interdependencies that result from different forms of institutional embeddedness in territories (Lawhon and Murphy, 2012). Coenen et al. (2012) add that it is essential to examine more closely the socio-spatial struggles that lead a regime or niche to spread beyond its initial boundaries. In the same perspective, the processes of scale articulation and trans-scalar relations (relations and interdependence between actors located at different scales, circulation of models, transnational networks), which could allow us to understand how these scales trigger or prevent transitions of sociotechnical regimes, are little addressed. According to Lawhon and Murphy (2012), MLP would benefit not only from being more sensitive to the role of geographical factors but also from being more responsible by recognising the power relationships factor as very important in guiding or hindering transition dynamics.

These criticisms have led to recent studies in the geography of sustainability transitions (Raven et al., 2012; Hansen & Coenen, 2015; Longhurst, 2015; Murphy, 2015; Truffer et al., 2015; Binz et al., 2020), which seem to pursue the issues of coexistence of models without, however, naming it as such. This is an emerging field in which the geographical dimension of transitions is addressed through a research effort on three key elements (Truffer et al., 2015): the socio-spatial anchoring of transitions, their multi-scalarity, and the integration of power relations. The socio-spatial anchoring of transitions is aimed at identifying the territorial conditions that are favourable (or unfavourable) to processes of transition towards sustainability. In particular, it is a matter of understanding the territorial inequalities associated with the transition processes (which spaces are favoured and which are disadvantaged). Taking the multi-scalarity (i.e. the articulation between different geographical scales and organisational levels) into account makes it possible to see how innovations emerge in different spaces, how these spaces interconnect and how actors situated at several different scales interact to disseminate these same innovations. Finally, these two dimensions lead to a third, which has to do with the unequal power relations in sustainability transition processes. According to the authors, the effects of these processes must necessarily be considered. This implies paying attention not only to the 'losing' and 'winning' actors, and to the interacting models, but also to the voices and interests of the actors who are part of these models, i.e. to the modalities of coexistence of different models resulting from these transition processes.

These authors' research perspectives are currently centred on geographical inequalities and the spatial variability of transition trajectories and their impact. They focus on two contexts in particular: urban transitions and transitions in developing countries but do not yet address the transition processes of agricultural and food models (Binz et al., 2020). However, these different contributions point to the need to analyse the territorial conditions and factors of these transition processes towards sustainability and their effects on a plurality of territories by verifying how the transition processes of food systems produce new modalities of coexistence of these same models at different scales. To further this reflection, we pose two working hypotheses that we examine in the light of the four contributions in this section. The first is that the coexistence of agricultural and food models can be the condition and the result of transition dynamics at work in food systems. What factors are the triggers for these transitions? What are the relationships between actors that drive or hinder these transitions? What territorial conditions encourage or constrain these processes? What are the future horizons expected by the different coexisting agricultural and food models? What paradigms, values and standards set them apart? Our second hypothesis is that, given that transition processes vary according to territories, their scales, their social and spatial configurations and their trajectories, the coexistence of models can be understood and governed at the territorial level. What are the effects of transitions on the conditions of interaction between agricultural and food systems in a territory? What are the new forms of coexistence produced and at what scale? Which actors are involved and what is the nature of their links and/or interactions (passive co-presence, tensions, synergy, complementarity, etc.)? What are the forms of public action, governance and support that enable a diversity of actors and systems to be committed to the same territorial development horizon, while respecting their singularities?

Transition Processes and Coexistence of Agricultural and Food Models in Territories: The Case Studies

We introduce here the three case studies that make up this part and indicate their common and complementary features. A fourth, panoramic chapter provides an original analysis for understanding transition trajectories of agroecological models, especially from the point of view of their diversity and operationality.

In Chap. 11, Claire Lamine considers the hypothesis that coexistence, understood as the presence of different agricultural models, both 'alternative' and 'conventional', in the same territory, produces processes of hybridisation and controversies. They contribute to the legitimisation of ecologised models and, consequently, to the processes of ecological transition, insofar as the transformation of visions, norms and relations between actors is concerned.

Her analysis of the south of France's Ardèche department is based on the coexistence of conventional and agricultural models, the coexistence of different rationales within agricultural activity, and the territorial coexistence of initiatives within the 'territorial agrifood system', an analytical category that allows the author to examine the territorial conditions of the transition processes and the ecologisation of agriculture. Her work highlights farmers' individual trajectories towards organic farming according to three rationales, all of which show forms of combination and hybridisation in the exercise of agricultural activity. These combinations and hybridisations are observed at the level of production methods (organic, non-organic), production choices (diversification or not) and marketing channels (short and long). They are all part of forms of collective functioning (traditional, new, informal), with farm viability as their goal. Her approach through the different categories of actors and their initiatives reveals the conditions for the emergence of new development models and the recomposition of the agrifood system. This system is the result of a plurality of individual and collective projects of agricultural and non-agricultural actors in a territory, concerning both specific products and more ordinary food products. There are conventional agricultural actors who invest in projects to qualify and structure agri-chains, which illustrate the processes of recomposition and re-differentiation within conventional models of deriving value from local production. For their part, alternative agricultural networks advocate and implement other collective initiatives (e.g collective sales points in short chains). Finally, other initiatives originate from new actors, such as local authorities and civil society actors, who choose to address agricultural and food issues in their territory. Their objective is to offer healthy and local food to all, including to the most vulnerable. All these projects and actors contribute to the recomposition of a territorial agrifood system.

Claire Lamine's chapter shows how the coexistence of agricultural models can trigger transition processes through those of recomposition and internal re-differentiation. It highlights, in particular, how territorial conditions (e.g. territorial identity or local food consumption habits) are levers of differentiation of transition processes (put in perspective with respect to the dynamics of another territory studied by the author: Biovallée).

In Chap. 12, Emmanuelle Cheyns and Nora Daoud analyse the transition of food systems and the coexistence of models through the fine grain of citizen participation in local purchasing groups by studying the daily practices of their members and their consequences on interactions with agriculture. These authors' proposal complements our analysis of the coexistence of agricultural and food models by exploring the modalities of collective action and solidarity, located on the fringes of the State's sphere of influence and at a distance from market instruments.

These two authors suggest that behind each agricultural and food model, patterns of engagement can be identified at the fine scale of individuals and collectives. The latter help explain the mechanisms of the coexistence of agricultural models, which take different forms: tensions, associations, and new ways of 'doing things together'. The authors invite us to explore the geography of everyday practices and to reflect upon radical breaks and modalities. For some of these citizen groups, the issue is no longer of simply revamping forms of supply but of positioning themselves through breaks with the market and by building or 'making' communities. The coexistence of agricultural and food models then seems to become difficult, insofar as coexisting would mean recognising other contested models and tolerating their rationale and validity. From an MLP perspective of transition, these purchasing groups can be understood and analysed as spaces of innovation, and the authors seek to determine the changes induced by these collective approaches. In their chapter, the engagement regime concept mobilised to address the functioning of purchasing groups goes beyond the simple description of shared regimes and values by underlining the tensions they generate and the different modalities of coexistence and solidarities within proximity spaces they lead to. The approach through these transitions, away from the official arenas, contributes in its own way to a transformation and a transition that takes into account a vulnerable public and producers who are sometimes outside the ambit of support mechanisms and the current agricultural models.

In Chap. 13, Guillaume Duteurtre and his colleagues respond to the dual hypothesis formulated above: transition processes generate situations of coexistence of agricultural and food models, and these very situations of coexistence, if we analyse them through the prism of long trajectories, themselves induce transition processes. Territorial, socio-political and economic conditions shape and orient these trajectories. This chapter sheds light on the modalities of governance of these transition processes and of the situations of coexistence, in the case of Vietnam, of agricultural models associated with dairy farming and its industrialisation.

The authors use the MLP framework to explain the multiplicity of the trajectories of this agricultural system, in which several models exist due to transition processes that span the long term. The abandonment of the collectivist economy in the country and subsequent farming reforms supported the development of a family farming model in the 1990s. But the melamine crisis in 2008 and the emergence of social demand for healthy and safe products triggered reforms that, this time around, supported more intensive and industrialised forms of agriculture, giving rise to commercial farms and mega-farms. The transition processes induce, as we hypothesise, not only new models but also a plurality of models and trajectories that imply forms of coexistence in a territory. Importantly, this chapter not only offers initial insights on the governance of these transitions (through national reforms) but above all it sheds light on the explicit willingness of public actors to recognise the issues of coexistence of the agricultural models that local authorities are trying to 'manage'. Land is used as a lever to govern this coexistence (access to land is controlled by the State through redistribution mechanisms), as are construction and sometimes the imposition of local partnerships between farmers, firms and local authorities, and the production of standards and conventions. Compromise, as a form of coexistence, goes hand in hand with the production of sense around the usefulness, necessity and importance of the agro-industrial model (provision of material resources, knowledge production, creation of employment in traditional dairy basins). These forms of coexistence also result in tensions, which highlights the changing nature of forms of coexistence when economic or health crises strike. This probably reflects the fragile and eminently political nature of the governance of coexistence models when it is carried out by local authorities. As the authors note, the issue of the drivers and mechanisms of this coexistence within local territories still needs to be addressed through a detailed analysis of the dynamics of land and financial capital, and their implications for the terms of this coexistence and its governance.

To conclude this part, Philippe Baret and Clémentine Antier propose an analytical and methodological approach to reflect on the effects of transitions and their operationality. Using agroecological transitions as a starting point, the authors defend the importance of taking the diversity of transition trajectories into account through a constructive critique of the MLP framework. Their proposal has the merit of better situating the diversity of transition situations, refining the characteristics of the different possible trajectories and their real-world implications. Starting from a model that seems to be unified (agroecology), they propose to translate it into four 'agroecological proposals' according to a dual characterisation: the extent of changes (scales, degree of integration of actors) and the modalities of this change (radical, incremental). It is a matter of clarifying and making explicit the political choices adopted when actors formulate transition projects for models, i.e. of thinking about the transition not only in terms of technical choices but also by paying attention to social, economic and cultural conditions. The authors stress in particular the need to adopt complementary, multidisciplinary and systemic approaches, while developing, at the same time, the critical and reflective dimension.

The three case studies in this part are characterised by the diversity of analytical scales used (a national scale, a meso-scale of a French institutional territory, and the 'micro'-scales of citizen collectives), and by transition modalities inscribed in differentiated historical, territorial, collective and individual trajectories. These case studies explore both the diversity of scales and that of the territorial anchorage of transition processes and their articulations. They show that the pathways of transition are not always linear, as shown by studies on transitions, and reveal, explicitly or implicitly, the challenges of coexistence of action regimes associated with agricultural models (Chap. 11, Lamine), of individual and/or collective engagement regimes associated with food consumption (Chap. 12, Cheyns and Daoud), or even, more broadly, of choices of governance of agricultural transitions at a national scale (Chap. 13, Duteurtre et al.). The theoretical and analytical frameworks used are based, on the one hand, on multilevel perspectives, and, on the other, on engagement regimes and justification theories. They each illustrate, from a different but complementary scalar perspective, how transition processes at different scales induce situations of coexistence of models that are driven by values, actors and spaces, which in turn participate in formulating transition goals. For its part, the panoramic chapter (Baret and Antier) makes a conceptual and methodological proposal, defending the importance of thinking about transition trajectories not only from the point of view of desired goals but also from the point of view of the choices adopted at the grassroots level, while pointing out the shortcomings of sociotechnical regime frameworks. It is a chapter that uses an innovative way to show the importance of reflexive, critical and engaged analyses.

Conclusion

Reflecting on the transition in terms of the coexistence of agricultural and food models has led us to formulate a dual hypothesis on the links between transition and coexistence of models, in particular the place of territories in these processes of change. Each of the case studies sheds light on a particular dimension of the territorial conditions of a production of situations of coexistence of models. They show the factors that trigger the transitions in question, the relationships between actors situated at different scales, and lead to reflections on territorial conditions that stimulate or hinder these transition processes. However, the case studies still do not address the question of the trajectories created by these dynamics. The chapter by Duteurtre et al. is quite enlightening in this respect. The panoramic chapter by Baret and Antier also revisits the necessity of shedding light on the political visions associated with these models, which would set out the terms of governance for the coexistence of agricultural and food models.

These contributions thus open up, to varying degrees, the issue of this governance of the coexistence of models within territories from a threefold perspective.

Perspective 1: *around the spatial scales of transition processes and the production of forms of coexistence.* The scalar issue calls for an exploration of the circulation of norms, values and contents of models resulting from transition processes. The effect of the articulation of these scales on the modalities of coexistence and their governance has still to be examined. Certain scales can be mobilised to consolidate, establish and legitimise certain innovations that create tension and conflict at other scales. In this case, we speak of trans-scalar connections (Cerdan et al., 2012; Peralta et al., 2014).

Perspective 2: *around actors and the understanding of their strategies and rationales of access to resources.* The contribution of micro-level analyses is very instructive in this context for thinking about the ways in which coexistence processes are constructed in local and remote territories. The processes of domination and power relations are often poorly explored in studies on the transition of models, and, as a result, the forms of coexistence and their social, spatial and political implications are little understood. Analyses of the governance arrangements for coexistence show us that we have to look at the renewal (or reproduction) of relations between the State, the market and civil society actors at territorial levels, and at the reproduction of structural inequalities.

Perspective 3: *around territorial trajectories in order, on the one hand, to investigate the issue of differentiated temporalities and spatialities, and, on the other, not only to grasp the effects of territorial contexts in all their complexity, but also the way they condition forms of governance of this coexistence of models.* In this way, the analysis of territorial trajectories allows us to move closer to genericity using comparative approaches.

These three perspectives inform the analysis and the understanding of the ways in which agricultural and food models coexist. The current context of health and climate crises makes it incumbent upon us to heed Baret and Antier's call to adopt systemic and multidisciplinary approaches for understanding these transition processes, approaches that are more reflexive, more engaged and politically situated. The chapters in this part invite us to do so more than ever.

References

Binz, C., Coenen, L., Murphy, J., & Truffer, B. (2020). Geographies of transition. From topical concerns to theoretical engagement: A commentary on the transitions research agenda. *Environmental Innovation and Societal Transitions, 34*, 1–3.

Bui, S. (2015). Pour une approche territoriale des transitions écologiques. Analyse de la transition vers l'agroécologie dans la Biovallée. Doctorate dissertation presented to the Institut des sciences et industries du vivant et de l'environnement (AgroParisTech). Speciality: Social Sciences, AgroParisTech, Paris.

Bui, S., Cardona, A., Lamine, C., & Cerf, M. (2016). Sustainability transitions: Insights on processes of niche-regime interaction and regime reconfiguration in agri-food systems. *Journal of Rural Studies, 48*, 92–103.

Coenen, L., Benneworth, P., & Truffer, B. (2012). Toward a spatial perspective on sustainability transitions. *Research Policy, 41*(6), 968–979.

Dumont, A., Gasselin, P., & Baret, P. V. (2020). Transitions in agriculture: Three frameworks highlighting coexistence between a new agroecological configuration and an old, organic and conventional configuration of vegetable production in Wallonia (Belgium). *Geoforum, 108*, 98–109.

El Bilali, H., Callenius, C., Strassner, C., & Probst, L. (2018). Food and nutrition security and sustainability transitions in food systems. *Food and Energy Security, 8*(2), 1–20.

Geels, F. W. (2002). Technological transitions as evolutionary reconfiguration processes: A multi-level perspective and a case-study. *Research Policy, 31*(8), 1257–1274.

Geels, F., & Raven, R. (2006). Non-linearity and expectations in niche-development trajectories: Ups and downs in Dutch biogas development (1973–2003). *Technology Analysis and Strategic Management, 18*(3–4), 375–392.

Geels, F. W., & Schot, J. (2007). Typology of sociotechnical transition pathways. *Research Policy, 36*(3), 399–417.

Grin, J., Rotmans, J., & Schot, J. (2010). *Transitions to sustainable development: New directions in the study of long-term transformative change*. Routledge.

Hansen, T., & Coenen, L. (2015). The geography of sustainability transitions: Review, synthesis and reflections on an emergent research field. *Environmental Innovation and Societal Transitions, 17*, 92–109.

Hinrichs, C. C. (2014). Transitions to sustainability: A change in thinking about food systems change? *Agriculture and Human Values, 31*(1), 143–155.

Hölscher, K., Wittmayer, J. M., & Loorbach D. (2018). Transition *versus* transformation: What's the difference? *Environmental Innovation and Societal Transitions, 27*, 1–3.

Ingram, J. (2015). Framing niche-regime linkage as adaptation: An analysis of learning and innovation networks for sustainable agriculture across Europe. *Journal of Rural Studies, 40*, 59–75.

Köhler, J., Geels, F., Kern, F., Markard, J., Wieczorek, A., Alkemade, F., Avelino, F., Bergek, A., Boons, F., Fünfschilling, L., Hess, D., Holtz, G., Hyysalo, S., Jenkins, K., Kivimaa, P., Martiskainen, M., McMeekin, A., Mühlemeier, M., Nykvist, B., Onsongo, E., Pel, B., Raven, R., Rohracher, H., Sandén, B., Schot, J., Sovacool, B., Turnheim, B., Welch, D., & Wells P. (2019). An agenda for sustainability transitions research: State of the art and future directions. *Environmental Innovation and Societal Transitions, 31*, 1–32.

Lamine, C. (2012). «Changer de système»: Une analyse des transitions vers l'agriculture biologique à l'échelle des systèmes agri-alimentaires territoriaux. *Terrains et Travaux, 20*(1), 139–156.

Lawhon, M., & Murphy, J. T. (2012). Socio-technical regimes and sustainability transitions: Insights from political ecology. *Progress in Human Geography*, 36(3), 354–378.

Levidow, L. (2015). European transitions towards a corporate-environmental food regime: Agroecological incorporation or contestation? *Journal of Rural Studies, 40*, 76–89.

Longhurst, N. (2015). Towards an 'alternative' geography of innovation: Alternative milieu, socio-cognitive protection and sustainability experimentation. *Environmental Innovation and Societal Transitions, 17*, 183–198.

Loorbach, D. (2010). Transition management for sustainable development: A prescriptive, complexity-based governance framework. *Governance, 23*(1), 161–183.

Markard, J., Hekkert, M., & Jacobsson, S. (2015). The technological innovation systems framework: Response to six criticisms. *Environmental Innovation and Societal Transitions, 16*, 76–86.

Meynard, J.-M., Messéan, A., Charlier, A., Farès, M., Bail, M. L., Magrini, M.-B., et al. (2013). *Freins et leviers à la diversification des cultures. Étude au niveau des exploitations agricoles et des filières.* Summary of the study report, Paris, Inra, Delegation of Collective Scientific Expertise, Future Prospects and Studies.

Murphy, J. T. (2015). Human geography and socio-technical transition studies: Promising intersections. *Environmental Innovation and Societal Transitions, 17*, 73–91.

Negro, S. O., & Hekkert, M. P. (2008). Explaining the success of emerging technologies by innovation system functioning: The case of biomass digestion in Germany. *Technology Analysis and Strategic Management, 20*(4), 465–482.

Raven, R., Schot, J., & Berkhout, F. (2012). Space and scale in socio-technical transitions. *Environmental Innovation and Societal Transitions, 4*, 63–78.

Rotmans, J., Kemp, R., & Van Asselt, M. (2001). More evolution than revolution: Transition management in public policy. *Foresight, The Journal of Future Studies, Strategic Thinking and Policy, 3*(1), 15–31.

Schot, J., & Geels, F. W. (2008). Strategic niche management and sustainable innovation journeys: Theory, findings, research agenda, and policy. *Technology Analysis and Strategic Management, 20*(5), 537–554.

Smith, A., Voß, J. P., & Grin, J. (2010). Innovation studies and sustainability transitions: The allure of the multi-level perspective and its challenges. *Research Policy, 39*(4), 435–448.

Spaargaren, G., Oosterveer, P., & Loeber, A. (Eds.) (2013). *Food practices in transition: Changing food consumption, retail and production in the age of reflexive modernity*, Routledge.

Truffer, B., Murphy, J. T., & Raven, R. (2015). The geography of sustainability transitions: Contours of an emerging theme. *Environmental Innovation and Societal Transitions, 17*, 63–72.

Van Gameren, V., Ruwet, C., & Bauler, T. (2015). Towards a governance of sustainable consumption transitions: How institutional factors influence emerging local food systems in Belgium. *Local Environment*, *20*(8), 874–891.

Vittersø, G., & Tangeland, T. (2015). The role of consumers in transitions towards sustainable food consumption. The case of organic food in Norway. *Journal of Cleaner Production*, *92*, 91–99.

Chapter 11
The Role of Interactions Between Organic and Conventional Farming in the Ecological Transition of a Territorial Food System

Claire Lamine

The territorial scale is increasingly being recognised as appropriate for addressing the ecological transition of food systems, both at the international level (IFPRI, 2015; IPES-Food, 2018) as well as at the national level (territorial food projects[1]). Several social science studies have explored the mechanisms that slow down or even prevent ecological transition processes or, on the other hand, facilitate them, whether at the level of agricultural or food practices (Cowan & Gunby, 1996), agri-chains (Lamine et al., 2009), alternative food systems (Brunori et al., 2011) or, more recently, territorial food systems (Lamine et al., 2015; Bui et al., 2016).

These studies, however, often focus on so-called 'alternative' networks, on farmers' networks or on networks that link farmers and consumers, and sometimes other actors in the agri-chains and territories. They are not overly concerned with interactions of these networks with local actors involved in more 'conventional' production, processing, retailing, and consumption practices. However, recent studies suggest that the combined presence in a given territory of networks and actors participating respectively in alternative and conventional agricultural and food models produces hybridisation processes, as well as evokes criticisms and generates controversies that contribute, over time, to the affirmation and legitimisation of ecologised models, and thus more broadly to the processes of ecological transition (Cardona & Lamine, 2014; Lamine, 2012, 2017).

This chapter discusses a case study in the southern part of the French department of Ardèche, where an agroecological transition trajectory, mainly based on conversion to organic farming, and also on other forms of ecologisation of practices and of renewal of food distribution chains and of agricultural development, was analysed for over a decade. We use the notion of territorial agrifood system, an analytical

[1] Law no. 2014–1170 of 13 October 2014 on the Future of Agriculture, Food and Forestry.

C. Lamine (✉)
Ecodevelopment Research Unit, INRAE, Avignon, France
e-mail: claire.lamine@inrae.fr

P. Gasselin et al. (eds.), *Coexistence and Confrontation of Agricultural and Food Models*,
https://doi.org/10.1007/978-94-024-2178-1_11

category that encompasses not only local actors in production, processing and distribution agri-chains, but also technical advice, territorial public policies, consumers and local civil society and, consequently, the various mechanisms and networks that link production, marketing and consumption. Thus, this analytical category not only makes it possible to integrate 'alternative' actors and networks, but also those that are part of more 'conventional' models.[2] In this chapter, we will explore, more specifically, the processes of interaction between alternative and conventional networks (hybridisation, as also criticism, controversy and redifferentiation). The underlying hypothesis is that these processes act as a driving force for an ecological transition at the scale of the territorial agrifood system by contributing to the affirmation and legitimisation of ecologising 'narratives' over time, and thus to the fostering of changes in visions and practices, and sometimes also of changes in the power relationships between alternative and conventional actors.

Southern Ardèche is a sparsely populated rural area alternating between largely cultivated valleys, terraces and mountains. It is dotted with towns of varying sizes, the largest of which is Aubenas, with a population of around 55,000, more than a third of the area's total population (140,000). Ardèche is often associated with 'neo-ruralists'.[3] In fact, in the 1970s, and later too, the region was host to the 'back-to-the-land' movement (Rouvière, 2015). At the beginning, these neo-ruralists, whose influx was in stark contrast to the great rural exodus, settled in abandoned areas such as mountain terraces. They restored these lands and some even started agricultural activities, including goat breeding, at least according to what popular perception would have us believe ('Settling in Ardèche to raise goats'). While many left, the others gradually integrated into the local community, actively participating in the social, cultural and economic life of the villages, often even developing innovative systems of collective organisation to support their activities, as we shall see later. Southern Ardèche is also an area which, from the 1950s to the 1980s, was marked by the golden age of fruit cultivation, with production being destined for outside the territory in a highly organised manner. At its peak, in the mid-1980s, the local fruit cooperative had over 2500 members and accounted for 95% of the fruit production in southern Ardèche, with combined annual fruits sales (mainly peaches, apples, pears, cherries) reaching 25,000 to 30,000 tonnes, making it a major player at the national and European levels. However, starting in the early 1990s, in a context of the increasing sway of large supermarket chains, local production became less competitive than in less isolated areas, and it became more difficult to rationalise the functioning of the rather small farming structures here than those in emerging and competing regions in southern France and in other countries. Fruit production has consequently experienced a period of continuous decline over the last two decades (Lamine et al., 2015). The cooperative currently manages to sell only around 2000 tonnes of fruits across all species (i.e. about 15 times less than in its heyday). Chestnuts, a traditional crop on sloping areas, is now the primary production.

---

[2] What qualifies as 'conventional' or 'alternative' is, of course, always relative and changing (Lamine, 2017).

[3] We should remember that this is a fairly heterogeneous category.

Our initial approach was to start from the trajectories of farmers who had initiated a transition to organic farming, regardless of whether it led to a partial or total conversion, or whether it was started at the initial installation stage or later.[4] This approach has made it possible to retrace these trajectories' various dimensions and interactions. These trajectories were, in fact, those of relatively formal groups (producer groups or shops, associations that support peasant agriculture (AMAPs[5]), cooperatives), and these farmers were involved in a variety of interactions with other organic and conventional farmers, and other actors of the territorial agrifood system. As a second step, we retraced, at the scale of this territorial agrifood system, the diversity of dynamics that have emerged since the 1990s in conventional agriculture as well as in organic and rural agricultural networks.[6] The aim is to understand how these different trajectories and dynamics are linked, or not linked, in a process of transition of the territorial agrifood system.

1 Farmers' Trajectories Towards Organic Farming: Combination of Models and Interactions with Conventional Actors

Our analysis of farmers' trajectories towards organic farming focused on two types of agricultural productions: market gardening and fruit cultivation. Despite their different dynamics—the first experiencing a certain revival in the territory, while the latter being in continuous decline for over two decades—they share the common characteristic that their producers use a wide variety of marketing channels. The aim of the study was to trace the evolution of producers and their farms through several components (production method and marketing, as also networks, learnings, and technical conceptions) from the time of installation of the agricultural activity to the present, including analysing how a producer's background shapes his or her decisions. The objective was to identify the farms' logics of evolution, underpinned by specific modes of interaction between production and marketing and by the particular links that producers have with various local networks, including conventional ones. Notwithstanding the uniqueness of the producers' trajectories, three principle farm development rationales can be identified.

The first logic of evolution is that of organic market gardeners choosing to continue with a highly diversified production system. This system aims to optimally leverage these very diverse productions, resulting in combined adjustments to the production and marketing systems in order to optimise the whole. These adjustments concern not only the surface area (which remains very limited), the quantities produced, and diversity, but also the quality of the production: the choice of an alternative production

[4] Survey of 15 farmers conducted in 2009 (Cambien, 2009).

[5] French: *Associations pour le maintien de l'agriculture paysanne (AMAP).*

[6] A long-term survey was conducted between 2009 and 2018. During this period, around 50 interviews were conducted, and dozens of events and meetings on these subjects were analysed.

method (e.g. the choice of varieties) leads to very diverse but often irregular products. But specific marketing chains (markets or 'food boxes' in particular) manage to derive value from this diversity and irregularity. Diversity, for some producers, is also tied from the very outset to an objective of self-subsistence and self-reliance.

The second logic of evolution is that of organic market gardeners gradually moving towards a 'rationalisation' of their production in order to supply greater quantities. While the farmers mentioned above are often new farmers who don't have a local agricultural background, these latter are more firmly rooted in this background (some even taking over the family farm). They have the same starting point as the previous group (limited surface area), but they progressively increase their surface area and production, which they rationalise through increasing mechanisation and a combination of short and long supply chains in pursuit of improved economic viability. These diverse outlets play a complementary role in these strategies. Wholesaling allows the producer to maximise production while saving on production costs, and direct sales constitute an outlet for products that are not valued in other chains, thus acting as a buffer and nudging average prices higher.

Finally, the third logic is that of conventional fruit and wine growers forced to diversify their production and marketing system in order to maintain their farms, which they do by introducing new crops and effecting a partial transition to organic farming. Vegetable production represents an important avenue and source of diversification. It can be set up rapidly and requires little investment. The development of direct sales or short supply chains that accompany these changes makes it necessary to diversify the range of vegetables produced.

In these three types of trajectories, it is, above all, the viability of a new installation or an existing farm that is the issue: optimisation of a very constrained farming structure; viability of the activity following a few years of hyperdiversification (production and marketing) that is difficult to sustain over the long term (Dupré et al., 2017); or maintenance of the farm in a context of sectoral crises. All these transitions are based on a *combination* of outlets (between different short supply chains, and often between short and long supply chains), productions, and even production methods (organic and conventional). These trajectories and forms of combination of outlets are part of collective modes of functioning, whether those that are already well established in the territory and institutionalised like local cooperatives, or collectives set up to supply certain chains such as AMAPs or other types of 'food boxes', collective farmers' shops, school food procurement, or informal networks of organic producers set up to share or exchange equipment and material (Lamine & Cambien, 2011).

2 A Diversity of Initiatives Originating from Both the Conventional World and the Organic and Alternative One

The initiatives that have emerged from the 1990s to the present day can be grouped into three categories: those originating from conventional agricultural actors (cooperatives, producer groups, Chambers of Agriculture, etc.), those originating from alternative rural and agricultural development actors (e.g. organic farmers' networks or CIVAM[7]), and those originating, often more recently, from local authorities and/or civil society actors. These different types of initiatives are contributing, over time, to a wider reconfiguration of the territorial agrifood system (Fig. 1).

Among the initiatives originating from conventional agricultural actors were those to qualify certain products specific to the region through geographical indications. Thus there are indeed Protected Designations of Origin (PDO) for wines, as also for picodon (goat cheese) since 1983, for chestnuts since 2006, and for *Fin gras du Mézenc* (beef). In the case of wine, the local cooperatives worked from the 1990s on a revival project supported by the setting up in 1994 of a union of cooperatives from across southern Ardèche. This union adopted a strategy based on economies of scale (which resulted, more recently, in the disappearance of a number of local cooperatives) for coordinating the development and marketing of local wines, and finally for segmentation into different vintages, including organic ones. At the same time, many producers have gradually chosen to engage in individual winemaking, developing new strategies aimed at differentiating their wines, and sometimes organising themselves collectively for certain processes (e.g. with a CUMA[8] bottling plant). In this way, many winegrowers have now joined the 'natural' wine movement (Barrey & Teil, 2011), while others have gone back to using ancient and rustic grape varieties such as *Chatus*, thus illustrating the process of recomposition and redifferentiation that is taking place between more conventional and more alternative forms of deriving value from local production.

The chestnut is a traditional crop of the territory and historically played a key role in the local diets, up until the beginning of the twentieth century. In addition, the chestnut is undergoing a process of revival that is receiving strong support from publicly funded agricultural programmes (Dupré, 2002). For the purpose of deriving greater value, producers and processing companies—some of which are more than a hundred years old—supported the process of creating a Chestnut of Ardèche PDO, which met with success in 2006. As for technical support, a programme to 'Restore the chestnut grove' was instituted in 2013 at the urging of the professional union, and with the support of the Monts d'Ardèche Regional Natural Park and the Chamber of Agriculture. This programme has been instrumental in the restoration of many abandoned or poorly maintained chestnut groves (Demené & Audibert, 2017). Here

---

[7] Centres for Initiatives to Promote Agriculture and Rural Areas. French: *Centres d'initiatives pour valoriser l'agriculture et le milieu rural (CIVAM)*.

[8] Farm machinery cooperatives. French: *Coopérative d'utilisation de matériel agricole (CUMA)*.

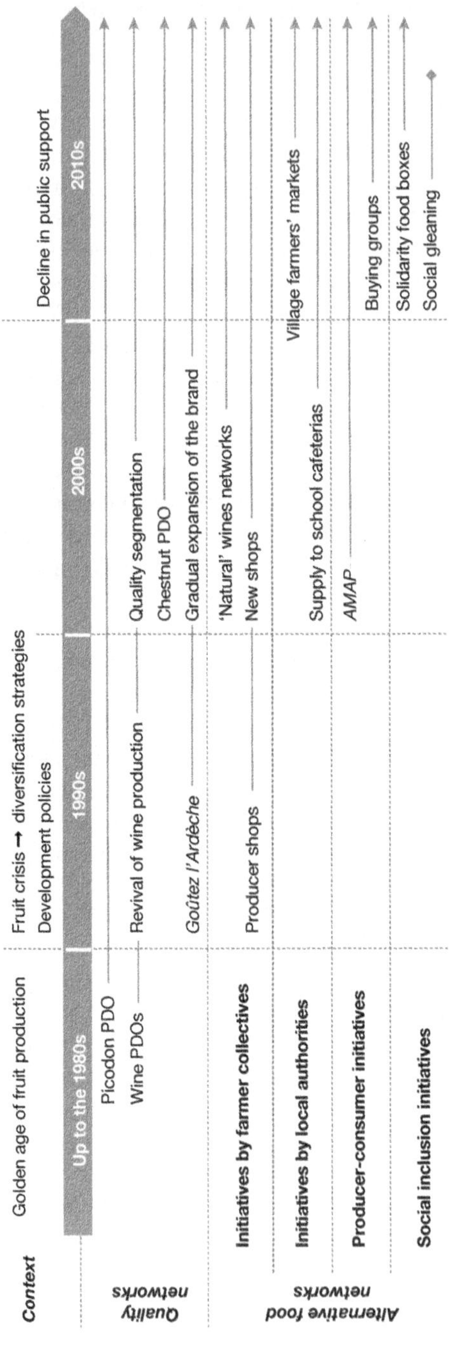

Fig. 1 Trajectory of various initiatives

too, as in the wine sector, a process of recomposition of the sector and redifferentiation is taking place, resulting in numerous small-scale, individual and collective value-adding initiatives, for example around collective processing units.

These approaches for qualification of products and the organisation of agri-chains have focused on specific products (wine and chestnuts), which are not or no longer key in the daily diets of the region's inhabitants. They have nevertheless contributed to an overall re-development of local agriculture. Wine is an emblematic example, since the wines from Ardèche, which were previously considered to be of very modest quality, are now sought after for local consumption, as well as by some restaurants, wine cellars and bars in big cities.

At the same time, a variety of initiatives have emerged concerning more common products such as fruits, vegetables, meat, bread, etc. Spanning both specific products (often with official quality labels) and common ones is the Goûtez l'Ardèche ('Come and taste Ardèche') umbrella brand created in 1994 by the three Chambers of agriculture, of commerce and industry, of trades and crafts. It currently encompasses about 400 product categories ranging from meats to preserves, wine to biscuits and a variety of other products, which are sold in a range of retail outlets, from supermarkets to shops in tourist areas. It also features on the menu cards of local restaurants that are partners of the brand.

Farmers belonging to certain alternative rural development networks and rural agriculture and/or organic farming networks have developed their own set of initiatives. The most symbolic of these is probably the producer shops. One of the first was opened in 1997 in Aubenas, the largest town in the area. It was created on the initiative of a coordinator of a local training programme focused on pluriactivity and the integration of activities of production, processing, marketing and services, which supported many young farmers and others in their projects during the 1990s and 2000s.[9] The initial idea was to support young farmers by involving them in marketing operations so that they could see how to best organise their farm and production based on the shop's requirements. The initiative was, and still is, based on the collective, the solidarity, and the direct relationship with consumers, since all the producers associated with this initiative have also to commit to devote some time in the shop, half a day a week, so that they learn how to promote not only their own products, but also those of other producers. Our initial surveys, conducted in 2009, showed that generational transmission seemed difficult as some older members regretted that the potential new members did not always easily incorporate the founding principles and were more 'consumers of the collective' than truly committed. Ten years later, the baton has been passed on, and other producer shops that have since opened seem to have found their public both in terms of committed producers and of consumers. During this period, different types of consumer have begun to show interest. This is reflected by the shop locations (the first shop was opened in the centre of the main

---

[9] The training programme was initially designed in Isère by the *Peuple et Culture* popular education associations network, with the support of Pierre Müller, a researcher at CNRS who was at the forefront of work on pluriactivity and, in fact, of the concept of the rural farmer (Müller et al., 1989), and was later replicated in Ardèche.

town of Aubenas but the newest one, in another municipality, opened in 2018 in the commercial zone on a highway); by the diversity of the range of products offered (in conjunction with a less binding system that does not require suppliers to devote time in the shop); by their communications strategy; and, finally, by probably a growing general interest in local products.

As for the organic shops, some of them have begun procuring local products (e.g. Biocoop in Aubenas) in accordance with the stated priority of the Biocoop network.[10] Finally, among the initiatives originating from civil society, rural agriculture and development networks, we also find here, as elsewhere, AMAPs, organic-products buying groups, and events such as 'farm to farm' weekends organised by CIVAM since 1999 that, although held only occasionally, reach a greater diversity of consumers.

The third category of initiatives, which involves local residents and ordinary everyday food items, is driven by local authorities and/or local associations. The first 'Country Bistros' (*Bistrot de pays*) opened in Ardèche in 2008. They are meant to offer visitors and also residents a restaurant that is open all year round (in a highly touristic region where many establishments are seasonal) and to provide a public service (in the broadest sense) which could, for example, be postal services, the provision of school meals, or a grocery shop. Farmers' markets have also flourished in the villages of the area in recent years. They are seasonal and are usually started by local authorities and associations. They have operating charters that accord preference to local producers and require them to commit to being present all through the season. These markets create opportunities for exchanges to take place between producers, residents (e.g. through the involvement of parents' associations of village schools) and tourists, and contribute to the liveliness and attractiveness of the villages. Finally, here as elsewhere, collective catering has gradually reoriented itself towards local supply, particularly in schools in the main town of Aubenas (Cambien, 2009) and even in smaller municipalities, through various strategies ranging from direct purchase from producers for directly managed kitchens to the replacement of the usual meal providers (large specialised companies) by local restaurants, for which this outlet provides a regular all-year activity that is complementary to tourism-driven summer activity.

Other initiatives have been launched by local civil society networks, primarily to reach populations—farmers or consumers—they consider are being neglected by agricultural and/or public actors. For farmers, a support system has been put in place for helping new farmers who want to start out based on a mentoring system that links these farmers with more experienced ones. On the consumer side, in addition to local versions of national schemes, such as Jardins de Cocagne 'solidarity food boxes', associations have developed a series of original initiatives for social accessibility to quality food as well as for fostering interactions and cross-learning between producers and consumers. For example, 'social gleaning' workcamps have

---

[10] Biocoop Case Study Report. Healthygrowth Project, 28 p. (French summary: Lamine C., Rousselle E., Étude de cas Biocoop, 10 p.).

been started, involving a public social centre, families and local farmers in collective activities of harvesting and processing of fruit and vegetables.[11]

3 The Difficulty of Forging an Agrifood Project that Involves All Actors

These very diverse initiatives ('quality' networks of conventional actors, alternative agricultural networks, initiatives by public actors and associations) have contributed to a reconfiguration of the territorial agrifood system over the last three decades. Indeed, they have influenced different types of farmers, supply chains and products, and have involved different links within this system (production, processing, distribution, public policies). While this reconfiguration is visible in the growing diversity of existing initiatives, it is also reflected in changes in some indicators. Thus, the share of organic farming in local agriculture (although this partly concerns crops exported out of the territory) is now around 25%, well above the national and regional averages. As for the share of local consumption, our study suggests that it has risen in recent times due to a more pronounced presence and visibility of these initiatives.[12]

However, as of 2018 there was no project that involved and was shared by all the actors in the territorial agrifood system.[13] Our initial surveys in 2009 revealed that some local actors wanted to explore the feasibility of a platform-type tool to facilitate the access by producers in Ardèche to collective catering and other outlets (distant wholesalers, supermarket chains, local shops, etc.). Our analysis of several collective approaches and their successes or difficulties[14] led us not only to insist on a (re)definition of the challenges of the sharing of risks and responsibilities but also, more broadly, to question what form a shared project could take. Collective catering seemed to be the best suited to serve as a 'bridge' between the conventional and alternative professional worlds, as well as between farmers and other residents of the territory, because all farmers consider this to be a legitimate 'chain', unlike others that might be considered elitist or marginal (Lamine & Cambien, 2011). In the meantime, in 2014, a platform called 'd'Ardèche et de saison'[15] was created, but improving the agri-chains' collective and territorial structuring, particularly with a view to supply collective catering, remains a strong challenge.

---

[11] http://civamardeche.org/Glanage-social (retrieved on 7 December 2021).

[12] A more detailed evaluation is still to be carried out on this point.

[13] In 2019, an action research project was actually launched with this aim. See https://www.assiette-territoire.com/.

[14] Notably that of an attempt at collective organisation around the production of organic vegetables, associating the local fruit cooperative, the chamber of agriculture and an organic wholesaler (Cambien, 2009; Lamine, 2012).

[15] A rough translation is 'From Ardèche and in season'.

In the same year, the French law on the future of agriculture institutionalised and defined, albeit in a somewhat vague manner, the notion of a territorial food project.[16] Three actors of the territory, the Chamber of Agriculture, the Agrifood Development Centre which owns the 'Ardèche le Goût' ('The Taste of Ardèche') brand, and the southern Ardèche region, set about defining such a project in 2016 by responding to the National Food Programme's (PNA) call for projects. This project focused on three main issues: raising awareness on reducing waste in school cafeterias, maintaining agricultural land, and labelling products for the markets. The project, called 'Food and territorial development', was based more on economic development and convergence around food than on ecologisation. This may seem surprising, in view of the territory's achievements and strengths, but it is understandable, given the choice of partners involved or left out. Indeed, this project did not directly involve any of the alternative actors in rural agriculture, organic farming and rural development, which led to a feeling of both under-representation and co-optation among some of the area's pioneering actors. This underlines the challenge of identifying and involving the diversity of actors who contribute, as our analysis has shown, to the on-going reconfiguration of the territorial agrifood system.

4 The Diversity of Interaction Processes: Combination, Influence, Redifferentiation

The three types of farmers' trajectories identified above partly overlap the boundaries between organic and conventional farming, between native and neo-rural farmers, and between short and long supply chains. Thus, producers who might have been considered opposites in their production and marketing approach, between diversified organic market-garden farms selling their products in short supply chains and conventional arboriculture farms initially very specialised and in long retail chains, are now converging somewhat in their strategies, practices and visions. This partial convergence is largely due to the increasing legitimisation and credibility of organic farming in the local agricultural landscape.

These partial convergences are also due to the existence of 'crossing points', in the form of actors, networks or places, between these professional worlds. In terms of actors, some producers appear to be mediators between the organic and conventional worlds, as well as between the world of short and long supply chains (Lamine & Cambien, 2011). In terms of networks, those that were initially more alternative or conventional now involve very diverse profiles. For example, innovative structures that were created by neo-rurals, such as producer shops or farm machinery cooperatives (CUMA) for processing, also now count non-neo-rural local producers

[16] The Regional Food Projects (French: *Projets alimentaires territoriaux, PAT*) are meant to bring together producers, processors, distributors, local authorities and consumers, and to develop territorial agriculture and food quality. They are mainly supported by the National Food Programme (French: *Programme national de l'alimentation, PNA*).

as members. Conversely, informal mutual aid networks initiated by 'native' farmers also now include neo-rural farmers. Both types of networks provide social spaces for intermingling and meeting in the territory. Finally, in terms of places, producer shops or village farmers' markets allow organic and conventional farm products and their producers to appear alongside each other.

Our surveys also show the extent to which the changes experienced and observed in the territory evoke criticism and generate new controversies. For example, while organic farmers are no longer considered by non-organic farmers as unreasonable or eccentric, organic farming is still criticised for certain aspects that are seen as inconsistent. At the environmental level, the most common criticism is directed at the use of copper. At the social level, questions are raised about the workload organic farming imposes on farmers in order to feed the rich consumers who can afford to buy these products (Lamine, 2017). This also leads to processes that can be described as a 'redifferentiation' within organic farming itself. For some, changes in practices do not occur without a profound change in the conception of 'good' organic farming. This is the case for some producers who are moving, even if gradually, from a highly diversified system oriented solely towards direct sales, to a system that refocuses production and turns partially to long retail chains. Thus, having started out with the 'utopian notion' of small-scale, diversified production based on short supply chains, one of these producers now considers that 'this vision of the large vegetable garden' is not consistent with the idea of 'producing'. He views his own journey as being part of an increasingly logical process that leads him to produce more, to be more profitable and incur lower production costs, which enables him to offer more people the possibility of healthy food at fairly low costs. For him, delivering small volumes to a small number of customers is less environmentally sound than delivering high volumes to a greater number of customers. We thus see the emergence of different visions of organic farming, with other farmers fearing a 'drift' in organic farming similar to that of conventional agriculture, and supporting the development of peasant organic farming, similar to that observed in peasant agroecology (Lamine, 2017). Not only do these 'redifferentiation' processes play out in the different visions, they also form part of the practices and initiatives themselves. This is how we can understand the movement of recomposition observed in 'food boxes' systems. At the same time as the latter were diversifying with the development of new economic models initiated by new types of intermediaries, the organic networks themselves tried to respond to consumers' need for flexibility and 'practicality' while continuing to be part of an associative model by setting up, in 2017, an online ordering system with delivery assured by the producers.

5 Conclusion

The case studied here raises the issue of coexistence, not so much between agricultural models as between different initiatives acting on different links of the system, while inducing processes of interaction between organic and conventional farming that also contribute to the observed transition. However, this study underscores the difficulties of constructing a transition project at the scale of the entire territorial agrifood system, largely because such a project was carried out, in its institutionalised form, by conventional actors while excluding the more alternative actors. It was the opposite in other cases studied, such as of the Drôme valley, where it was mainly initiated and supported by actors who were very focused on undertaking a radical transition towards alternative systems (organic, local, etc.), with a tendency to exclude, in an opposite of the Ardèche case, conventional professional actors (Bui, 2015). However, in order for such a process to integrate all the actors concerned, it appears that a solid foundation such as a strong territorial identity or a coherence between local production and local food practices would help. The challenge is to recreate or strengthen not only social links and a shared identity (around specific products such as wine, chestnuts or goat's cheese), but also organic interdependencies in the biological sense (consuming local products, living in the same territory, sharing the same landscapes), whereas the contemporary world, even the rural world, has progressively freed itself from organic interdependencies, in the sociological sense of the term (when the farmer depended on various other trades in the village, and vice versa) (Rémy, 1987).

References

Barrey, S., & Teil, G. (2011). Faire la preuve de l'« authenticité » du patrimoine alimentaire. *Anthropology of Food, 8*.

Brunori, G., Rossi, A., & Malandrin, L. (2011). Co-producing transition: Innovation processes in farms adhering to solidarity-based purchase groups (GAS) in Tuscany, Italy. *International Journal of Sociology of Agriculture and Food, 18*(1), 28–53.

Bui, S., Cardona, A., Lamine, C., & Cerf, M. (2016). Sustainability transitions: Insights on processes of niche-regime interaction and regime reconfiguration in agri-food systems. *Journal of Rural Studies, 48*, 92–103.

Bui, S. (2015). *Pour une approche territoriale des transitions écologiques. Analyse de la transition vers l'agroécologie dans la Biovallée* (Doctoral thesis, Sociology speciality). AgroParisTech.

Cambien, L. (2009). Les dynamiques territoriales de production et de transition vers l'agriculture biologique en Ardèche méridionale: des formes mixtes au profit d'une multiplicité d'acteurs. Mémoire de mastère EMTS, Museum-AgroParisTech.

Cardona, A., & Lamine, C. (2014). Liens forts et liens faibles en agriculture: l'influence des modes d'insertion socio-professionnelle sur les changements de pratiques. In A. Bernard de Raymond & F. Goulet (Eds.), *Sociologie des grandes cultures. Au cœur du modèle industriel agricole* (pp. 97–113). éditions Quæ.

Cowan, R., & Gunby, P. (1996). Sprayed to death: Path dependence, lock-in and pest control. *Economic Journal, 106*(436), 521–543.

Demené, C., & Audibert, O. (2017). Promouvoir l'agriculture dans le projet de territoire pour faciliter la mise à disposition du foncier: le cas de la châtaigneraie ardéchoise. *VertigO – La revue électronique en sciences de l'environnement, 17*, 1.

Dupré, L., Lamine, C., & Navarrete, M. (2017). Short food supply chains, long working days: Active work and the construction of professional satisfaction in French diversified organic market gardening. *Sociologia Ruralis, 57*(3), 396–414.

Dupré, L. (2002). *Du marron à la châtaigne d'Ardèche. La relance d'un produit régional* (335 p). Éditions du CTHS.

IFPRI. (2015). Global nutrition report 2015: actions and accountability to advance nutrition and sustainable development. International Food Policy Research Institute. www.fao.org/fileadmin/user_upload/raf/uploads/files/129654.pdf

IPES-Food. (2018). *Breaking away from industrial food and farming systems: Seven case studies of agroecological transition* (110 p). International Panel of Experts on Sustainable Food Systems.

Lamine, C. (2012). « Changer de système »: Une analyse des transitions vers l'agriculture biologique à l'échelle des systèmes agri-alimentaires territoriaux. *Terrains Et Travaux, 20*(1), 139–156.

Lamine, C. (2017). *La fabrique sociale de l'écologisation de l'agriculture* (225 p). Éditions La Discussion.

Lamine, C., Bui, S., & Ollivier, G. (2015). Pour une approche systémique et pragmatique de la transition écologique des systèmes agri-alimentaires. *Cahiers De Recherche Sociologique, 58*, 95–117.

Lamine, C., & Cambien, L. (2011). Les transitions vers l'agriculture biologique: une approche à l'échelle d'un système agri-alimentaire territorial. In *Écologisation des politiques et pratiques agricoles*, March 16–18, 2011, Avignon, Écodéveloppement Inra.

Lamine, C., Meynard, J.-M., Perrot, N., & Bellon, S. (2009). Analyse des formes de transition vers des agricultures plus écologiques: Les cas de l'agriculture biologique et de la protection intégrée. *Innovations Agronomiques, 4*, 483–493.

Müller, P., Faure, A., & Gerbaux, F. (1989). Les entrepreneurs ruraux. Agriculteurs, artisans, commerçants, élus locaux. *Économie rurale, 194*(1), 53.

Rémy, J. (1987). La crise de la professionnalisation en agriculture: Les enjeux de la lutte pour le contrôle du titre d'agriculteur. *Sociologie Du Travail, 29*(4), 415–441.

Rouvière, C. (2015). *Retourner à la terre. L'utopie néo-rurale en Ardèche depuis les années 1960* (502 p). Presses universitaires de Rennes.

Chapter 12
Contesting and Caring: Forms of Solidarity in Local Buying Groups

Emmanuelle Cheyns and Nora Daoud

Various forms of buying groups have emerged in recent years, such as the Food Buying Clubs in the United States, the Organic Buying Groups in the United Kingdom, the *Groupements d'achats communs* in Belgium, the *Gruppi di Acquisto Solidale* in Italy and the *Groupements d'achats locaux* in France. These buying groups are groups of consumers who buy in bulk directly from various producers. They are organised around a common principle: a commitment to solidarity, which was already present as a fundamental concept in their historical forms as consumption cooperatives. These cooperatives appeared in the nineteenth century in the wake of the utopian socialism of Charles Fourier and Robert Owen, and were intended to provide poor families with access to quality foodstuffs, to build up collective savings or to buy at a fair price in order to remunerate the producer's work (De Boyve, 1889; Guillaume, 2007a).

While these consumption cooperatives declined with the advent of mass retailing in the 1970s, a new wave of alternative consumer buying groups in the last two decades has developed with similar motivations, albeit renewed (De Munck, 2011). Like the associations that support peasant agriculture (AMAPs[1]) and short supply chains, they are based additionally on ecological concerns, in a new context of globalisation and the health crises of the 1990s, which call into question the domination of food supply chains by large-scale distribution systems (Chiffoleau, 2008; De Munck, 2011).

[1] French: *Association pour le maintien d'une agriculture paysanne (AMAP).*

E. Cheyns (✉)
Cirad, UMR Moisa, Montpellier, France
e-mail: emmanuelle.cheyns@cirad.fr

N. Daoud
Consorcio Andaluz de Impulso Social (CAIS), Seville, Spain
e-mail: noradaoud@hotmail.fr

P. Gasselin et al. (eds.), *Coexistence and Confrontation of Agricultural and Food Models*,
https://doi.org/10.1007/978-94-024-2178-1_12

The buying groups are spaces for experimenting with forms of solidarity built outside the State's ambit and influence and at a distance from market instruments, in a broader context of criticism of public and private forms of solidarity (Tremblay, 2007). Indeed, the support policies of the post-war welfare state have been steadily diluted by programmes to reduce social expenditure and to target benefits, anchored in new principles of social protection that accord value to efficiency: the New Public Management. Forms of solidarity provided by the market[2] are similarly called into question because of the reduction of common goods into certified properties (Cheyns & Thévenot, 2019) in a market of passive and atomised consumers 'making choices' (Hubaux, 2011).

By coming together in groups, consumers explore alternative ways of living. This is especially true for buying groups that lay emphasis on self-management and participation, which require investment of time and effort in a collective whose aim is to 'make a community'. Do these buying groups bring about social change or a transition? If yes, what kind of change or transition? What forms of solidarity do they engage in, between contesting conventional agriculture and caring for vulnerable people? In this chapter, we discuss the different 'regimes of engagement' (Thévenot, 2006, 2015) and tensions between members of buying groups, which reflect various forms of solidarity, in particular two forms: a solidarity in familiarity and caring for others, and a public civic solidarity more 'at a distance' from the producer.

1 'Making a Community': A Survey of Buying Groups

In order to characterise the diversity of buying groups, our survey[3] first focused on 26 groups in the Languedoc-Roussillon region[4] in France. A series of interviews with these groups' members led us to distinguish five types of buying groups, two of which were of particular interest to us because of the high level of investment of members in their groups. The first of these two types are activist groups, self-managed, characterised by their decision to remain independent from institutions (they are not registered under the 1901 French law of associations, refuse State financial aid, etc.) and by a critical stance towards the market, and in particular towards large-scale distribution systems and supermarket chains. The second type is based on an intimacy between close persons, goodwill and a desire for togetherness.

[2] For example, the engagement of firms through corporate social responsibility (CSR), voluntary certifications and sustainability standards, etc., which aim for equity, protection of vulnerable people, respect for fundamental rights at work, etc.

[3] This research was carried out in 2011 with the support of the PSDR project 'Coxinel' (Short agriculture and agrifood marketing circuits: innovations for regional development, French: *Circuits courts de commercialisation en agriculture et agro-alimentaire: des innovations pour le développement régional*), funded by the Languedoc-Roussillon region, INRA, Cemagref, CIRAD and SupAgro (2007–2011).

[4] Former French region consisting of the following departments: Aude, Gard, Hérault, Lozère and Pyrénées-Orientales.

For this reason, the number of members in this type of group is often restricted in order to maintain the ease provided by a 'family size'. The other three types of groups, which we do not discuss here, are characterised by a centralisation of decision-making that is oriented towards efficiency and/or a low investment in the group (no meetings or collective decision-making) (Daoud, 2011).

In more than half of the 26 groups, decision-making is a collective process and participation is voluntary. Members are tacitly expected to participate in the group's activities by attending regular meetings and by getting involved in logistics (contact, purchasing, delivery, etc.). But the ways of making a community in these groups, in which a lot of personal investment is required, differ. In a second step, we thus studied two buying groups[5] marked by a strong investment in the community, corresponding to the first two types identified above.

The first is an activist group, the Self-Managed Socio-Ecological (SEMSE)[6] buying group, in Montpellier. We can observe a political engagement of its members based on a denunciation of the capitalist system and on a self-managed collective organisation leading to a strong requirement for public 'civic justification' (Boltanski & Thévenot, 1991). This group undertakes investigations and subjects producers to an interview-test in which they have to justify their social and environmental conditions of production. These activities have come in for criticism by some of the group's members, who decry the group's lack of care towards producers. The second buying group, Yummy-Yum, allows us to observe a completely different kind of engagement, in the 'familiar' and in the 'close', by adjusting to the environment and context in order to achieve a certain level of ease (Thévenot, 2006). What results is a solidarity 'in familiarity' and affection and a concern of taking care of others (*ibid.*). This regime of engagement is put under strain by some members who criticise the group for its inefficiency and a lack of political engagement.

These regimes of engagement and the tensions they generate reflect different forms of solidarity (Thévenot, 2006, 2015), in particular a solidarity in familiarity and affection and of caring for others, and a more distant public civic solidarity. These forms of solidarity indicate differences in the ways of forming a group and making a community as well as a coexistence of forms of support that these groups offer to local and peasant agriculture and intend to leverage to transform society.

---

[5] We interviewed 20 members of these two groups, including founder-members, and 10 producers supplying them. We attended the groups' monthly meetings and participated in other get-togethers organised by these groups. We also met people who decided to leave these groups or decided not to join them (5 individuals). Finally, our study also relies on the written documents of these two groups: minutes of meetings, planning or logistical tools, charters and e-mails.

[6] We use pseudonyms for the names of groups and individuals.

2 Contesting, Arguing, Caring: A Diversity of Solidarities

2.1 Caring in What Is Familiar

The Yummy-Yum buying group consisted of 17 individuals in 2011. They were all residents of a few neighbouring and peri-urban villages near Montpellier, but were not originally from these villages. This buying group accords value to the sense of ease and conviviality that can be created by the proximity of friendly persons, in the sense that they become closely tied through friendship or personal commitments and familiarisation processes (Thévenot, 2006). While most of the members did not know each other before the creation of the buying group, they now consider themselves friends or emphasise 'a feeling of friendship' (interview with a member).

The monthly meetings of the members take place in a residence, most often in the kitchen or living room of one of the members (in turn), with family photos and other signs of private and intimate life all around. The members share a meal there, in the comfort of a familiar place. While discussing group-related matters, members intersperse conversations about their lives and possible mutual aid. It is not uncommon at the beginning of a meeting to ask about each other's families as well as about those who are absent, and thus to take some time to greet each other.

Engagements of care—attention, solicitousness, and concern[7]—are at the core of the relationships. They allow to consider positively a relationship based on vulnerability (Garrau & Le Goff, 2009) or a way of being that reveals the vulnerability and dependencies of the human being (Centemeri, 2015). At the time of this survey, several of this group's members were in vulnerable situations, financially (precarious jobs, bankruptcy of the family artisanal business) or emotionally (bereavement, loss of property). Therefore, a central concern for these members was to share the comforting familiar with others, to develop a community of familiarity, made up of ease and solicitousness for others. This ease provides a reassuring foundation for the individual, which is crucial for exploring new things—or even for gradually building up autonomy (Centemeri, 2015).

Familiar engagement is also present in the relationships that group members have with producers. The consumer members are called 'godfathers' or 'godmothers' of the products, instead of the person 'responsible' for the product, indicating a good-natured and familiar relationship of accompaniment. The members seek above all a personal relationship with the producers, whom they choose more 'through acquaintance' and word of mouth than on the basis of a debated charter, perceived in the group as 'too theoretical' or 'intellectual' (interview with members, see below). The group frequently goes to meet the producers on their farms, to discuss their difficulties (material, access to land, farming set-up, etc.) and to share a meal, sometimes 'between two rows of vegetables'. These forms of meetings encourage a growing concern on what may affect the producers. In case a producer is experiencing personal difficulties, group members show their sympathy through personal gestures, such as

[7] See for example Paperman and Laugier (2005), Tronto (1993).

by writing a letter of condolence on the group's behalf. In addition, some producers are themselves consumer members of the group, and attend the group's meetings in the same spirit of togetherness. These meetings and farm visits have made it possible to develop ease and personal links little by little.

2.2 *Contesting and Arguing in Public for Solidarities*

The SEMSE group from Montpellier is much larger[8] and its engagement is based on expressing indignation towards capitalism.

This buying group, which historically grew out of an anarchist-inspired group in the mid-2000s, defines itself as a self-managed group, in a rejection of forms of domination and hierarchy. Unlike the Yummy-Yum group, the members meet (every month) in a public place (an activist hangout). Most of them are involved in global movements such as anti-globalisation, support for the autonomy of Zapatista communities, etc., about which they exchange information, linking the local and concrete level of their engagement to more global causes (see also Louviaux, 2011). More political than a simple charter, their manifesto sets out the group's higher principles, based on a 'market' critique of capitalism and a 'civic' engagement (Boltanski & Thévenot, 1991). This engagement accords importance to solidarity with producers (e.g. help in setting up peasant agriculture) and to independence from the current dominant economic system, especially from supermarket chains 'which exploit the land as well as the people' (buying group manifesto, 2011).

This engagement to build a fairer and more just world is manifested in a public test: an open, frank and vigorous discussion by the participants of all arguments for and against each proposal. Participants require a high degree of emotional detachment since disagreements are very publicly exposed. Decisions by the group are taken 'by consensus', i.e. without a vote[9] and, above all, by the unanimous agreement of those present (without proxies). The members have to be substantially invested: apart from the fact that one has to be present to be able to influence a decision, obtaining the unanimity of those present is based on a presentation and discussion of everyone's arguments, with debates concluding with a final 'going around the table'. The capacities required to take part in this public test of 'qualifying the common good' (Boltanski & Thévenot, 1991) can make some participants uncomfortable.

The process also requires an investment over a long period of time that is not well defined in advance, far removed from an efficiency-oriented managerial framework: 'We are not in a hurry […], we are not going to pursue profitability […] so we are going to take the time we need to discuss' (Véronique, one of the founding members).

---

[8] About 200 members in 2011, although attendance at monthly meetings consists of only between 5 and 40 members.

[9] Voting is eschewed because it tends to cut short the debate ('Voting means giving up discussion'), imposes a majority point of view and because the decision, which becomes less reversible, lends itself less to reflective examination.

But at the same time the discussion process is a key aspect that most of the members enjoy. Indeed, for members the discussion is accorded more importance than the arrival at its resolution—which can even be reviewed in a subsequent reflective moment. However, this reflective attitude makes the process demanding, with some participants even experiencing it as tiring (see also Louviaux, 2011).

Some members express their capacity of critical distancing as humour, self-mockery and irony. Some use irony, not only towards the system they are criticising, but also sometimes to express disagreements within the buying group. These moments of 'implied criticism' can be seen as a fumbling for criticism (Daucé, 2017; Thévenot et al., 2017), and also a dissidence which, if not taken to its conclusion, becomes part of a movement of emancipation of thought (Géraud, 1999). The use of nicknames or the inclusion of cultural quotations[10] by some members in their signature blocks in written exchanges demonstrates their commitment to a critical, but also playful, public life,[11] which distances them from their everyday life so that they can act differently (Legout, 2003). This resonates with the inventive energy of the 'humorous utopia' of the activists of 19th-century consumption cooperatives (Guillaume, 2007b[12]). Humour, which is milder than irony, is present in a number of oral and written exchanges, for example in the form of critical puns or comic poetic prose, such as a meeting report entirely in rhyme, which also allows for a certain reflective lucidity.[13]

The participants express their concern about creating social links, in a political conception which is rooted in individual freedom and not in attachment. In contrast with the Yummy-Yum buying group, 'If a person has not been coming for a while, we are not concerned, we don't ask questions.' Not everyone knows each other's first names (or even their nicknames). Most of the participants have a stable professional and social situation and/or a political posture that values individual autonomy (teachers, civil servants, activist members of multiple networks or of anarchist culture, etc.). The spirit of autonomy, in the sense of freedom of choice and independence from near and dear ones (Pattaroni, 2007), leads members to oppose any centralisation and specialisation of tasks (e.g. they take turns to write minutes of meetings). The members keep the ties of proximity that bind them at a distance, as illustrated by this statement from Thomas: 'We all find it hard to come forward when we are in a difficult situation. We are too afraid to ask for help from others, society has taught us to manage on our own.'

Nor do members get very close to the producers even after several months. Only the person in charge of a product is in contact with the producer, and theirs is not

---

[10] From literature, notable thinkers, cult films, etc.

[11] Criticism of institutions and hierarchies, such as the family and school (which transmitted and assigned value to a first name): 'There comes a moment when you become an adult, you make yourself; I was made for something other than school' (a member who gave himself a nickname).

[12] In reference to Gallus, *La Marmite libératrice ou le commerce transformé. Simple entretien* (1865). Preface by Henri Desroche, Paris, Balland, Bibliothèque des utopies, 1978.

[13] One of the lines, 'Even if, as always, no consensus emerges', underlines, for example, the constraint that surrounds the deliberation. Another, 'The prophetic tribe of pains-in-the-backside rice eaters' refers to a heated discussion of disagreements (meeting report, 2014).

necessarily an enduring relationship. Almost half of the producers are geographically distant. Farm visits are much rarer than for the Yummy-Yum buying group.[14]

Finally, in the manner of the *Ligue Sociale d'Acheteurs* (LSA) of the early 1900s (Chessel, 2017), the producers are chosen and 'validated' after an 'investigation' by the group's members. This investigation, in the form of an interview-test, is sometimes undertaken collectively, during a meeting of the group to which the candidate producer is invited to answer specific questions on the social and environmental conditions of his production. Each proposal for a new product is an opportunity for the group to take a reflective look at its own practices and values, and potentially a source of tension.

3 Tensions in the Buying Groups Between Regimes of Engagement

As places of collective social experimentation, these buying groups also experience tensions. These tensions have the advantage of constantly reminding the members of—or redefining—what is important, and, in this case, of redefining solidarity.

3.1 The 'Investigation' of Production Conditions as a Source of Tensions

The internal tensions in the SEMSE buying group in 2011 were largely crystallised around the process of the 'investigation', and in particular the interview process for approval of producers by the group, in the members' presence. This interview requires the producer to call on his skills of public presentation and justification in front of a group that behaves like an informed jury. In many cases, the producers felt they emerged more worthy after having passed the test and from being fully recognised for their often isolated activity. 'It was a bit intimidating at times [...], but on a narcissistic level, it was great. I felt good that everyone was interested in me, which is not often the case' (producer). But some producers experience a certain amount of anxiety during this public test, similar to that felt during an oral exam. This public experience proved to be very trying for some of the producers and consequently for some members of the group, who criticised what they saw as a lack of care towards the invited producers. For example, a producer who was invited to present his product to the entire group and answer a series of the members' questions mentioned that his trees, affected by a disease, had to be treated with a synthetic pyrethrum. He was then subjected to a stream of criticism from members, who urged him to 'convert to organic'. The producer finally got angry, reminding the members of his

[14] This was true at the time of the survey. The buying group later tried to overcome internal tensions (mentioned below) by getting closer to the producers and visiting their farms more often.

financial dependence on and his links to his cooperative. One of the group's members then interrupted him, loudly reminding him of his freedom of choice and individual responsibility: 'It's your decision, too! [...] I say: don't make excuses. Farmers have gotten into the subsidy business because it makes them money.' Several members of the group then proffered advice to the producer by giving him information and contacts of associations that help with the 'conversion to organic'. This encounter provoked a series of reactions during subsequent discussions between members. One of them expressed her unease, taking offence at the emotionally violent ordeal: 'I find that people [producers] are put in unacceptable situations... I'm sorry, but this is intolerable! [...] He knows what organic is, he wasn't born yesterday.' This opinion was not shared by all, with other members responding that such exchanges are also the purpose of the process: 'Being in front of a group can make him think and admit that he can still change his production method.' Ultimately, this producer's products were not 'validated' by the group, and tensions eased after the reminder of the necessity for a pre-investigation before inviting a candidate to appear before the group. The internal criticism revealed by this episode highlights the modalities of exclusion inherent in a model of critical deliberation, given the constraint placed on individuals unprepared for these tests, especially vulnerable individuals (Young, 2000; Charles, 2012).

Furthermore, disagreements between the group's members also sometimes stray from the ideal model of general and formal deliberation, and slide into personal attacks and exchanges some experience as particularly aggressive. This has led some members to leave the buying group. The lengthy debates of indefinite duration, disagreements that sometimes appear to be irreconcilable[15] and the unevenly applied boundaries of a self-framing by the members sometimes lead to exasperation and the escalation into more hostile exchanges.

Solidarity is also tested by the freedom of choice, valued in a pursuit of individual autonomy (see also Pleyers, 2011). One producer, who announced an increase in the price of a product without any justification, faced an immediate reaction from members: they stopped buying from him. He then complained to the group about a lack of solidarity. He finally explained the significant losses he was making on his crop, forcing him to review his cost price. He also regretted the lack of response from the group to his various requests for non-monetary aid. At a subsequent meeting, faced with this price increase, some members argued that 'purchasing is a matter of individual choice', and finally noted that 'there are limits to solidarity'. In this particular case, the group also discussed the capacities of the producers that it had validated in this way: 'Does the producer have to be a good communicator [i.e. able to justify or clearly explain] in order to bring about solidarity?' The group finally recognised a need to develop closer relations with the producers in their own

[15] For example, many members expressed support for undocumented migrants whereas a new participant was against 'mass immigration'. During discussions, both sides' positions hardened, and their written exchanges degenerated into personal attacks and insults ('ignorant fools', 'your argument is most stupid', 'cultural colonist', etc.).

environments, through farm visits by its members, and thus made an effort to get closer to producers.

These tensions highlight the pressure that the modalities of constant investigation (Chessel, 2017) and individual autonomy exert on the regime of familiarity and affection and on care.

3.2 *Yummy-Yum: Pressure on the Regime of Familiarity*

In the Yummy-Yum buying group, it is, in contrast, the engagement in the regime of familiarity and affection that has invited criticism internally by some members because it reduces 'political engagement' in the choices of products and producers. Three members in particular lamented the fact that instead of an investigation into the social and environmental conditions of production, it is 'love at first sight, the interpersonal relationship with the producers [living in or near these villages] or the price negotiation that prevails.' They suggested introducing, as in the SEMSE group, the process of investigations and debates, and drafting a charter to define the group's main principles. But this charter was rejected by the group's core members:

> 'I thought it was too much, compared to what we were doing, it was much too intellectual.' 'I think that the system still works because we know each other well, we see each other [in other contexts, e.g. meeting by chance in the market or in the village, etc.].'

During a meeting, Virginie, one of the aforementioned three members, criticised the others for not having asked sufficiently probing questions to a trout producer proposed by the group. She questioned the choice of this producer, who had not been surveyed about his ecological footprint, the density of his farm or the anthropisation of the water. But the group's core chooses to maintain a distance from the expert knowledge required to engage in an investigation, and instead accords value to social ties and proximity:

> 'We went to visit his farm. He said he wasn't organic, that's true, but compared to the others, we knew that he was better. Afterwards, it's true, I didn't ask him how many trout per cubic metre of water he had. I'm not expert enough in trout farming to tell him how he should do it' (statement at a meeting).

The regime of familiarity and affection was also called into question when these three group members regretted the 'lack of formalism' which would 'improve efficiency', noting the absence of minutes of meetings, order forms, and product data sheets, necessary in their opinion for organising and even developing the group's activities. In response, almost all the members expressed their fears that this focus on efficiency would undermine the kinship and ease that prevail in familiar places and ways. Luc preferred the 'efficiency of his hands' over managerial efficiency. Laura feared losing 'feelings of togetherness' and pointed out that if the group were to grow, the (family) conditions for meetings would be jeopardised, and Agnès added that she does not feel 'very comfortable in large groups'.

The same members were also very unenthusiastic about opening up the group to new members. While all agreed with the idea of opening it up to a moderate level, most were uncomfortable with Virginie's proposal to communicate publicly at a farmers' market and to offer an open sign-up list in order to bring in new people. One member (also a producer) tried to counter this proposal by suggesting that they should instead rely on word of mouth, so that personal relationships between individuals could be maintained. However, one participant finally suggested that the next meeting, in preparation for this opening up of the group, should not be held in a familiar place, but in a public hall in the village. Finally, at the meeting following the farmers' market, organised in the public hall, the buying group's members discovered that none of them had taken responsibility for contacting the people on the list. The arrival of potential new members creates uncertainty on the pursuit of engagements of care and familiarity.

Generally speaking, criticism in meetings from a few members about the group's lack of effectiveness, its reluctance to take in new members or the methods of selecting producers are not followed by a debate, nor by clearly expressed opposition from other members. The latter, when they are challenged, prefer to consider splitting the group. Some critical members have also already left the group and one of them complained: 'There are no debates, no decisions because people know each other and when there are debates, it is not to clarify a decision (a disagreement), it fails immediately.' Another critical member sees this avoidance of decision-making and debate as a difficulty in voicing disagreement. The ease of familiarity and the care for each other developed between members leaves little room for public dispute, which could be detrimental to the personal relationships that the members maintain. The consequences of a possible break-up go beyond the simple loss of a source of quality food products. For some members in fragile professional or family situations, it is the continuity of familiar ties and the assurance they provide (Thévenot, 2015) that is more important than an informed choice by consumers regarding production conditions.

These tensions reveal what the members of the group are keen to preserve: instead of a deliberative model, an exchange that is based on an emphasis on solicitousness and attention, and which aims 'not at the independence of individuals, but at an attitude of mutual concern open to the always specific forms of vulnerability'[16] (Garrau & Le Goff, 2009).

4 Solidarity with Producers

These two buying groups both aim to support peasant agriculture, which encompasses not only varied production methods, but also very specific (and unconventional) ones. They re-ascribe value to local ecologies in contrast to the specialisation of labour and land found in the agri-chains of globalised and integrated markets. Through the products they choose, they support mobility of agriculture (transhumance of animals and beehives, nomadism), the association of several species on the same plot of land, pluriactivity, biodynamics, collection of produce, animal husbandry on natural meadows, and local hardy breeds. Some of the producers they support do not yet have a formal status of farmer when they start or are *cotisant solidaires*.[17]

These groups also promote products recognised for their dietary virtues (e.g. spirulina, old varieties) and local and/or organic farming, with or without certification (the investigation or familiar engagement with producers already allows consumers to ensure environmental protection).

However, these two groups do not engage in the same forms of solidarity with producers. The specific modalities of their collective actions lead us to distinguish between two different forms of support.

4.1 Civic Solidarity 'at a Distance'

SEMSE members engage in solidarity through the expression of indignation based on a principle of 'civic justice' (Boltanski & Thévenot, 1991). Their financial support, relatively 'at a distance', is aimed at societal change.

Members choose to help low-income producers and/or those newly starting out, on the margins of conventional agri-chains, by committing themselves to financial support. This can be done through sponsorship by pre-financing part of the production, contributions to participatory funding (interest-free loans), or accepting the prices proposed by these producers without any negotiation. One beekeeper, for

[16] Translation by the author.

[17] Formal status accords full protection under social security, but requires full contributions. *Cotisant solidaires* (joint contributors) are those who make limited contributions to the social security system and, in return, enjoy accident insurance but no health insurance or retirement benefits. Most farmers embarking on the profession with small surface areas at the beginning choose this status until their activities increase.

example, obtained half of the funding for her hives from the group's members. They provided more than 2000 euros in total in exchange for the delivery of 3 kg of honey per year to each funder. This support for starting out was critical, especially since the producer did not yet have a formal status. In another case, members of the group encouraged a producer to revise his prices upwards in order to incorporate missing elements into his production costing and thus make a better living from his work. Price negotiation is eschewed, as it is seen as unfair in the case of a producer who has no marketing skills, and as a way of opposing the relationships of domination and power observed in globalised food chains.[18] Proposals for attractive products have also been refused on the grounds that the producer did not need the group's financial support, as revealed for example 'very clearly by the communications budget invested on his website'.

The group supports producers who explicitly engage in the same criticism of conventional agriculture—and even of institutions—as the members, in a form of political convergence. By adopting them, the group allows them to deploy their critical postures and alternative practices through improved visibility and networking. This is the case of the 'Zapatista rebel coffee', bought from a cooperative in Chiapas in Mexico through an association that supports the coffee farmers' demands for autonomy. Another producer, supplier to the group, presented itself as 'a collective enterprise that functions without ever having asked for or received a cent of public aid', in the same distancing from the State as that of the group's members. The investigation process also selects candidates on this basis (via questions on the support that the producers have requested or received 'from Europe', for example).

Finally, these producers usually have the same critical capacities and individual autonomy as the members,[19] and are endowed by some intellectual and/or financial capital. The producers we met were either in the process of changing careers to farming after leaving a desk job or even long studies, or are embarking on structural changes in the family farm, envisaged as creations or projects. Here too, the validation process filters towards these capacities, if only because the prospective producers have to handle and pass the test of the interview (see above), argue about their project and spell out their commitment. Although they emerge from this process feeling more worthy, these tests are more difficult for vulnerable producers.

4.2 *Solidarity in Familiarity and Affection*

Some of the producers in the Yummy-Yum group are less endowed with intellectual or financial capital, even though they too may have changed careers (after working as a labourer, for example). Some do not own land. They are all geographically close to the group's members (the two most distant producers are 60 km away, the others

[18] Even though these principles are challenged by the principle of freedom of choice (see above).

[19] Recourse to humour and literary quotations (high cultural capital) are also part of the repertoire of some of the group's producers (on their flyers for example).

on average 15 km). This is in contrast with the SEMSE buying group, almost half of whose producers are from outside the Hérault and Gard departments.

Solidarity in this buying group is expressed by a concern and a solicitousness for what affects the other, from major stressful events to the small details of everyday life. It is also expressed through support for close producers, for example by coming to help on the farm in case of difficulties, including personal ones. During a meeting, Valentine, a goat cheese producer and also member of the group as a consumer, spoke of difficulties concerning her farm, which she could no longer manage on her own, following the departure of her husband. The group heard her out very carefully, and some members offered to organise mutual aid days (to fence her plot, etc.). One of the workcamps involved refurbishing her mobile homes, in which she was living alone with her children, repairing the roof and hooking them up to running water. Aid concerned not only her professional life, but also her personal life; the group took care of her daughters.

Listening to turbulent and distressing life stories of producers and taking care to welcome them in tactfully created conditions of ease, sometimes involving listening to a third party who not only knows the person well, but also has experienced these difficulties himself or herself, complements this moral and emotional support. Valentine refers to this group's unusual ability to care about the difficulties that others may be experiencing and their vulnerability: 'This group has a dynamic that is quite unusual, and yet it has experience of some hard times, because there is me, all right, but there are others who have lived through difficult times' (referring also to non-producer members).

This group is more welcoming of producers in vulnerable situations, or who have a very small production that is very unsystematic and does not allow them to develop a real market. Maintaining the link is also crucial. One of the producers, for example, always delivers his products, even though not everyone thinks they are of very good quality: 'She invites Gérard (producer) into her house to eat something even if his products are not great. It's good to have people like that, who leave the door open. Gerard is reassured and encouraged, rather than being turned away; […] she brings this confidence to say… everyone does what they can!'.

5 Conclusion: Transitioners[20] but Towards What Type of Solidarities?

Our analysis of the regimes of engagement specific to each of the two groups indicates two quite different movements to support peasant agriculture in territories. The first supports producers who do not conform to a model of industrial agriculture integrated with the large-scale distribution of supermarket chains, a model which they contest. It thus accords value to local agroecologies, embedded in a project to (re)qualify the common good. This regime of engagement facilitates and makes visible a social and

[20] Term used in a buying group.

environmental criticism of production models and supports alternatives to them, as evidenced by the SEMSE group.

The second regime of engagement has the capacity to support producers in vulnerable situations through care and solicitousness. It proceeds from an engagement with the familiar, which favours relationships between members and accommodates the environment to achieve a certain familiarity. For example, in the case of Yummy-Yum, meetings around shared meals, organised at members' homes or at farms, allow individuals to establish links with each other and with producers. The purpose of these meetings is far from deliberative; it is instead to experience an emotional communication, which accords primacy to narratives and greetings for example, in order to make a community (Young, 2000, quoted by Garrau & Le Goff, 2009; Thévenot, 2015). This comparison makes it possible to place the transition, in this second case, in a wider space than that of the visible public space, whether critical or technical. A more silent transition (Lucas et al., 2020), although present, does not necessarily involve the formulation of a project or a protest. Other ways of making a community, less visible, develop in a familiar, benevolent engagement.

These two groups are vehicles for a transition that is very different from those of transformation projects driven by sustainability standards, technical indicators and objectives, which are now favoured by the market and public policies (from certified product properties to performance contracts). Groups such as SEMSE are the proponents of a transition underpinned by an explicitly critical political project, an alternative to a technical democracy that renounces the qualification of the common good. They help raise general awareness of the issues at stake and are catalysts for critical positions (Hubaux, 2011). However, due to their mistrust of the State, official arenas remain oblivious to these developments. These groups are largely linked in a spirit of 'convergence of struggles' and aim at a social transformation that would take place through 'swarming', following the example of the cooperatives of the nineteenth century (Guillaume, 2007a), forming a 'politics of small steps' (Louviaux, 2011). For their part, groups anchored in the regime of familiarity and affection (such as Yummy-Yum) are the proponents of a transition that envisages relations of vulnerability as potentially positive and are capable of welcoming them. They are open to producers who do not necessarily have the critical capacities and individual autonomy that are necessary in the previous case.

The originality of these buying groups is that they support critical capacities that emphasise civic engagement and the capacities to embrace a vulnerable public, which are no longer necessarily the focus of public solidarity policies, let alone the market. Indeed, the types of producers supported by these groups are not generally the most visible to or targeted by support policies.

References

Boltanski, L., & Thévenot, L. (1991). *De la justification. Les économies de la grandeur* (1st ed., 483 p, 1987), Gallimard [English translation: 2006. *On justification: Economies of worth* (400 p)]. Princeton University Press.

Centemeri, L. (2015). L'apport d'une sociologie des attachements pour penser la catastrophe environnementale. Hal-01163221.

Charles, J. (2012). Les charges de la participation. *SociologieS*. First texts.

Chessel, M. E. (2017). Consumers' leagues in France: A transatlantic perspective. In *The Expert Consumer* (pp. 53–69). Routledge.

Cheyns, E., & Thévenot, L. (2019). Le gouvernement par standards de certification, consentement et plaintes des communautés affectées. *La Revue des droits de l'homme. Revue du Centre de recherches et d'études sur les droits fondamentaux, 16*.

Chiffoleau, Y. (2008). Les circuits courts de commercialisation en agriculture: diversité et enjeux pour le développement durable. In G. Maréchal G. (Coord.), *Les circuits courts alimentaires: bien manger sur les territoires* (213 p). Educagri Éditions.

Daoud, N. (2011). *Régimes d'engagements, tensions et compromis dans la création de groupements d'achat* (135 p). Cirad, UM3 Paul Valéry.

Daucé, F. (2017). Éprouver le politique dans un média russe. *Revue D'études Comparatives Est-Ouest, 3*, 159–182.

De Boyve. (1889). *Histoire de la coopération à Nîmes et son influence sur le mouvement coopératif en France* (120 p). Guillaumin et Compagnie Éditeurs.

De Munck, J. (2011). Alterconsommation: la reconfiguration d'une critique. In G. Pleyers (Coord.), *La consommation critique. Mouvements pour une alimentation responsable et solidaire* (pp. 17–44). Desclée de Brouwer.

Garrau, M., & Le Goff, A. (2009). Vulnérabilité, non-domination et autonomie: vers une critique du néorépublicanisme. *Astérion. Philosophie, histoire des idées, pensée politique*, (6).

Géraud, V. (1999). L'ironie au siècle des Lumières. *L'information Grammaticale, 83*, 3–8.

Guillaume, C. (2007a). Les coopératives de consommation. *Cahiers Charles Fourier, 18*.

Guillaume, C. (2007b). La Marmite libératrice, ou l'association de consommation comme rêve de transformation sociale. *Cahiers Charles Fourier, 18*.

Hubaux, S. (2011). Le Groupe d'achats communs de Louvain-la-Neuve : convivialité et engagement. In G. Pleyers (Coord.), *La consommation critique. Mouvements pour une alimentation responsable et solidaire*. Desclée de Brouwer.

Legout, M. C. (2003). Modes de régulation des pratiques ludiques en salle de jeux réseau. *Les Cahiers Du Numérique, 4*(2), 135–148.

Louviaux, M. (2011). Le Groupe d'achats communs de Barricade (Liège): à petits pas vers un autre monde. In G. Pleyers (Coord.), *La consommation critique. Mouvements pour une alimentation responsable et solidaire*. Desclée de Brouwer.

Lucas, V., Gasselin, P., Barbier, J. -M., Pignal, A. -C., Cittadini, R., Thomas, F., & de Tourdonnet, S. (2020). Une agroécologie silencieuse au sein de l'agriculture française. In C. Bosc & M. Arrignon (Coord.), *La transition agroécologique en France ou les conditions du changement agricole* (pp. 147–160). coll. Territoires, Presses universitaires Blaise-Pascal, Clermont-Ferrand.

Paperman, P., & Laugier, S. (dir.). (2005). Le souci des autres. Ethique et politique du care. Raisons Pratiques, *16* (éd EHESS).

Pattaroni, L. (2007). Le sujet en l'individu : la promesse d'autonomie du travail social au risque d'une colonisation par le proche. In F. Cantelli , J. -L. Genard (Eds.), *Action publique et subjectivité* (P. 45). coll. Droit et Société, LGDJ.

Pleyers, G. (2011). Consommation critique: des promesses de la modernité aux enjeux de l'âge global. In G. Pleyers (Coord.), *La consommation critique. Mouvements pour une alimentation responsable et solidaire* (pp. 17–44). Desclée de Brouwer.

Thévenot, L. (2006). *L'action au pluriel. Sociologie des régimes d'engagement* (310 p). La Découverte.

Thévenot, L. (2015). Making commonality in the plural, on the basis of binding engagements. *Social bonds as freedom: Revising the dichotomy of the universal and the particular* (pp. 82–108). Berghahn.

Thévenot, L., Tsinman, J., & Zambiras, A. (2017). En commun, en différend. *Revue D'études Comparatives Est-Ouest, 3*, 45–93.

Tremblay, P.-A. (2007). Le sens de la solidarité privée et publique dans la lutte contre l'insécurité alimentaire. *Économie Et Solidarités, 38*(1), 41–57.

Tronto, J. C. (1993). *Moral boundaries: A political argument for an ethic of care*. Routledge.

Young, I. M. (2000). *Inclusion and democracy* (p. 320). Oxford University Press.

Chapter 13
Governing the Coexistence in a Transition Economy: Trade-Offs Between Smallholders and Mega Farms in the Vietnamese Dairy Sector

Guillaume Duteurtre, Pascal Bonnet, Nathalie Hostiou, Nguyen Mai Huong, Pham Duy Khanh, Jean-Daniel Cesaro, and Emmanuel Pannier

Vietnam has been undertaking a transition towards a 'socialist-oriented market economy' for the past 30 years.[1] What impact is this transition having on the diversity of the forms of agricultural production? How does this experience shed light on the coexistence of agricultural and agrifood development models?

[1] Parts of this chapter have been published in Duteurtre G. et al., 2021. 'Economic Reforms and the Rise of Milk Mega Farms in Vietnam: Governing the Post-socialist Transition'. *European Journal of Development Research*. https://doi.org/10.1057/s41287-021-00456-3.

G. Duteurtre (✉) · J.-D. Cesaro
UMR Selmet, Cirad, Montpellier, France
e-mail: guillaume.duteurtre@cirad.fr

J.-D. Cesaro
e-mail: jean-daniel.cesaro@cirad.fr

P. Bonnet
DGDRS, Environment and Societies Department, Cirad, Montpellier, France
e-mail: pascal.bonnet@cirad.fr

N. Hostiou
UMR Territoires, INRAE, Clermont-Ferrand, France
e-mail: nathalie.hostiou@inrae.fr

N. Mai Huong · P. Duy Khanh
Rudec-Ipsard, Hanoi, Vietnam
e-mail: maihuong.nguyenmh@gmail.com

P. Duy Khanh
e-mail: khanh.rudec@gmail.com

E. Pannier
UMR Paloc, IRD, Paris, France
e-mail: emmanuel.pannier@ird.fr

P. Gasselin et al. (eds.), *Coexistence and Confrontation of Agricultural and Food Models*,
https://doi.org/10.1007/978-94-024-2178-1_13

Table 1 Changes in annual milk production in six Asian countries (tonnes)

Year	1990	2000	2010	2017	Ratio 2017/1990
Bangladesh	1,593,503	1,507,310	2,035,550	2,005,405	**1.3**
Indonesia	599,155	1,009,289	1,492,848	1,540,200	**2.6**
Thailand	130,278	520,115	911,000	421,961	**3.2**
India	53,678,000	79,661,000	121,847,000	176,272,357	**3.3**
China	6,820,400	11,986,000	40,803,769	34,469,224	**5.1**
Vietnam	60,471	84,525	338,662	909,103	**15.0**

Source FaoStat (2019)

As in most countries of the former communist bloc, any transition corresponds to a radical change in political orientation, especially in economic matters. Vietnam witnessed profound reforms following the rolling out in 1986 of the *Đổi Mới* ('Renovation/Innovation') policies adopted by the Vietnamese Communist Party. This transition to a 'post-socialist' Vietnam is still ongoing (Fortier & Trang, 2013). It is profound, gradual and concerns many sectors. In addition to the economic transition, there are several other transitions taking place, pertaining to demography, food, technology and agriculture, even if these stylised processes are, in fact, part of a more complex reality (Lagrée, 2010).

The analytical framework provided by the multi-level perspective on sustainable transitions allows these multi-dimensional changes to be considered as a 'sociotechnical transition' (Geels, 2004; Geels & Schot, 2007). This framework is especially suitable for analysing long-term transitions in the agricultural sector and their impact on sustainable development (Darnhofer, 2015). In particular, it is important 'not to consider the transition only as a comparison between two situations isolated in time, but to understand what happens during the transition: the state of change' (de Terssac et al., 2014).

The multi-level perspective suggests taking into account three components (or 'levels') that determine the dynamics of change. First, the sociotechnical regime is defined as a coherent set of practices, techniques and social rules. The qualifier 'dominant' is sometimes used to express the pre-eminence of one type of regime at a given point. Second, niche innovations reflect a radical departure from the dominant regime, sometimes at the local level, and have the ability to challenge the dominant regime. And third, the sociotechnical landscape determines the general context, the conditions outside the regime, such as overall demographic and environmental trends, political orientations, social values, etc. This landscape evolves as a result of decisions or shocks, or under a general influence of trends (Geels, 2004; Geels & Schot, 2007).

We propose to address the transition taking place in Vietnamese agriculture through a case study of its dairy sector. This sector is interesting because of the rapid pace at which changes have taken place. Between 1990 and 2017, milk production in Vietnam increased 15-fold, making it the highest growth rate in the dairy sector in Asia (Table 1).

This chapter is a synthesis of multi-disciplinary studies conducted between 2014 and 2016 on the evolution of the dairy sector in several Vietnamese regions. It encompasses field surveys with different stakeholders (livestock farmers, milk processors, milk collectors, and local and national policymakers). The studies analysed farm trajectories (Khanh et al., 2016) and the transformation of agri-chains and territories (Duteurtre et al., 2015, 2017; Huong et al., 2017).

1 The *Đổi Mới* Policies and Support for the Peasant Farming Model

The development of the peasant farming model was a result of the *Đổi Mới* reforms, whose aim was to give wings to individual initiatives at the expense of collectivist organisations, which were deemed inefficient.

*1.1 Peasant Farms at the Heart of the Post-*Đổi Mới *Dairy Economy*

Up until the advent of *Đổi Mới*, dairy farming remained the exclusive domain of 'State farms' (*Nông lâm trường*), some of which were the result of the nationalisation of erstwhile colonial farms (Duteurtre et al., 2015). The re-emergence of household farms (*hộ nông nghiệp*), and the official recognition of their role in the Vietnamese dairy sector, was made possible due to the gradual rollout of the political, economic and land reforms of *Đổi Mới*. The continued existence of 'home gardens' under the collectivist system had allowed a significant residual peasant economy to survive from as early as the 1970s and 1980s. Dairy farming, however, was yet to penetrate this domestic economy (Brocheux, 2009).

It was the 6th National Congress of the Communist Party of Vietnam, held in 1986, that approved a change in policy orientation. In 1988, resolution no. 10 of the Communist Party recognised family farming as the principle model for agricultural production, and allowed peasants (*nông hộ*) to market their produce directly. The Land Law of 1993 established the term 'private land use right', defined around limited duration land leases. Certificates of land use rights (*giấy chứng nhận quyền sử dụng đất*), also known as 'red books' (*sổ đỏ*), were issued for a renewable period while the land remained State property. This land tenure system allowed for the redistribution of part of the collective land to families 'in proportion to the number of eligible persons per household' (Gironde, 2008). These land reforms led to a rapid development of private agricultural production. In 2006, there were 10.46 million agricultural households farming an average area of 0.9 ha (GSO, 2018).

In the dairy sector, these reforms resulted in the growth of peasant dairy farms within the boundaries or on the peripheries of former State farms. These government farms, which had experienced considerable management difficulties, were converted into research and development centres (case of Ba Vi farm) or into semi-private firms (e.g. Moc Chau farm). The farms' cows were given to former workers or newly settled peasants. A small number of government farms were sold to private entities (e.g. Son Dong farm).

In order to support this movement, the National Dairy Development Plan (NDDP), launched in 2001, placed the development of rural families at the heart of its strategy. This plan[2] succeeded in strengthening public structures responsible for supporting farms: credit to purchase heifers, technical training, and subsidies for equipment and inputs. At the same time, several public investment programmes sought to strengthen rural infrastructure.

This post-*Đổi Mới* regulatory context allowed the development of individual peasant projects, and was accompanied by major investments in the dairy sector by domestic private firms (such as Vinamilk and IDP) and by international ones (such as Nestlé and Dutch Lady). These firms bought the milk while providing industrial feed and credit. This association between peasants, firms and local authorities helped increase the national milk production fivefold between 1990 and 2010. In 2010, there were 20,000 dairy farms each with an average of 6 cows. These farms produced a total of 328,000 tonnes of milk per year (Fig. 1).

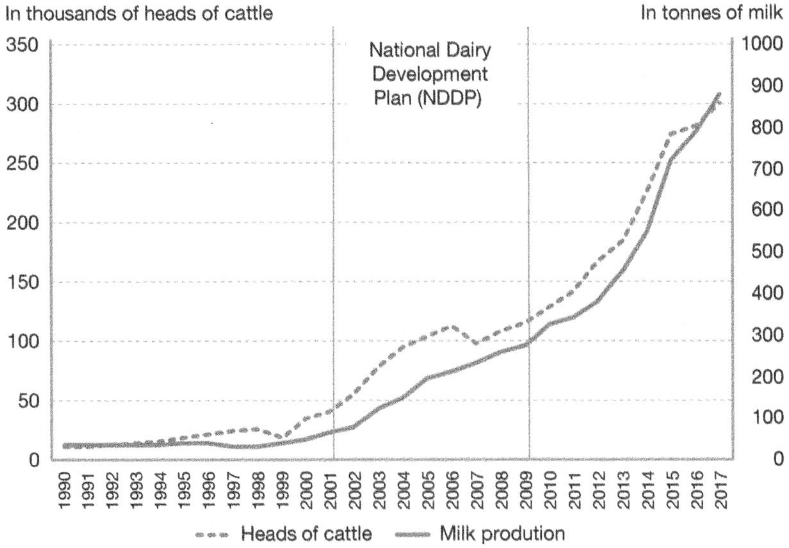

Fig. 1 Development of the national dairy herd and milk production in Vietnam from 1990 to 2017. *Source* GSO (2019)

[2] Ratified by decision no. 167/2001 of the Prime Minister.

This expansion of the peasant dairy sector was based on technical solutions adapted to the local constraints of very small farms of less than a hectare. Stable rearing of Holstein crossbred cows that were trough fed with a mixture of industrial concentrates and green fodder formed the basis of this highly labour-intensive system. The practice of cultivating elephant grass (*Pennisetum purpurem*), which can provide high yields on very small areas, quickly spread, thanks in particular to the development of a highly productive hybrid variety (VA06), soon followed by the adoption of fodder maize cultivation. The growth of small farms was accompanied by the emergence of a territorial network of upstream and downstream service companies that made this agricultural development possible: milk processing units, dairy industries, feed concentrate manufacturers, collectors and traders. The emergence of this local private agrifood sector also benefited greatly from the complementary provision of local public services to livestock farmers, with the NDDP and other extension projects facilitating training, access to credit and equipment for livestock farmers in the main milksheds (Duteurtre et al., 2015).

1.2 A Period Marked by a 'Peasant' Sociotechnical Regime

Thus, from 1993 to 2008, dairy farming in Vietnam was dominated by a sociotechnical regime that can be described as 'peasant'. This regime was characterised by a combination of a coherent set of practices, techniques and social rules (Table 2). This period also corresponded to a 'rehabilitation of the household economy' (Gironde, 2008) or, in other words, to changes in collective norms and values. A new model of agricultural development emerged, based on a social conception of agriculture's role and on a new demand for diversified foods that were synonymous with health and modernity, such as dairy products. These changes in values were reflected in several regulatory changes and the implementation of public policies in favour of individual dairy farms.

The emergence of this rural regime was a response to a radical change in the sociotechnical landscape (crisis of the collectivist economy, advent of a market economy) that led to the implementation of new regulations and policies favouring household farms and private trade. The emergence of this regime, during the 1990 and 2000 decades, can also be interpreted as the result of several niche innovations that appeared during the crisis of the collectivist system: the continued existence of 'home gardens' from the 1970s onwards, and the emergence of 'production contracts' in the early 1980s (Fig. 2).

Table 2 Characterisation of the two sociotechnical regimes in the dairy sector from 1986 to 2019

Domains	Sociotechnical components of the 'peasant' regime (1986–2008)	Sociotechnical components of the 'corporate' regime (2008–2019)
Livestock practices	**Intensive production practices** based on in-stable rearing of crossbred dairy cows, purchase of industrial feed, intensive cultivation of green fodder	Intensive **family-run commercial farms** and **industrial mega farms** based on in-stable rearing of purebred Holstein cows, fed with a mixture of concentrate and silage
Organisation of the economy	**Liberalisation of domestic markets**, regular sales of collected milk to industry **Construction of a mixed private–public economic fabric** consisting of State enterprises, technical services, private firms, SMEs **Development of a mass retailing system** (shops, supermarkets) and the appearance of new consumption practices	**Opening of markets to foreign competition** through trade agreements and the establishment of health standards. Price-driven competition **Construction of dairy oligopolies** made up of large dairy firms, some of which integrate all upstream and downstream activities **Development of corporate capitalism** based on the Hanoi and Ho Chi Minh City stock exchanges
Technology and know-how, research and development	Practice of **artificial insemination** allowing cross-breeding, control of the **sanitary** environment of livestock farms, know-how and innovations in **fodder crops** and milk **collection and processing**	**High labour productivity** systems based on the **mechanisation** of most livestock practices and **a high level of dairy technology and capital** (precision livestock husbandry) **Industrialisation of processing**
Cultural values and food and social norms	**Social issues**: milk production as a driver for the development for rural families **Health issues**: milk = health, growth and modernity	**Social issues**: appeal of modern technology, of technological gigantism **Health issues**: industrial milk = symbol of safety, health and modernity

Source Our surveys

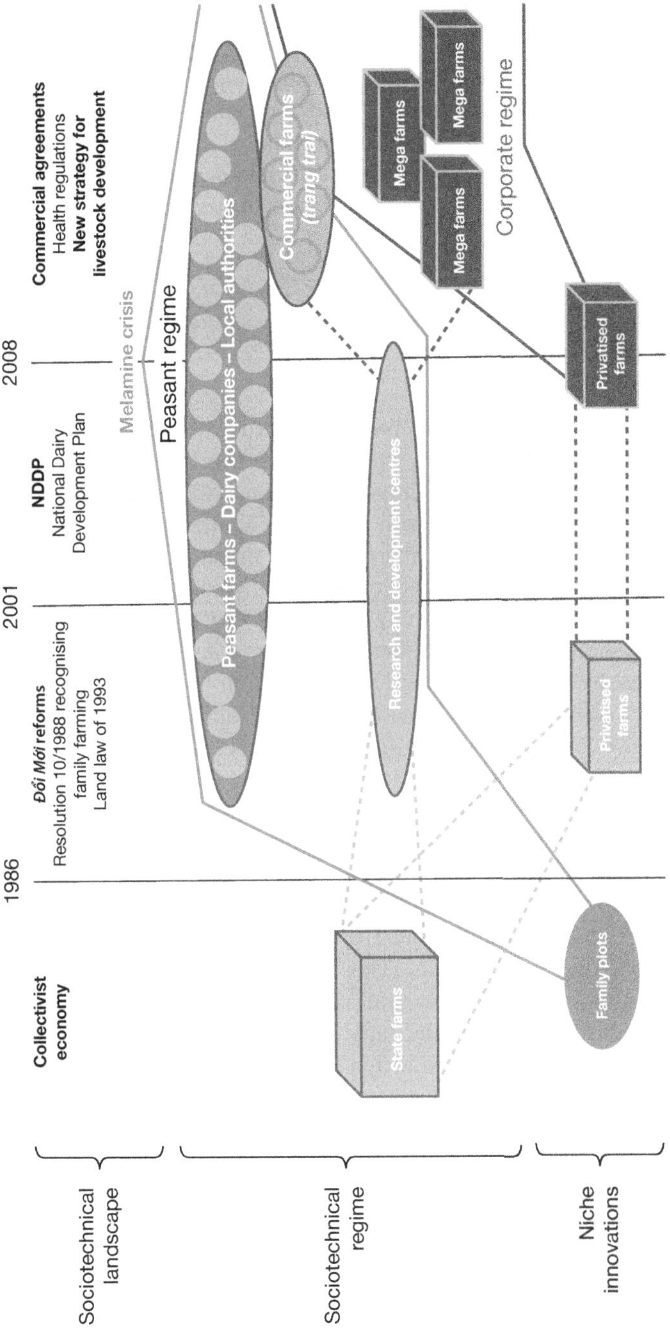

Fig. 2 Sociotechnical trajectory of the Vietnamese dairy sector

2 The Emergence of Mega Farms and the Establishment of a 'Corporate' Regime

From 2008 onwards, the Vietnamese dairy sector underwent major changes that resulted in a change in the sociotechnical landscape and the emergence of a new regime. Public policies were directed towards the industrialisation of the dairy sector. The aim was to promote large farms, reduce the trade deficit, promote new health standards and strengthen the sector's competitiveness to face the challenges of foreign competition.

2.1 Challenges to the 'Peasant' Regime

The Vietnamese dairy sector was initially very affected by the melamine contamination crisis. In October 2008, at a time when Vietnam was importing the equivalent of 618,000 tonnes of milk annually, amounting to about 80% of national consumption, the country was forced to deal with imports of adulterated milk powder from China. The presence of melamine in a large number of batches of infant formula in China led to the hospitalisation of tens of thousands of young children, and the death of six babies. Vietnam reacted by halting all imports of Chinese milk, and closing down some local industries using adulterated milk powder. This crisis resulted in a renewed interest by industries in the local production of milk, in a context of a severe erosion of consumer confidence in the dairy industry. A number of these industries then invested in systems for supporting domestic dairy producers and in the establishment of specialised industrial dairy farms.

The melamine crisis occurred at the same time as domestic demand was growing for 'healthy' and 'safe' products (*thực phẩm sạch*) that met industry standards or were part of guarantee systems. Vietnam increased the number of its trade agreements following its accession to the WTO in 2007. As a result of these developments, the Food Safety Law,[3] which was ratified in 2010, gave rise in the ensuing years to several implementation decrees and circulars. This led to the emergence of certification systems in other sectors, like meat or 'safe' vegetables. These health safety policies encouraged the industrialisation of the dairy sector, with a greater concentration of production, and an integration of production by the industries themselves.

At the same time, in the livestock sector, a new kind of policy emerged in 2008 to address the need to reduce the country's dependence on imports. This shift was initiated by the 'Strategy on animal breeding development up to 2020' launched in 2008.[4] This strategy's primary objective was to create favourable conditions for the emergence of household farms with intensive production and of large industrial

[3] Law no. 55/2010/QH12 of 17 June 2010.

[4] Decision no. 10/2008 of the Prime Minister.

farms. This regulation was followed, in 2014, by a new decision[5] of the Ministry of Agriculture and Rural Development, which aimed to improve value addition by the livestock sector, while ensuring the principles of sustainable development. This decision confirmed the orientations of the horizon 2020 strategy, but included environmental safeguards. Accordingly, Vietnam adopted a livestock law in 2018[6] that emphasised the need for cooperation between actors in the livestock sector, and which promoted production areas with stringent health safety requirements and the maintenance of coexistence of different livestock farming models. This explicit inclusion of the term 'coexistence' in the law showed the State's concern at the strong challenge posed to the peasant system by the rapid growth of larger farms.

2.2 The Promotion of 'Commercial Farms' and 'Companies'

The emergence of large individual farms was first encouraged by the certification of family farms as part of the *trang trại* label, which we translate here as 'commercial farms'. The aim was to define criteria to help local authorities at the district level register larger family farms in order to orient certain aid programmes towards these farms. The criteria for the certification of the *trang trại* farms were first defined in 2000 by a Ministry of Agriculture circular. They were revised upward in 2011 via Circular no. 27 from the same Ministry. The turnover threshold for the livestock sector increased to 1 billion dong. For a dairy farm, this represented a herd of about 25 adult dairy cows each producing 3000 L per year.

There were at total of 33,500 'new criteria' commercial farms in Vietnam in 2016, of which 21,060 were in the livestock sector. Although these commercial farms accounted for a mere 0.35% of the total number of farms in the country, their number had seen an increase by 67% since 2011. According to the 2016 census, this 'farm economy' (*kinh tế trang trại*) accounted for 135,500 permanent workers, 44% of whom were from within the family, and 56% were salaried employees. In the same period, between 2006 and 2016, the number of farm households decreased from 10.5 million to 9.3 million (GSO, 2018). In the dairy sector, this change reflected an increase in the number of farms with 20 or more cows.

At the same time, several regulations allowed the establishment of 'private firms' (*doanh nghiệp*) in agriculture and agro-industry. Between 2006 and 2016, the number of firms approximately doubled from 2136 to 3846 nationwide (GSO, 2018). In the dairy sector, these companies mainly involved themselves in processing, supply of inputs and marketing of dairy products.

[5] Decision no. 984/2014 of the Ministry of Agriculture and Rural Development (MARD).

[6] Law no. 32/2018 on livestock farming passed in the National Assembly.

2.3 The Emergence of the 'Mega Farm' Model

However, it was primarily the advent of mega dairy farms that completely transformed the organisation of the dairy sector in Vietnam. Consumer interest in locally sourced milk products (following the melamine crisis), the support of the authorities for this type of project, and the anticipated benefits of an integrated industrial organisation led to the construction of a large number of mega farms. Vinamilk, a private industrial group which had been collecting milk mainly from peasant dairy farms, set up five industrial dairy farms between 2007 and 2014, each with about 1000 milch cows. In 2017, this company opened a 500-head certified organic dairy farm in Dalat. In 2018, it started a new 4000-head mega farm in Thanh Hoa. In 2019, it announced the launch of two new mega farms: one in Tay Ninh (8000 heads) and one outside Vietnam, in Laos (24,000 heads).

At around the same time, the TH Milk company started what would go on to become the largest private mega farm in Vietnam by 2009. By 2014, this farm, located in Nghe An province, already had 44,000 dairy cows. In 2017, the company announced the launch of a new 10,000-head mega farm project in Ha Giang province, followed by another 5000-head project in Phu Yen. In 2019, TH Milk further announced that it was planning a 10,000-head farm in Thanh Hoa and another of 20,000 heads in Soc Trang. We must also mention Future-Milk's farm of close to 1000 cows, set up in 2008 on the former State farm of Son Dong, and Moc Chau Dairy company's three industrial farms, set up between 2010 and 2015.

These mega farms constitute a niche innovation that has generated renewed interest in technology and capital intensive systems, to the detriment of the more labour intensive family systems. The mega farms are based on the in-stable rearing of Holstein cows supplied with an automated feeding system consisting of a mix of silage and industrial feed. Forage cultivation practices, feed distribution and milking are mechanised, resulting in a higher labour productivity than in household dairy farming. They are based on the provision by the authorities of large-scale land holdings, most of which were part of erstwhile State farms.

In 2018, mega farms accounted for 32% of the national bovine herd, compared to 25% in 2014, and less than 5% in 2008. The rest is owned by family farms.

This period also saw the rise of international-scale dairy processing industries. Following the privatisation of the erstwhile government-owned dairy Vinamilk in the 2000s, the company became the third largest private company in Vietnam in 2017, accounting for half of the dairy sector's market share. Also worth noting is the stock market listing of TH Milk (the 166th largest private company in Vietnam in 2018), the purchase of the IDP dairy company by a Japanese investment fund in 2015, and the complete privatisation of the Moc Chau Dairy company in 2018.

2.4 A Period Marked by a 'Corporate' Sociotechnical Regime

We choose to describe the sociotechnical regime as 'corporate' since it resulted from changes in the Vietnamese sociotechnical landscape, insofar as this new regime aimed to promote productive investments in agro-industry, to the detriment of the complementarity between firms and peasants that had prevailed until then (Table 2).

Today, this reversal of the sociotechnical regime appears to be both a change in the development model and a strategic adaptation of actors to the new sociotechnical landscape. Livestock farmers are taking advantage of market opportunities to grow in size, thanks to credit obtained from firms. Firms decide to invest where the opportunities are most favourable, whether in terms of access to land or of milk collection. And local authorities favour the setting up of industrial systems to create local employment and income for the district,[7] and to meet the rapidly growing domestic demand for local milk. These strategic decisions lead to the establishment of new collective norms, and to configurations in agri-chains that reflect 'compromises' between local actors, industries and local authorities. While these compromises serve as the basis for 'coexistence' between different models, their overall balance in terms of competition and access to resources remains very fragile.

3 The Coexistence of Dairy Models: Between Pragmatic Trade-Offs and the Dynamics of Capitalism

The trade-offs made by authorities to orient this coexistence are illustrated in particular by land management and the promotion of local partnerships.

3.1 Land Management, a Prerogative of the State to Orient the Transition

The issue of land has emerged as a central element of the sociotechnical landscape. Since State services are in charge of land matters, they could orient the outcome of the transition in a definite matter. From the early 1990s, the 'distributions' to peasant families following the *Đổi Mới* reforms favoured a small peasant farming model. This distribution of resources was particularly equitable, with an average of 0.9 ha available per household. This redistribution, confirmed through the issue of 'red books' to peasant families, led to the rise of very land- and labour-intensive diversified farming systems (Khanh et al., 2016).

---

[7] The 63 provinces of Vietnam are divided into districts. Each district is further divided into communes.

However, from 2008 onwards, the authorities favoured the consolidation of land to enable the gradual emergence of larger farms that could meet the stated requirements of the livestock farming development strategy. The number of 'commercial farms' increased, while the number of farm households decreased. At the same time, local authorities encouraged land deals that favoured the emergence of agro-industrial activities by allocating land that had remained under direct State control. This trend was particularly evident in the dairy sector with the emergence of the numerous mega farms mentioned above. Of course, the vast land holdings of the former State farms played a significant role in this land transition.

3.2 Local Partnerships: A Factor in the Structuring of Milksheds

The importance of local authorities in managing national policy priorities was also reflected in the emergence of local partnerships to support the dynamics of creating milksheds. When mega farm projects were launched in areas where peasant dairy farms already existed, the partnerships between local authorities, private investors and peasants led to compromise situations. The mega farms were presented either as demonstration farms (as in the case of the Ba Vi industrial farm) or as units that provided quality heifers for small livestock farmers. Mega farms also promised to create local jobs, buy fodder maize from neighbouring peasants and sell manure to the farms.

Local authorities acted as 'guarantors' of this coexistence to ensure the success of these adjustments and compromises. This was reflected, for example, in the establishment of agreements between the firms and local authorities. In Ba Vi, for example, the IDP company signed an agreement with district authorities in 2012 to support the development of local livestock farms. The local authorities also encouraged the establishment of certification labels or local geographical indications to promote the collection of fresh milk, such as the Moc Chau Milk brand or the Ba Vi Fresh Milk certification. With this in mind, the Ministry of Agriculture put out a circular in 2017 to promote the establishment of private–public partnerships to manage agricultural investments.[8]

3.3 The Limits of the Logic of the Agri-chain

However, the coexistence of different forms of production in the same territory was not always taken for granted. In these local partnerships, the logic of the agri-chain sometimes prevailed over the objectives of seeking complementarity. In 2015, for instance, the milk crisis in the Hanoi region led several firms to limit their milk

[8] MARD circular no. 14/2017 on 'partnerships' within value chains.

collection to the largest livestock farmers to reduce supply costs, a move that forced peasant livestock farmers to diversify their activities. During the milk price crisis in 2016, Cu Chi district, in the south of the country, sought to develop its own certification brand for locally produced milk. The district's identity was, however, not strong enough to support its development.

Sometimes the lack of compromise can even leads to local tensions. This was the case, for example, in 2014 with the TH Milk farm in Nghia Dan, following the pollution of the watercourses of neighbouring villages by the farm's manure slurry. The authorities attempt in such situations to encourage modifications in the firms' strategies towards more sustainable trajectories. In concrete terms, the firms have invested in projects to support local communities, or to offer school scholarship programmes, but such projects have met with varying degrees of success. These experiences show that the dynamics of the evolution of milksheds in Vietnam play out in collaboration between the State, the firms and the peasants. Because of the possibility and variability of local trade-offs, the outcome of the agrarian transition remains uncertain.

4 Conclusion

Our observations of the livestock transition in Vietnam highlight the pluralistic nature of the dimensions of sociotechnical change. The transition appears to be a gradual process in which individual, collective and cognitive dimensions interact to produce differentiated trajectories (de Terssac et al., 2014). Rather than the replacement of certain forms by others, transition leads to parallel trajectories, i.e. the superposition of several regimes whose relative importance varies according to local trade-offs. These developments shed light on the coexistence of different forms of production in the same territory. Indeed, this coexistence appears to be the product of political orientations, and therefore of power relations, while at the same time being part of market dynamics driven by demand, techniques, investments and cognitive models. Finally, trade-offs by public authorities, which reinforce or weaken coexistence, turn out to be 'pragmatic', insofar as they respond to adjustments in the face of socio-economic contexts that are undergoing profound change.

The coexistence of peasant farms with industrial farms illustrates these pragmatic adjustments. For the mega farms, it is a matter of increasing their purchases of fodder maize from neighbouring farms, or of emphasising the impact of reselling good quality heifers to small livestock farmers. For industrial dairies, it is a matter of highlighting their role of collecting milk from peasant livestock farmers. More generally, private dairy farms clearly understand the need to follow 'inclusive' trajectories with respect to household farms (MARD, 2019).

Our approach allows us to better identify the temporal dynamics of innovations. In particular, the Vietnamese case turns the classical perspective of the conception of the agroecological transition on its head. The intensive and productivist model described as 'conventional' in Europe is considered in Vietnam as a form to be promoted, an outcome expected from the transition. The new regime is also based on a social construction of new food models concerned with 'health security' which leave little room for the incorporation of environmental issues and local specificities. As a result, the mobilisation of the concept of the agroecological transition is likely to run up against the local context. Further studies are needed to better qualify the sustainable forms of agricultural production that meet the challenges of emerging Vietnam. Research on farm trajectories, 'medium-sized' farms and 'commercial farms' could help identify agroecological solutions that are locally relevant. Similarly, it will be of interest to examine the trajectories of a return to more integrated forms of agricultural production.

Finally, it seems essential to link the coexistence approach to a more in-depth analysis of the dynamics of agricultural capital and of land. Indeed, the emergence of mega farms may seem to constitute a return to concentrated forms of production that are very similar to the State farms that were set up during the collectivist economy era. But what is different this time is the increasing importance of financial capital in the transformation of these economies. As De Koninck (2010) notes about the agrarian transition, we are witnessing a 'shift from a society characterised by accumulation in agriculture to one in which accumulation takes place in industry.' Such research, focusing on the dynamics of 'agrarian capitalism' in Vietnam, could lead to a better understanding of the social issues of coexistence. The aim would be to shed light on the social impact of the transition and to analyse the evolution of rural workers' livelihoods.

References

Brocheux, P. (2009). *Une histoire économique du Viet Nam: 1850–2007* (p. 257). Les Indes savantes.

Darnhofer, I. (2015). Socio-technical transitions in farming: Key concepts. In Sutherland et al. (Eds.), *Transition pathways towards sustainability in European agriculture. Case studies from Europe* (pp. 17–31). CAB International.

de Terssac, G., Truong A. Q., & Catlla, M. (dir.). (2014). *Viêt-Nam en transitions* (295 p). ENS Éditions

De Koninck, R. (2010). La transition agraire. In S. Lagrée (Ed.), *Transitions décrétées, transitions vécues* (pp. 45–58). AFD.

Duteurtre, G., Khanh, P. D., & Cesaro, J. D. (2015). Bassin laitier de Ba Vì: un territoire d'élevage façonné par les politiques publiques, entre logiques industrielles et soutien à la paysannerie. In Napoléone, Corniaux & Leclerc (Eds.), *Voies lactées: dynamique des bassins laitiers entre globalisation et territorialisation* (pp. 67–87). Cardère Éditeur.

Duteurtre, G., Sautier, D., Pannier, E., & Huong, N. M., et al. (2017). Alliances between agri-chain actors for a sustainable development of territories in Vietnam. In Biénabe (Ed.), *Sustainable development and tropical agri-chains* (pp. 55–69). Springer.

FaoStat. (2019). Country data from http://www.fao.org/faostat/en/#home

Fortier, F., & Trang, T. T. T. (2013). Agricultural modernization and climate change in Vietnam post-socialist transition. *Development and Change, 44*(1), 81–99.

Geels, F. W. (2004). From sectoral systems of innovation to socio-technical systems: Insights about dynamics and change from sociology and institutional theory. *Research Policy, 33*, 897–920.

Geels, F. W., & Schot, J. (2007). Typology of sociotechnical transition pathways. *Research Policy, 36*, 399–417.

Gironde, C. (2008). Grandes réformes et petits arrangements dans les campagnes vietnamiennes. *Autrepart, 48*, 113–128.

GSO. (2018). *Results of the rural, agricultural and fishery census 2016* (680 p). General Statistical Office (GSO), Edited by Statistical Publishing House, Hanoi.

GSO. (2019). *Data on the milk production sector in Vietnam, Hanoi* (1 p).

Huong, N. M., Duteurtre, G., & Moustier, P. (2017). What shapes the governance of the dairy value chain in Vietnam? Insights from Ba-Vì milkshed (Hanoi). *World Food Policy, 4*(1), 57–81.

Khanh, P. D., Duteurtre, G., Cournut, S., Messad, S., & Hostiou, N. (2016). Caractérisation de la diversité et de la durabilité des exploitations laitières familiales au Vietnam : Une étude de cas en zone péri-urbaine de Hanoï. *Revue D'élevage Et De Médecine Vétérinaire Des Pays Tropicaux, 69*(4), 131–141.

Lagrée, S. (Ed.). (2010). Transitions décrétées, transitions vécues: du global au local, approches méthodologiques, transversales et critiques. In *Actes de l'université d'été régionale en sciences sociales «Les journées de Tam Dao»* (411 p), July 2009, Vietnam.

MARD. (2019). *Report of a Workshop on "Livestock Outlook and PPP Dialogue: Challenges and Potential to Develop Dairy Farming and Dairy Products"* held in on 24th September, 2019 in Pan Pacific Hotel, Hanoi, organised by Partnership for Sustainable Agriculture in Vietnam (PSAV) Secretariat, Ministry of Agriculture and Rural Development (MARD), Hanoi.

Chapter 14
Considering the Diversity of Transition Trajectories

Philippe V. Baret and Clémentine Antier

1 The Necessity of an Agroecological Transition

Highly productive but environmentally unsustainable agricultural production models are now running up against the limits of our planet, resulting in demands for a quick and comprehensive transition of agricultural models (De Schutter, 2010). This transition—commonly understood as a passage from one state to another—is highly complex in reality when we consider current agricultural systems, which are fully or partially embedded in often globalised food systems. Indeed, the diversity of actors, practices and norms in the agricultural and food sector make up a multitude of interacting sociotechnical systems. A change in farmer or consumer behaviour can contribute to a process of transition, but the extent of the change that is necessary and desired requires a strategy thought out on a large scale within a constructed theoretical framework. And this framework will need to incorporate the multi-actor and multi-scale dimension of the transition.

According to the model of Geels and Schot (2007), it is the interactions between the actors at the heart of an existing system and those situated on this system's periphery which will initiate the process of transition. On the one hand, within the industrialised agrifood system—which is dominant in Europe and even worldwide—standards define what is acceptable and desirable, actors often share a long common history, and these actors coordinate around practices. This is known as an organised 'sociotechnical regime'. The actors in this dominant regime interact to improve the system according to and consistent with their own criteria and, in this way, maintain it.

Philippe V. Baret (✉) · C. Antier
Sytra (Transition of Food Systems), Université Catholique de Louvain, Louvain-la-Neuve, Belgium
e-mail: philippe.baret@uclouvain.be

C. Antier
e-mail: clementine.antier@uclouvain.be

P. Gasselin et al. (eds.), *Coexistence and Confrontation of Agricultural and Food Models*,
https://doi.org/10.1007/978-94-024-2178-1_14

Globalised milk production and processing systems[1] or soy-feed based industrial pig farming are two good examples of this type of dynamics of improvement without any fundamental change. On the other hand, outside the dominant regime, niche innovators propose, often on a small scale, other ways of doing and thinking about agriculture and food. Finally, influential factors (media, consumers, citizen dynamics, cultural changes, etc.) determine the long-term trends of the 'landscape' in which the actors evolve. These trends can have a positive or negative effect on the dynamics of transition.

At the same time as this theorisation of transition processes proposed by Geels and Schot (2007), a new innovation paradigm[2] gained in importance at the beginning of the twenty-first century: agroecology gradually emerged as an alternative to the trajectory of technical mastery that marked the great modernisation of agriculture after the Second World War (Wezel et al., 2009; De Schutter, 2010; Holt-Giménez & Altieri, 2013). Inspired by the concept of the ecosystem, agroecology as defined by Altieri (1987) looks beyond the plants and animals themselves to improve the efficiency and sustainability at the farm and food system levels. Agroecology seeks to optimise the agricultural system on the basis of synergies with natural processes and aims at an independence from synthetic inputs (pesticides, chemical fertilisers). Furthermore, an agroecological farming system is farmer-driven, which implies that farmers regain their decision-making autonomy and socio-economic principles are implemented (Dumont et al., 2016). This initial definition of agroecology has, moreover, since been extended to the scale of the entire food system by including the economic, sociological and political dimensions within agri-chains and national and international governance systems (Francis et al., 2003; Gliessman & Tittonell, 2015).

In a context of transition, an agroecological pathway can emerge either from the dominant regime through a process of insularisation (Vankeerberghen et al., 2014) or the development of an innovation niche. These niches can have older or younger historical roots: from organic farming, which has significant historical background (Bellon & Penvern, 2014), to the more recent emergence of models built on the principles of permaculture (Ferguson & Lovell, 2014).

Still a novel proposition in Europe in the early 2000s, agroecology is now presented as a credible alternative to agricultural systems that are more dependent on synthetic inputs and based on a mastery over nature (HLPE, 2019). As the agroecological movement grows, new questions arise. Does it contribute to a radical break with the models inspired by the Green Revolution, as proposed by Altieri (1987) as early as the 1980s, or does it constitute a gamut of new technical proposals that will allow the current regime to evolve from within (Conway & Toenniessen, 1999)?

[1] *Das System Milch* (2017), documentary film by Andreas Pichler, www.dassystemmilch.de.

[2] We recall that the concept of innovation paradigm refers to the technical approach chosen to respond to a question or a problem (Vanloqueren & Baret, 2009). For example, when confronted by a fungal cereal disease, conventional farmers will choose to use chemical solutions (fungicides) whereas organic farmers will use biopesticides or shift to a variety or varietal mixture with higher resistance to diseases (Vanloqueren & Baret, 2008).

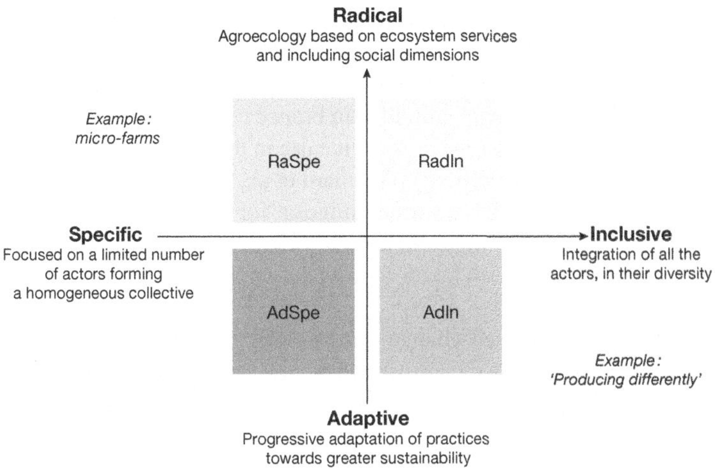

Fig. 1 Two axes and four quadrants to situate transition approaches in agriculture. The axes make it possible to situate agricultural initiatives and approaches in four contrasting quadrants: RaSpe (radical and specific approach), AdSpe (adaptive and specific approach), RadIn (radical and inclusive approach) and AdIn (adaptive and inclusive approach)

2 A Two-Dimensional Framework to Situate Transition Approaches

To address this question, we propose a framework that aims to situate agroecological proposals in a two-dimensional space: on the one hand, the scope of the proposal, and on the other, its degree of radicality (Fig. 1). By analogy with the concept of the ecological niche (Chase & Leibold, 2003), our hypothesis is that the different spaces defined by these two axes gather initiatives with diverse or specific properties and behaviour.

Agroecological proposals range from that of Stéphane Le Foll for French agriculture (Le Foll, 2012; Pluvinage, 2013; MAAF, 2015) to small-scale enthusiastic ones for microfarms (Morel, 2016). When we look at how collective organisations are stuctured, we can distinguish, on the one hand, to the left of the ordinate axis, the proposals that are aimed at a particular specific group of actors, usually of committed and already convinced persons. Most often, these are small-scale proposals, even though these collectives may subscribe to a broader ambition for change. In general, the actors participating in these initiatives form a relatively homogeneous group whose intention is to bring about change by gradually expanding from one person to the next.[3] On the other hand, to the right of the ordinate axis are inclusive proposals that aim, from the outset, to modify the entire agricultural system by integrating all existing forms of agriculture at the scale of a region (Antier et al., 2017), a country

[3] This notion of proximity has now gone beyond simple geographical proximity through the 'magic' of the Internet.

(Solagro et al., 2016), Europe (Poux & Aubert, 2018) or the entire world (Dorin et al., 2011). The target groups of these initiatives are heterogeneous in nature, as they encompass the diversity of an entire agricultural sector (milk, meat, cereal production, etc.). The Ecophyto[4] initiative in France corresponds to this second type of proposal: the goal being to reduce pesticide use in the country by involving all the actors in French agricultural systems (Guichard et al., 2017). The intention of inclusion does not by itself exclude a strong ambition for change. It takes into account the agricultural system as a whole and aims to change the behaviour of all its actors. In so doing, it opens up a much larger field of action than those of the smaller-scale proposals.

The nature of the expected change varies along the vertical axis in Fig. 1. A proposal above the abscissa axis corresponds to the requirement of a radical change most often driven by a logic of a break with the existing dominant system. Those below the abscissa axis have an adaptation objective in which the systems' actors have to embark on a trajectory of gradual and adaptive transformation.

At the technical level, the proponents of radical change advocate a comprehensive reconfiguration of systems (relinking of crop cultivation and livestock husbandry, agroforestry, etc.). In a more progressive vision, adaptive change is based on a pursuit of efficiency (more sparing use of pesticides or fertilisers, precision farming, etc.). Substitution approaches (e.g. organic farming model without synthetic inputs) are intermediate between these two types (Hill & MacRae, 1995).

The intersection of these two axes creates four quadrants in which agroecological initiatives and approaches can be situated: RaSpe (radical and specific), AdSpe (adaptive and specific), RadIn (radical and inclusive) and AdIn (adaptive and inclusive). Situating empirical initiatives in these quadrants makes it possible to highlight their strategy to contribute to an agroecological transition. This positioning is not normative and is not meant for comparisons between different initiatives. It does, however, make it possible to discuss, from the moment an initiative emerges, a specific trajectory favourable to the agroecological transition.

In the RaSpe (radical-specific) quadrant, we find, for example, small-scale and specific initiatives such as permaculture (Ferguson & Lovell, 2014), micro-farms and urban farming. These models are characterised by the small size of the areas cultivated (often less than 5 ha) and the specificity of the actors (most often farmers with no mainstream agricultural background). These initiatives are often managed by local collectives involving farmers and consumers. At the technical level, the initiatives in the RaSpe quadrant are often focused on diversified horticulture, but their goal can extend to a model that imagines a territory entirely covered by micro-farms.[5] These initiatives are not inclusive (in terms of integrating a wide range of actors and production methods), as each of them advocates a specific type of model. They

---

[4] https://agriculture.gouv.fr/ecophyto.

[5] On this subject, see Simon Gouin's article (in French) published in Bastamag.net on 18 June 2014, titled '*Bienvenue dans l'agriculture de demain, libérée des pesticides et du pétrole, et créatrice de dizaines de milliers d'emplois*' ('Welcome to the agriculture of tomorrow, free of pesticides and fossil fuels, and creator of tens of thousands of jobs') (https://www.bastamag.net/Bienvenue-dans-l-agriculture-de).

are deployed on a small scale, but they could reach a significant global production capacity if they were replicated widely (Morel, 2016).

In the RadIn (radical-inclusive) quadrant, we find initiatives such as conservation agriculture or farmer seed networks (Demeulenaere & Bonneuil, 2010). A large-scale change is expected, but one that remains anchored to the existing agricultural network. In France, the proposals of the Centres for Initiatives to Promote Agriculture and Rural Areas (CIVAM[6]) are also located in this quadrant. In these approaches, agroecology is not always identified as the innovation paradigm being mobilised. As for conservation agriculture, there exist differing views of its degree of radicality. Some authors believe that its practices place it instead in the AdIn (adaptive-inclusive) quadrant (Landel, 2015). On the one hand, conservation agriculture appears to be a radical change of vision, favouring ecosystem services and biodiversity (Chabert & Sarthou, 2020). The concept of 'living soil' is mobilised, as opposed to a soil that has lost its biological activity after years of chemical-based agriculture (Lemieux, 1996). These aspects imply that conservation agriculture can be thought of as a technical model radically different from that of tillage-based agriculture inherited from history, and hence it clearly belongs in the RadIn quadrant. On the other hand, the fact that conservation agriculture is still very dependent on glyphosate brings it closer to the AdIn quadrant, as an inclusive adaptation strategy, but without breaking with the dominant 'conventional farming' model (Ferdinand et al., 2020).

The purpose of proposing a classification into four quadrants (and the illustrative examples mentioned above) is to stimulate such debates on the intentions and strategies of transition initiatives. Indeed, the aim is not to judge the respective qualities of different transition proposals in a normative way, but instead to foster a debate on the possible options and the scope of their change strategy, and to understand the possible synergies or potential antagonisms between proposals. An analysis of the positioning of the different actors, in a dynamic and comparative logic, will open up the possibility of a coordinated vision of the agroecological transition. The aim is not to build a consensus—a consensus that that would anyway be unlikely to be reached—but to help find complementarity between the approaches and clearly understand the horizon and potential impacts of each of them. Furthermore, the classification we propose is not set in stone; it could evolve over time and be modified appropriately when new types of initiatives emerge.

In the AdIn quadrant, we can include initiatives such as pesticide reduction schemes (Ecophyto), policies in favour of diversification (Meynard et al., 2013) and territorial approaches aimed at preserving water catchments (Becerra & Roussary, 2008). The objective of this category of initiatives is to help a large group of farmers, or even all the farmers in a country, change and improve their practices. Even if these proposals are based on small-scale initiatives, sometimes quite different from the dominant farming models and farm types, for example the Ecophyto demonstration farms (Cerf et al., 2015; Guichard et al., 2017), the objective is for everyone to embrace the proposed change. It must be noted that the trade-off between inclusion and radicalness in these AdIn initiatives can lead to failures, results that are slow to

[6] French: *Centres d'initiatives pour valoriser l'agriculture et le milieu rural (CIVAM).*

appear (Écophyto), dubious justifications (glyphosate vs. climate) or even outright greenwashing. As AdIn initiatives are conceived to be implemented on a large scale, they should provide the ideal framework for a process of generalisation of new practices and a potential reconfiguration of the existing regime. Given the slow pace of these changes, it is in these AdIn systems that the 'lock ins' and roadblocks to transition have most often been studied (Cowan & Gunby, 1996; Vanloqueren & Baret, 2008; Meynard et al., 2013).

Finally, it is difficult to attribute initiatives to the AdSpe (adaptive-specific) quadrant since adaptive processes usually have a broad scope. Regional PDO (Protected Designation of Origin) differentiation initiatives such as the structuring of the Comté cheese sector could probably correspond to this quadrant (Jeanneaux & Perrier-Cornet, 2011).

3 Mapping Initiatives to Help Reflect on an Agroecological Transition Pathway

A canonical reading of Geels and Schot's (2007) transition theory leads to a horizon where the initial regime, under the influence of the sociotechnical landscape and niches, reaches a new state. However, other horizons of a transition process can be imagined. One other possible outcome is the emergence of an alternative regime alongside a dominant regime that has itself changed (Dumont et al., 2020). The coexistence of these two regimes (a regime inherited from the dominant regime and a regime emerging from the convergence of innovation niches) can manifest in different ways: competition between regimes, cooperation between regimes or each regime ignoring the other, with each of them developing its own value chain and targeting a different type of consumer. Such a coexistence of regimes would lead to market segmentation across the entire sociotechnical context, i.e. not only in economic dimensions, but also in norms, relationships between actors and practices.

The development of a specific Limousin cattle agri-chain in Wallonia (southern Belgium), in parallel with the still dominant Belgian Blue cattle agri-chain, corresponds to this situation. Breeders have developed this new agri-chain as an alternative niche, most often undertaken and marketed as organic farming. It has developed gradually, initially relying on marketing through already developed French networks before setting up its own marketing and distribution network (the first auction was organised only in 2014; Buron et al., 2014). Today, in Wallonia, 80% of beef cattle belongs to the conventional breed, the Belgian Blue, and the remaining 20% belongs to French breeds (mainly Limousin, Charolais and Blonde d'Aquitaine).

It is difficult to discern when a developing niche becomes a regime—either by substituting the dominant regime or by coexisting with an already existing regime (Fig. 2). This is all the more true for our example since other niches have developed in Wallonia following a similar pattern: Charolais, Blonde d'Aquitaine, Angus, etc.

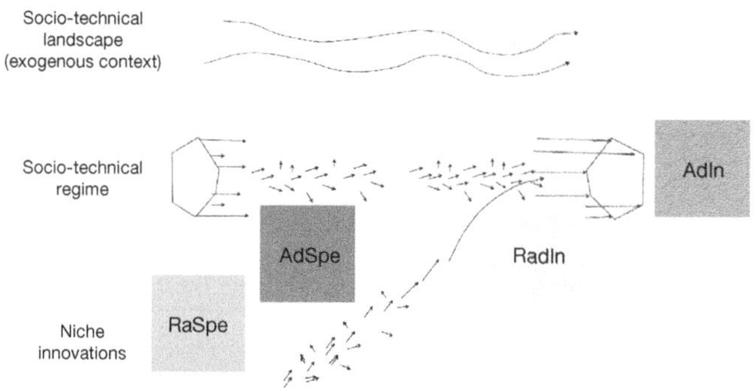

Fig. 2 Positioning of the different quadrants in the Geels and Schot model. The axes make it possible to situate agricultural initiatives and approaches in four contrasting quadrants: RaSpe (radical and specific approach), AdSpe (adaptive and specific approach), RadIn (radical and inclusive approach) and AdIn (adaptive and inclusive approach) (see Fig. 1). Adapted from Geels and Schot (2007)

The criteria to determine when a niche becomes a regime could be the extent of the niche's development, or the establishment of a distinct network of actors.

Is market size or market share sufficient to distinguish between niche and regime? Are the new initiatives the work of a very small number of actors, or do they now constitute a real value chain involving major players? Do they have 5, 10 or 20% market share? The objective is not to set a standard or a threshold, but rather to understand the role that each initiative can play in the transition of an entire sector, such as the beef sector in Wallonia. What status should be accorded, for example, to the micro-farms that are proliferating, to farmer seed networks that are being set up, or to the short supply chains that are developing? Even though these initiatives have a high media profile, they remain negligible compared to the global seed system or the global market.

This issue of the coexistence of two regimes can also be discussed from two perspectives, that of polarisation and that of the 'glass ceiling' (Fig. 3). When seen through a perspective of polarisation, the coexistence of two regimes implies a re-organisation around two contrasting poles, with the disappearance of intermediate models. This is the case in the milk sector, where two main models are perceived today as promising. The first is a model based on increasing the farm size and optimising processes through the use of automated milking robots, precision farming equipment and more intensive and mechanised feeding methods. The second is a model of adding value through on-farm processing, or in very short networks, and the development of close ties with consumers. At least at present, most farmers believe that the path to be followed is that of modernisation and expansion. And yet, this predominant choice leads to competition between farmers, and between dairies, and to a headlong rush into overproduction that requires new markets to be found (Pouch & Trouvé,

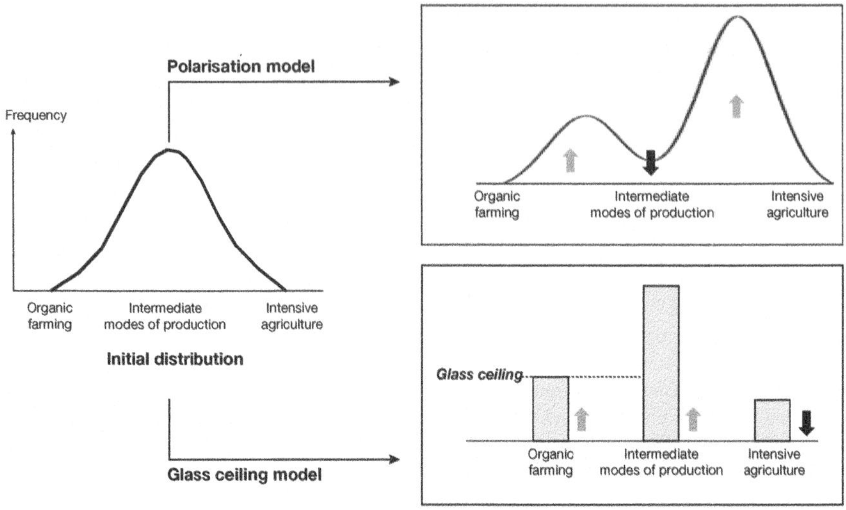

Fig. 3 Two models of changes in the distribution of existing systems

2018). A key element of this evolution, in a highly integrated sector like milk,[7] is the development of new governance models and related processing and/or marketing structures (De Herde et al., 2019).

The other perspective is that of a coexistence of regimes consisting of a niche's substantial development and, in parallel, a corresponding erosion of the dominant regime, but only up to a certain limit described as a 'glass ceiling'. The development of organic farming in Europe has led to this type of coexistence, alongside conventional farming and markets. Indeed, after an initial spurt (Darnhofer et al., 2019), the increase in the share of organic farming seems to slow down for no discernible reason (Willer et al., 2019).

These examples also illustrate the difficulty of setting a common horizon—i.e. a shared vision of the future and a set of associated objectives. While the work of the IPCC on climate change (Porter et al., 2017) and that of IPBES on biodiversity loss (Pascual et al., 2017) call for a rapid response and a reorganisation of our societies, the issue of new trajectories opens up the debate on the balance to be struck between rapid implementation and long-term planning. Should we act quickly at the risk of choosing ineffective or insufficient solutions, of ignoring the rebound effects or of creating polarisations between those convinced of the need for radical change and those in favour of gradual adaptation? Or should we, in contrast, plan, set objectives, give ourselves time to validate the various possible trajectories, convince the most reluctant amongst us, and ensure that there are no unanticipated negative effects? Moreover, it is not easy to allocate resources in terms of research priorities, citizen involvement and political debate in order to find a balance between implementing

[7] The milk sector, due to the very nature of the product and the way it is currently processed, requires a tight integration of all its actors, from collection of raw milk to distribution of final products.

available solutions and investigating desirable horizons. The bulk of the scientific literature on agroecology, which is now firmly on the political agenda, mainly focuses on agroecological practices and their implementation. Research on the roadblocks to agroecology and on its systemic dimensions is increasing but is still not very significant.

Given the ecological challenges of the twenty-first century and the significant role of the agricultural sector, one might expect that proposals for sociotechnical development trajectories for sectors such as dairy, meat or, more generally, agriculture as a whole would pay close attention to not exceeding planetary limits, and that trajectories that break with the past would be designed and followed. Such prospective approaches have been developed in the energy domain (Association négaWatt et al., 2012) and for agrifood systems (Paillard et al., 2010; Solagro et al., 2016; Poux & Aubert, 2018). In the energy sector, these trajectories now have legal backing (mandated increases in the share of renewable energy, etc.), come with consequences in case of deviations, and are integrated into corporate strategies. However, in the agrifood sector, foresight has so far had little impact on regulatory frameworks and company strategies. We think that such prospective approaches could actually contribute to the agroecological transition by defining a desired future horizon. This would enable the design and implementation of strategic changes that are more far-reaching than those of current trajectories stuck in business-as-usual ruts or which undertake only minor adaptive changes in response to short-term constraints.

4 Conclusion: The Challenge of Diversity

How can we reconcile the diversity of current agricultural and food systems with the necessity and rationales of transition? Can the academic and research world contribute to the evolution of our systems beyond merely alerting us to climate and biodiversity issues?

A geographical map does not tell the whole story, but it does allow us to situate ourselves and to understand the linkages between scales. In the absence of a typology, a plan, and foresight, the agricultural sector, in all its diversity, has difficulty finding a consistent response to the challenges of sustainability. It functions today as if it were following a GPS that shows the direction of the journey without the destination being clearly defined. Most of the proposals are technical and focused on the 'farm' system without taking into account the social and economic conditions for change. Ecologically intensive agriculture (Griffon, 2013), conservation agriculture (Kassam et al., 2019) and the '4 per 1000' initiative (Rumpel et al., 2019) all claim to be responses to the challenges of the twenty-first century. But they are characterised by an essentially technical bias that underestimates the part that cultural, social and economic dimensions have played in the past trajectories of agricultural systems and will certainly play in future ones. Emphasising the need to change farm-level practices (reducing pesticides, tillage, etc.) tends to make us forget the role that actor networks, political and economic choices, and macroscopic phenomena such as market globalisation have influenced and continue to influence the evolution of agricultural and food systems. Broadening the reflection to include these dimensions requires an awareness of the importance of these systemic factors, and the laying out of a broad and well-documented vision from which the trajectories to be undertaken can be negotiated together.

Situating oneself clearly in relation to a process of transition and an existing sociotechnical regime, as we propose here, should make it possible to foster debates that would move us away from a binary logic to construct credible and collective trajectories.

References

Altieri, M. A. (1987). *Agroecology: The scientific basis of alternative agriculture* (p. 246). Westview Press.

Antier, C., Petel, T., & Baret, P. V. (2017). *État des lieux et scénarios à l'horizon 2050 de la filière céréales en région wallonne* (p. 66). Sytra-Earth and Life Institute, Université catholique de Louvain.

Association négaWatt, Salomon, T., Jedlickza, M., & Marignac, Y. (2012). *Manifeste Négawatt. Réussir la transition énergétique* (376 p). Actes Sud.

Becerra, S., & Roussary, A. (2008). Gérer la vulnérabilité de l'eau potable: Une action publique désengagée ? *Natures Sciences Sociétés, 16*(3), 220–231.

Bellon, S., & Penvern, S. (2014). Organic food and farming as a prototype for sustainable agricultures. In S. Bellon & S. Penvern (Eds.), *Organic farming, prototype for sustainable agricultures* (pp. 1–19). Springer.

Buron, M.-H., Bouquiaux, J.-M., & Marsin, J.-M. (2014). *Blanc Bleu Belge, Blonde d'Aquitaine, Charolaise, Limousin : les quatre races viandeuses les plus répandues en Wallonie* (167 p). CER-SPW Éditions, Bilans et perspectives.

Cerf, M., Veiga, I., Prost, L., & Barcellini, F. (2015). Designing for transition in agriculture: addressing the gap between design and innovation. In *Proceedings 19th Triennial Congress of the IEA*, Melbourne (pp. 9–14).

Chabert, A., & Sarthou, J.-P. (2020). Conservation agriculture as a promising trade-off between conventional and organic agriculture in bundling ecosystem services. *Agriculture, Ecosystems and Environment, 292*, 106815.

Chase, J.M., & Leibold, M. A. (2003). *Ecological niches: Linking classical and contemporary approaches* (222 p). University of Chicago Press.

Conway, G., & Toenniessen, H. (1999). Feeding the world in the twenty-first century. *Nature, 402*, C55–C58.

Cowan, R., & Gunby, P. (1996). Sprayed to death: Path dependence, lock-in and pest control strategies. *The Economic Journal, 106*(436), 521–542.

Darnhofer, I., D'Amico, S., & Fouilleux, E. (2019). A relational perspective on the dynamics of the organic sector in Austria, Italy, and France. *Journal of Rural Studies, 68*, 200–212.

De Herde, V., Maréchal, K., & Baret, P. V. (2019). Lock-ins and agency: Towards an embedded approach of individual pathways in the Walloon dairy sector. *Sustainability, 11*(16), 4405.

De Schutter, O. (2010). *Agroecology and the right to food: report presented at the 16th session of the United Nations human rights council.* A/HRC/16/49, Geneva, Switzerland, United Nations Human Rights Council, 21 p.

Demeulenaere, E., & Bonneuil, C. (2010). Cultiver la biodiversité. Semences et identité paysanne. In B. Hervieu, N. Mayer, P. Muller, F. Purseigle, & J. Rémy (Eds), *Les mondes agricoles en politique. De la fin des paysans au retour de la question agricole* (pp. 73–92). Presses de Sciences Po.

Dorin, B., Treyer, S., & Paillard, S. (2011). *Agrimonde: Scenarios and challenges for feeding the world in 2050* (250 p). éditions Quæ.

Dumont, A. M., Gasselin, P., & Baret, P. V. (2020). Transitions in agriculture: Three frameworks highlighting coexistence between a new agroecological configuration and an old, organic and conventional configuration of vegetable production in Wallonia (Belgium). *Geoforum, 108*, 98–109.

Dumont, A. M., Vanloqueren, G., Stassart, P. M., & Baret, P. V. (2016). Clarifying the socioeconomic dimensions of agroecology: Between principles and practices. *Agroecology and Sustainable Food Systems, 40*(1), 24–47.

Ferdinand, M., Bertin, P., & Baret, P. V. (2020). Conservation agriculture and glyphosate: Strategies, lock-ins and diversity in the Walloon region. In *25th National Symposium for Applied Biological Sciences (NSABS)*, Gembloux, Belgium, 31 January 2020.

Ferguson, R. S., & Lovell, S. T. (2014). Permaculture for agroecology: Design, movement, practice, and worldview. A review. *Agronomy for Sustainable Development, 34*(2), 251–274.

Francis, C., Lieblein, G., Gliessman, S., Breland, T. A., Creamer, N., Harwood, R., Salomonsson, L., Helenius, J., Rickerl, D., & Salvador, R. (2003). Agroecology: The ecology of food systems. *Journal of Sustainable Agriculture, 22*(3), 99–118.

Geels, F. W., & Schot, J. (2007). Typology of sociotechnical transition pathways. *Research Policy, 36*(3), 399–417.

Gliessman, S., & Tittonell, P. (2015). Agroecology for food security and nutrition. *Agroecology and Sustainable Food Systems, 39*(2), 131–133.

Griffon, M. (2013). *Qu'est-ce que l'agriculture écologiquement intensive?* (224 p). éditions Quæ.

Guichard, L., Dedieu, F., Jeuffroy, M.-H., Meynard, J.-M., Reau, R., & Savini, I. (2017). Le plan Écophyto de réduction d'usage des pesticides en France: décryptage d'un échec et raisons d'espérer. *Cahiers Agricultures, 26*(1), 14002.

Hill, S. B., & MacRae, R. J. (1995). Conceptual framework for the transition from conventional to sustainable agriculture. *Journal of Sustainable Agriculture, 7*(1), 81–87.

HLPE. (2019). *Agroecological and other innovative approaches for sustainable agriculture and food systems that enhance food security and nutrition.* Report by The High Level Panel of Experts on Food Security and Nutrition, HLPE report 14, FAO, Rome, 162 p.

Holt-Giménez, E., & Altieri, M. A. (2013). Agroecology, food sovereignty, and the new green revolution. *Agroecology and Sustainable Food Systems, 37*(1), 90–102.

Jeanneaux, P., & Perrier-Cornet, P. (2011). Stratégie d'élévation des coûts des concurrents pour préserver un système productif agro-alimentaire. Le cas d'une filière fromagère d'appellation d'origine. *Revue D'économie Industrielle, 135*, 115–132.

Kassam, A., Friedrich, T., & Derpsch, R. (2019). Global spread of conservation agriculture. *International Journal of Environmental Studies, 76*(1), 29–51.

Landel, P. (2015). Réseaux d'action publique et accès aux connaissances pour la «transition écologique». *Économie Rurale, 347*, 59–78.

Le Foll, S. (2012). *Déclaration de M. Stéphane Le Foll, ministre de l'Agriculture, de l'Agroalimentaire et de la Forêt, sur le projet agroécologique de la France.* Conclusion de la journée «Produisons autrement» à Paris le 18 décembre 2012.

Lemieux, G. (1996). *Cet univers caché qui nous nourrit: le sol vivant* (p. 51). Université Laval.

MAAF. (2015). *Le projet agro-écologie en France* (4 p). Ministère de l'Agriculture, de l'Alimentation et de la Forêt.

Meynard, J.-M., Messéan, A., Charlier, A., Charrier, F., Le Bail, M., Magrini, M.-B., & Savini, I. (2013). Freins et leviers à la diversification des cultures: étude au niveau des exploitations agricoles et des filières. *OCL, 20*(4), D403.

Morel, K. (2016). Les microfermes participent à la transition agroécologique. *La Revue durable.*.

Paillard, S., Dorin, B., & Treyer, S. (2010). *Agrimonde: scénarios et défis pour nourrir le monde en 2050* (296 p). éditions Quæ.

Pascual, U., Balvanera, P., Díaz, S., Pataki, G., Roth, E., Stenseke, M., Watson, R. T., Dessane, E. B., Islar, M., & Kelemen, E. (2017). Valuing nature's contributions to people: The IPBES approach. *Current Opinion in Environmental Sustainability, 26*, 7–16.

Pluvinage, J. (2013). Agricultures. Produisons autrement. *Le Courrier de L'environnement de l'INRA, 63*, 141–145.

Porter, J. R., Howden, M., & Smith, P. (2017). Considering agriculture in IPCC assessments. *Nature Climate Change, 7*(10), 680.

Pouch, T., & Trouvé, A. (2018). Deregulation and the crisis of dairy markets in Europe: Facts for economic interpretation. *Studies in Political Economy, 99*(2), 194–212.

Poux, X., & Aubert, P.-M. (2018). *Une Europe agroécologique en 2050 : une agriculture multifonctionnelle pour une alimentation saine* (78 p).

Rumpel, C., Amiraslani, F., Chenu, C., Cardenas, M. G., Kaonga, M., Koutika, L.-S., Ladha, J., Madari, B., Shirato, Y., & Smith, P. (2019). The 4p1000 initiative: opportunities, limitations and challenges for implementing soil organic carbon sequestration as a sustainable development strategy. *Ambio*, 1–11.

Solagro, Couturier, C., Charru, M., Doublet, S., & Pointereau, P. (2016). *Afterres 2050: quelle utilisation des terres en 2050 en France?* (96 p).

Vankeerberghen, A., Dannevoye, B., & Stassart, P. M. (2014). L'insularisation comme mode de transition, le cas de l'agriculture de conservation en région wallonne. In R. Antoine, & F. Goulet (Eds.), *Sociologie des grandes cultures, au cœur du modèle industriel agricole*. éditions Quæ.

Vanloqueren, G., & Baret, P. V. (2008). Why are ecological, low-input, multi-resistant wheat cultivars slow to develop commercially? A Belgian agricultural 'lock-in' case study. *Ecological Economics, 66*(2–3), 436–446.

Vanloqueren, G., & Baret, P. V. (2009). How agricultural research systems shape a technological regime that develops genetic engineering but locks out agroecological innovations. *Research Policy, 38*(6), 971–983.

Wezel, A., Bellon, S., Doré, T., Francis, C., Vallod, D., & David, C. (2009). Agroecology as a science, a movement and a practice. A review. *Agronomy for Sustainable Development, 29*(4), 503–515.

Willer, H., Schaack, D., & Lernoud, J. (2019). Organic farming and market development in Europe and the European Union. In H. Willer & J. Lernoud (Eds.), *The world of organic agriculture. Statistics and emerging trends* (pp. 217–254). Research Institute of Organic Agriculture FiBL and IFOAM-Organics International.

Part V
A new paradigm of territorial development?

The Challenge of the Territorial Governance of Coexisting Models. Introduction to Part V

Pierre Gasselin, Sylvie Lardon, Claire Cerdan, Salma Loudiyi, and Denis Sautier

The first four parts of this book examine the situations of coexistence and confrontation of agricultural and food models according to the four dimensions of the territorial development analysis framework proposed in the general introduction: diversification/specialisation, innovation, adaptation and transition. This fifth and final part has a threefold ambition of an opening up, a more detached analysis and a conclusion. The first sub-section comprises the contributions of three researchers (Jérémie Forney, Kae Sekine and Gilles Allaire) whom we have invited to present new perspectives on the coexistence and confrontation of agricultural and food models based on their personal work. The second sub-section consists of chapters by Ronan Le Velly and Patrick Caron, whom we warmly thank for agreeing to share their personal and critical reflections on the contents of the entire book. We conclude by examining the title question of the book: Do the coexistence and confrontation of agricultural and food models open the way to a new paradigm of territorial development?

New Perspectives in Switzerland and Japan and in the 'Quality' Economy

The first three chapters of Part V extend the geography of the book's case studies with situations of coexistence in Switzerland and Japan and offer new theoretical perspectives for studying these situations.

In the first chapter, Jérémie Forney (Chap. 15) uses the diversity of forms of dairy production in Switzerland to question the relevance of the usual categories that are defined as models (plains and mountains, industrial and artisanal, conventional and organic, etc.). He emphasises that it is the local, legal and economic conditions that largely determine these models. They are interdependent, and their fates are 'inevitably linked', both in Switzerland and internationally (dependence of part of Swiss milk production on soya imports from Brazil). Furthermore, the dynamics of these models pose real analytical difficulties (e.g. when does one model start and end). Jérémie Forney, therefore, suggests considering models as an assemblage of elements

P. Gasselin et al. (eds.), *Coexistence and Confrontation of Agricultural and Food Models*,
https://doi.org/10.1007/978-94-024-2178-1

(production, processing, distribution, consumption) defined by the interactions that compose it and by its links to other models. In doing so, the boundaries of assemblage are blurred, weakening the notion of hybridisation, on the one hand, and calling for an examination of 'transformative forces' and the inconstancy of the model, on the other.

In the second chapter, Kae Sekine (Chap. 16) reports on the coexistence of contrasting agricultural models in Japan in the context of neoliberalisation of agricultural policies that has led to a decline in agricultural commodity prices and an economic crisis of the family farming model created in the context of post-Second World War land reforms. Kae Sekine examines the coexistence of family farming with, on the one hand, the multinational Dole Food Company, and, on the other, the US and Japanese multinationals involved in the process of reconstruction of the Fukushima region after the March 2011 tsunami. These two case studies reveal the agro-environmental, socio-economic and cultural tensions generated between agricultural models at the territorial level, which result in manifestations of resistance from local actors. Furthermore, there now exists a crisis of legitimacy of neoliberal agricultural policy.

In the third chapter, Gilles Allaire (Chap. 17) examines the notion of coexistence in the light of the 'quality turn' of capitalist development, which has been underway since the 1990s and is marked by the emergence of alternative production systems and value chains. He situates his thinking in regulation theory and analyses the competition (and cooperation) regime that is ensconced in systems of standardisation subject to 'the pressure of a conflicting plurality of visions of the future'. Gilles Allaire reminds us that quality 'is not the property of a thing'; it is instead a contextual judgement rooted in values and, at the same time, an institution based on doctrines (prevailing conceptions at a given moment of what is healthy, what is 'sustainable', etc.) and market standards backed by monitoring and sanctioning mechanisms. By illustrating this movement of qualification (of products, services, etc.), especially in organic farming, Gilles Allaire calls into question the coexistence of political projects. Alternative systems can, on the one hand, become conventionalised and lose their potential for radical change, and, on the other hand, renew themselves 'within the failures of the system'. The quality crisis, based on the questioning of doctrines, requires the mediatisation of criticism in a social movement in which the consumer becomes an actor in the debate (as, for example, in the case of mad cow disease). Gilles Allaire thus offers us a theoretical perspective based on institutional economics centred on quality, which has become 'an issue in the restructuring of activities and markets' and their coexistence.

These three chapters are based on theoretical propositions (assemblage for Jérémie Forney, compatibility/incompatibility between models for Kae Sekine, quality regimes for Gilles Allaire) that broaden and enrich the frameworks for analysing situations of coexistence that have been discussed in this book.

Benefits and Limitations of Models

In addition, these three chapters discuss the benefits as well as the limitations of the agricultural and food model as defined in this book's introduction. Let us remind ourselves here of its three acceptations: the model can be considered as an ideal type, an archetype of an observed reality (analytical representation); as a type of ideal (normative and programmatic representation, whether it be a desired or criticised future); or as a standard for action. This discussion is also a central element of the two chapters by Ronan Le Velly and Patrick Caron. While we do not present a summary of these chapters here, as they already put into perspective the main ideas developed in the book, we will cover a few of their central ideas.

The archetypal model requires the identification of regularities and polarities in the tumult of reality. Without an archetypal model, how can we account for the great diversity of actors' practices, discourses and positions? Our theories lead us to think about heterogeneity according to its various frameworks. To take an example from the field of agricultural studies, different currents of research have endeavoured to identify this heterogeneity, such as sustainable rural livelihoods (Chambers & Conway, 1991; Scoones, 1998; Farrington et al., 1999; Scoones, 2009), comparative agriculture (Mazoyer & Roudart, 1997; Cochet, 2011), farming styles (van der Ploeg, 2010; 2012), 'territorial pacts' (Albaladejo, 2009) and the sociology of agricultural worlds (Hervieu & Purseigle, 2013; 2015). Each of these analytical frameworks sheds light on the diversity of agricultural models with a focus on particular dimensions: labour, technical and economic performance, markets, relationship with nature, territorial integration, historical trajectories, etc. But it is also a matter of choosing scales of analysis and postures in the context of actions taking place. As Patrick Caron reminds us, 'Agreeing to look at diversity means recognising and grasping it, and this exercise in abstraction is closely tied to the intention to act'. Looking beyond this analytical diversity, we argue that the archetypal model should always be grasped in its temporality and its territorial embeddedness, and be confronted by the practical forms observed.

Indeed, several authors in this fifth part underscore the risk of cloistered thinking in terms of models that are likely to blur the complexity, diversity and dynamics of reality. Ronan Le Velly sums up the difficulty well: 'How can we not believe too much in agricultural and food models, but believe in them all the same?' The archetypal model encourages intellectual laziness or, worse, blindness in the belief of a world that is only represented by the model. The desired/criticised-future model, which organised actors use as the standard for their claims and projects, also entails the risk of obscuring the plurality of ideological currents that run through them and the practices that emanate from them. And finally, the standards-based model, such as that of organic farming, is likely to restrict social and technical transformation, hobble innovation and paralyse the capacity to adapt.

As we pointed out in the general introduction, models are often categorised by dual opposition (industrial vs artisanal, conventional vs alternative, modern vs traditional, etc.). This book is no exception to this tendency. This dualism is intrinsic to our intellectual, political and cultural heritage. We are subject to the dualism of biology (masculine and feminine), of certain religions (God and the devil), of moral conceptions (good and evil), of philosophies (the intelligible and the sensible), of currents of thought (Descartes's body and thought) and of political organisations (the right and the left in the legislature). This dualism, widely discussed since the Renaissance, does not, however, reduce the authors' analysis to a Manichean perspective. Ronan Le Velly agrees: 'The wide range of practices within each model also makes it impossible to continue to support dualistic reasoning'. Indeed, all the authors of this book emphasise the co-evolution and interfaces between these binary models, justifying the imperative necessity of understanding their coexistence and confrontations.

Given the risks of cognitive narrowing associated with a 'rigid-model' thinking, we also believe it is necessary to examine the extent to which the three acceptations of models (archetypes, desired/criticised futures, norms for action) interact closely, draw from each other, hybridise and even overlap. The actors who define a model as an ideal to be followed are strongly inspired by the analytical ideal types, and, for their part, the researchers produce archetypes inspired by the models under debate in society. This book invites us to explore more in depth this coexistence between analytical and normative models. Patrick Caron also shows that coexistence itself can be considered in its analytical dimension (taking note of reality) or in its normative dimension, noting that 'coexistence would be preferable to uniformity. In any case, it implies being able to exist in the first instance'. However, Gilles Allaire warns us: 'We are thus moving from the coexistence of normative goals in confrontation to a normative goal of coexistence'.

The Challenge of the Territorial Governance of Coexistence

Coexistence can also be a project, or rather a gamble, as Patrick Caron writes, to intentionally manage diversity and organise mediation. It is a gamble, but also a challenge. In economics, the notion of governance has its roots in the work of 'institutionalists' on the corporate world and corporate governance (Coase, 2007 [1937]). It was subsequently adopted in the field of urban governance and then by the international financial institutions, which defined 'good governance'. Governance has thus moved from the corporate level to the political field and its regulation and is now applied to nations, markets and territories (Gasselin, 2013). It presupposes the putting in place of adequate monitoring and sanctioning mechanisms (Foucault, 1975).

Applied to a territory, governance can be conceived as a process and instruments that together enable the maintenance or re-establishment of a cohesive collective and political action at the local level (Leloup et al., 2005; Lardon et al., 2008). Territorial governance brings together the processes, mechanisms and tools for coordinating various actors, social groups and institutions to achieve goals that have been collectively discussed and defined, including forms of public action. Patrick Caron prefers more voluntarist and explicit terms of management and mediation to that of governance. He also emphasises that this 'construction [...] presupposes that the terms of the confrontation between the elements present are clearly explained, whether they be actors, forms of organisation, actions, etc., and, in particular, of what may be controversial. [...] As coexistence is not self-evident, it is necessary to clarify the positions of each party and to establish or re-establish the conditions for dialogue between them, and to identify the obstacles that need to be overcome'. Looking beyond the nuances of the polysemous and controversial concept of governance (Torre & Chia, 2017), we note that the governance of the coexistence of agricultural and food models in a territory is therefore also that of their confrontation, or even their hybridisation, in order that new forms of organisation adapted to territorial development issues can be developed.

Without claiming to provide a recipe for the governance of coexistence, we return in the last chapter to the question in this book's title, identify three epistemological positions of the authors and summarise the approach that we propose for further research on this front. We hope that this proposal will be widely debated and thus enriched.

References

Albaladejo, C. (2009). Médiations territoriales locales et développement rural: Vers de nouvelles compétences d'accompagnement de l'activité agricole. Les agricultures familiales dans les transformations territoriales en Argentine, au Brésil et en France (p. 304). Accreditation to supervise research (HDR), University of Toulouse, Toulouse.

Chambers, R., & Conway, G. (1991). Sustainable rural livelihoods: Practical concepts for the 21st century (p. 33). IDS Discussion Paper, 296.

Coase, R. H. (2007 [1937]). The nature of the firm. *Economica, 4*(16), 386–405.

Cochet, H. (2011). L'agriculture compare (p. 160). coll. Indisciplines, éditions Quæ.

Farrington, J., Carney, D., Ashley, C., & Turton, C. (1999). Sustainable livelihood in practice: Early applications of concepts in rural areas. *ODI Natural Resources Perspectives*, (42), 15 p.

Foucault, M. (1975). *Surveiller et punir: Naissance de la prison*, (p. 360). Gallimard.

Gasselin, P. (2013). Gouvernance foncière et dynamiques territoriales, dans une perspective de sécurité alimentaire. In *Séminaire GloFoodS, travaux préparatoires du métaprogramme Inra « Étude des transitions pour la sécurité alimentaire mondiale »*, 12 and 13 February 2013, Serris, France, 10.

Hervieu, B., & Purseigle, F. (2013). *Sociologie des mondes agricoles* (p. 320). Armand Colin.

Hervieu, B., & Purseigle, F. (2015). The sociology of agricultural worlds: From a sociology of change to a sociology of coexistence. *Review of Agricultural and Environmental Studies, 96*(1), 59–90.

Lardon, S., Tonneau, J.-P., Raymond, R., Chia, E., & Caron, P. (2008). Dispositifs de gouvernance territoriale durable en agriculture. Analyse de trois situations en France et au Brésil. *Norois, 2008/4*(209), 17–36.

Leloup, F., Moyart, L., & Pecqueur, B. (2005). La gouvernance territoriale comme nouveau mode de coordination territoriale? *Géographie, économie, société, 7*(4), 321–332.

Mazoyer, M., & Roudart, L. (1997). *Histoire des agricultures du monde: Du néolithique à la crise contemporaine* (p. 505). Éditions du Seuil.

Scoones, I. (1998). Sustainable rural livelihoods: A framework for analysis, (p. 22). Working Paper 72, Institute for Development Studies, Brighton.

Scoones, I. (2009). Livelihoods perspectives and rural development. *Journal of Peasant Studies, 36*(1), 171–196.

Torre, A., & Chia, E. (2017). Nouvelles controverses du développement territorial: Quelle gouvernance et quelles innovations? *Canadian Journal of Regional Science/Revue canadienne des sciences régionales, 40*(2), 91–102.
van der Ploeg, J. D. (2010). Farming styles research: The state of the art. In *Keynote lecture for the Workshop on "Historicising Farming Styles"*, Melk, Austria, 21–23.
van der Ploeg, J. D. (2012). The genesis and further unfolding of farming styles research. *Historische Anthropologie, 20*(3), 427–439.

Chapter 15
Coexistence as Assemblage: The Multiplicity of Dairy Models in Switzerland

Jérémie Forney

During meetings with farmers, I often hear them say that each farm and each situation is unique and can only be understood in its irreducible distinctiveness: the specific soil, plot layout, heritage and history, outlet, expertise, passion or desire, etc. The ethnographer must take such an observation seriously. However, the human mind insists on looking for points of convergence and lines of demarcation to bring order to the mishmash of the diversity of reality. Whether we are farmers, agronomists, sociologists, civil servants or others, we mobilise the analytical tools familiar to us in an attempt to find unity in diversity. It is this very fundamental thought process that allows us in this book to mobilise the concepts of agricultural 'models' and their coexistence. In itself, the exercise makes sense and offers fertile ground for reflecting on the complexity of agriculture in a given territory and its articulation with the perspectives for this territory's future, as we see in this book's various chapters. However, every process of categorisation brings with it a risk: the fixation with reifying a concept, its slide from the status of a tool for dynamic thinking to that of a box in which the complexity of reality is locked up and concealed, as my interlocutors insist. The concept of an agricultural 'model' is not immune to this risk, whether it refers to an expert's ideal-type, to a project of committed actors (a desired ideal) or to a regulatory and normative standard.[1] When an ideal-type is accorded too much importance, it masks specificity and originality. An ideal model requires a specific translation for each application context. And it is essential to accept that a standard always offers only a partial and reductive vision. In contrast, once put into practice, a model becomes anchored in space and time, and its limits dissolve in the richness of reality.

[1] According to the triple definition of the agricultural 'model' proposed by this book's editors.

J. Forney (✉)
Institute of Anthropology, University of Neuchâtel, Neuchâtel, Switzerland
e-mail: jeremie.forney@unine.ch

P. Gasselin et al. (eds.), *Coexistence and Confrontation of Agricultural and Food Models*,
https://doi.org/10.1007/978-94-024-2178-1_15

In this reflective chapter, I wish to use the case of the Swiss dairy sector and what makes it internally diverse, particular and united at the same time to propose an approach to coexistence and to the concept of a model that does not divide reality into fixed analytical categories. To this end, I will start from categories commonly used to make sense of the diversity and complexity of agricultural worlds, and will discuss their value and limitations: lowland and mountain, milk and cereals, production and environment, industrial and artisanal (Protected Designation of Origin, PDO), conventional and organic, policies and market.

I propose to use this process to reflect on coexistence by shifting the focus away from agricultural models (without giving them up altogether) as central elements of the analysis and paying more attention to the multiplicity of modes of engagement and integration in an agricultural and territorial assemblage, and to the links between them. This approach draws on Deleuze and Guattari's (1980) concept of the French term 'agencement', which is usually translated as 'assemblage' in the international literature. Even more than the concepts themselves and a discussion about them, what interests us here are the possibilities offered by such an approach to think about the multiplicity and coexistence of agricultural models in a non-exclusive, dynamic way that is open to possible futures (Forney et al., 2018), in order to help make them more precise and effective tools for territorial and food governance. The notion of assemblage is thus offered as a compromise between a reification of models and a renunciation of the attempt of categorisation.[2]

1 Multiplicity and Uniqueness of an Agricultural Sector as a Starting Point

To think about and represent the multiplicity of social forms of agriculture and agricultural models (*who* produces?) is also to ask a series of questions about the *why* (what place in society?), the *for what* and *for whom* (with what partners and for what markets?) and the *how* (what techniques?). I will use these basic questions to trace a path through the Swiss dairy sector by calling into question certain categorisations that are commonly used to analyse the diversity of forms of dairy production.

1.1 Who? Producing Milk in Switzerland

Swiss dairy farming is a relatively homogeneous agricultural sector. In the vast majority of cases, it is represented by a highly capitalised and technologically advanced family farm, small in size by international standards (average of 25 ha). This relative homogeneity is the result of history and a strong and generally protectionist

[2] This chapter was written as part of the New Directions in Agri-Environmental Governance project, funded by the Swiss National Science Foundation, which I thank here for its support.

agricultural policy, which was able to alleviate certain economic pressures during the twentieth century while actively promoting productivist agricultural modernisation. Today, protection against imports through customs duties and production subsidies have largely given way to other instruments tied to environmental specifications that the vast majority of farms (nearly 98% across all sectors) conform to, most notably out of economic necessity. Thus, these 'Required Ecological Performances' (REPs) define a national environmental standard. These successive protections have however not prevented the sector's erosion, which has accelerated since the abolition of the milk quota system in 2009.[3] Falling prices and a lack of coordinated quantity controls have led to a phenomenon of restructuring and growth in production: fewer producers, but more milk per farm. In one decade, from 2007 to 2017, the dairy sector lost almost 30% of its producers.[4] To summarise, this political and economic context provides a shared framework that induces a certain uniformity in the sector, between the adoption of agri-environmental parameters and market pressures.

Milk production can also be described through its diversity. One of the main demarcation criteria generally used in public discourse (media, producer organisations, public services, etc.) is between the plains and the mountains. However, what underpins this demarcation has always varied depending on economic and political contexts. The specialisation of the mountain dairies in cheese production has historically taken place as a complement to the cereal orientation on the plains and because of the emergence of export markets. Competition from American wheat in the nineteenth century pushed the farms on the plains towards more intensive milk production, and mountain cheeses saw new competition with the development of cheese dairies on the plains (Ruffieux & Bodmer, 1972). Nevertheless, post-war food security policies have revitalised crop production (Moser, 2003) and provided strong support to mountain farming.[5] The story of dairy specialisation, partially presented here, is far from straightforward when put into a historical perspective. Dairy territories have been constructed and deconstructed according to different logics. Natural conditions interact with industrial history, the active promotion of agricultural techniques (such as silage), and the circulation of agricultural products, to name just a few factors. Today, the progressive liberalisation of the dairy sector is producing new forms of heterogeneity through differentiations in increasingly competitive markets. To summarise this initial overview, dairy farming is caught up in a set of forces, some of which have an obvious homogenising power, while others are leading to increasingly specific processes. This observation also reminds us of the fact that '*who* produces

---

[3] The milk quota system was introduced by the Swiss federal government in 1977. It allocated a maximum production level to each farmer. Subsequently, trading in these quotas (sale and rental) was allowed to impart more flexibility to the system. Its dismantling left the issue of quantity management in the hands of market forces.

[4] According to the website of the Swiss Milk Producers' Federation: https://www.swissmilk.ch/fr/producteurs-de-lait/marche-du-lait/faits-et-chiffres/graphiques-illustrations/chiffres-annuels/.

[5] This preferential support for mountain areas continues to this day in the form of the direct payment system: in 2017, mountain farms received an average of CHF 84,431 each in direct payments annually, as compared to CHF 66,344 for farms on the plains (according to the 2018 Agricultural Report of the Federal Office for Agriculture, https://www.agrarbericht.ch/fr).

milk' can only be understood by situating the question in a broad societal framework that combines technologies, markets, and national and international policies.

1.2 Why? The Opposition Between Production and the Environment

The dairy sector is the most important one in Swiss agriculture in terms of its actual size as well as of its symbolic significance. However, it costs more to produce milk in Switzerland than in other countries.[6] Looking beyond differences in calculation methods and their technical and political issues, we note that the factors behind higher production costs are the small size of the farms, the often difficult natural conditions (mountain farming) and the high cost of labour and equipment. In a context of progressive liberalisation, these costs pull down the farms' economic performance. However, several mechanisms still protect Swiss producers and Swiss markets from foreign competition. One example is the ban on the import of fresh milk for the Swiss market. (This is allowed, however, for processing into products for direct export.) The fact remains that milk has lost a significant part of its economic value and that this is perceived by many producers as a fundamental decline of the productive and food functions of agriculture.

But in the context of Swiss agricultural policy based on promoting multifunctionality, asking *why* milk is produced forces us to think beyond the roles of producers and food providers, and to confront an upsurge in societal expectations from agriculture and political objectives concerning it. Over the past 20 years, the Swiss federal government has implemented a policy of paying farmers for the public services they provide that are unremunerated by the markets. Thus, in 2017, each Swiss milk producer received on average the equivalent of about Euro 60,000 per year from the federal government[7] in the form of direct payments in return for complying with required ecological performances (REPs) and for participating in various programmes for animal welfare, promotion of biodiversity and sustainable use of natural resources.[8] Finally, the amount received through the direct payment system is on average equivalent to one third of the farm's income and exceeds the activity's final profit. In other words, the income from the sale of milk does not cover production costs, and the farmer survives on the money earned from providing environmental services. No doubt, averages gloss over huge differences, and the significance of direct payments in a farm's economic performance varies according

[6] For example, a study comparing Switzerland and Norway attributes the high cost of Swiss production primarily to structural costs (especially machinery and buildings), identifying in particular 'real investments in buildings that are 47–63% higher in Switzerland than in Norway, net of subsidies' (Gazzarin et al., 2014: p. 254).

[7] According to the 2018 Agricultural Report of the Federal Office for Agriculture, https://www.agrarbericht.ch/fr.

[8] This is a somewhat of a rough summary of a complex system. For further information, see the website of the Federal Office for Agriculture (www.blw.admin.ch).

to many factors (type of production, geographical location, size, strategic choices). Different farming models and production methods are inevitably constructed in interaction with and based on this policy framework. However, this same policy produces a fundamental distinction between production and the environment, which is the topic of public debates and finds expression in the actors' discourses. This separation (or decoupling) between the market (products) and environmental services (direct payments) is indeed this policy's fundamental principle. The result is a tension between a de facto overlap (subsidy instruments as determinants of agricultural practices) and a fragmentation of sense (the abstract separation between productive and environmental functions).

1.3 For What and for Whom? 'Quality' Agri-chains and Strategies

The Swiss dairy sector is usually divided into two main agri-chains according to the type of final product. First, the 'cheese milk' agri-chain represents about 43% of total volumes and is based on several PDOs, some of which are well-known.[9] Because of the restrictions imposed by cheese making (unpasteurised milk), this milk is produced without the use of fermented fodder (silage). The State compensates the farmers with a non-silage subsidy. Second, the 'industrial milk' agri-chain encompasses milk processed by dairy companies into fresh products for the national market—milk for drinking, yoghurt, cream, quark cheese and other products, which represent about 26% of the total volume—or into butter (15%) or other milk preserve products (10.8%) for the food industry or export. The cheese milk agri-chain is better protected against the difficulties that confront the industrial milk agri-chain. This is clearly shown by the significant difference in the prices paid to producers.[10] To complete this first binary overview, we note that the share of organic production continues to grow in the entire sector (cheese milk and industrial milk combined) and had reached about 7% of total volumes in 2018, offering more remunerative and stable prices to producers.[11] As expected, this seems to confirm the advantage of so-called 'quality' strategies (PDO and organic) over standardised industrial production. The idea that

[9] Gruyère PDO (341 million kg) and Emmentaler PDO (207 million kg) together account for more than half of the milk processed by the country's cheese dairies (1100 million kg).
(https://www.swissmilk.ch/fr/producteurs-de-lait/marche/acteurs-et-structure-du-marche/transformation-du-lait/, retrieved 25 November 2021).

[10] According to the October 2021 Milk Price Monitoring report, 77.67 CHF cents per kg for cheese milk versus 65.41 CHF cents per kg for 'industrial' milk, average over 12 months, 'effective ex-farm REP' price (https://api.swissmilk.ch/wp-content/uploads/2021/12/rapport-psl-prix-du-lait-surveillance-2021-octobre-2021-12-23-fr.pdf, retrieved 17 January 2022).

[11] According to the October 2021 Milk Price Monitoring report, 88.64 CHF cents per kg for organic 'industrial' milk and up to 92.23 CHF cents per kg for organic cheese milk (Gruyère PDO), 'effective ex-farm REP' price (https://api.swissmilk.ch/wp-content/uploads/2021/12/rapport-psl-prix-du-lait-surveillance-2021-octobre-2021-12-23-fr.pdf, retrieved 17 January 2022).

Swiss agriculture should strive for 'quality' in globalised and competitive markets makes for common sense in agricultural circles and in wider society. However, this notion of a quality strategy needs to be rethought, as does the separation between agri-chains. Indeed, the different agri-chains and strategies can also be analysed through what links them.

Let us take a few examples. The Vacherin Fribourgeois PDO, a small brother of the big, well-known PDOs, shares their link with historical cheese production in mountain pastures. Today, the largest producer of Vacherin PDO is Cremo, the second largest Swiss dairy company, which also specialises in the production of industrial milk preserves (butter, powder). The same 'quality strategy' thus brings together producers from the plains who deliver their milk to an industrial entity and mountain farmers who process their own milk in the mountains. Another example: when the Gruyère PDO interprofessional organisation decreed a reduction of 10% in the quantities produced in 2015, the industrial actors bought and incorporated part of the surplus of this cheese milk into their supply chains and thus played a welcome role of buffer for the cheese milk producers. However, such a role can lead to dissatisfaction on the part of industrial milk producers due to the additional pressures resulting from the arrival of new quantities of milk in an agri-chain already close to saturation.

The success of 'quality' approaches also deserves examination. The example of Emmental PDO shows that a protected designation does not guarantee success, even for a cheese with a world-famous name and well-established production structures. Production of this cheese has collapsed in recent years, with periods when the price paid to milk producers has fallen below the average price paid for industrial milk. This uncertainty of economic viability is also found in the industrial agri-chain, where some actors specialising in processing fresh products for Swiss consumers are in a very different situation from other processors whose significant part of the production is destined for other less favourable markets (agrifood industry, international markets, etc.). Furthermore, a discussion on quality-oriented strategies goes hand in hand with the issue of quantity, even if only pertaining to efforts to avoid oversupply or to maintain market share. The cut-backs in quantities imposed in 2015 by the Gruyère interprofessional organisation, for example, made it possible to anticipate and compensate for the drop in exports that an unfavourable exchange rate would lead to. More generally, the possibility for a milk producer to join a specialised agri-chain depends largely on access to production rights. Whether in the framework of a PDO or the organic sector, the strict control of quantities conditions and limits the access to these agri-chains by new producers. In essence, not all producers can adopt quality strategies even if they want to.

These few examples are an apt illustration of a key characteristic of the coexistence of dairy models in Switzerland: the broad agri-chain categorisations conceal a multiplicity of interdependencies and overlaps. These interdependencies are such that it is difficult to imagine the existence of one model without that of the others that complement it.

1.4 How? *Grasslands-Rich Country and Yet Fodder Imports*

Switzerland is a grasslands-rich country. The climatic and geomorphological conditions of Swiss agricultural territory confirm the reasoning that since grass is what grows best in this territory, livestock farming, especially dairy farming, is necessary in order to transform the grass into food for human consumption. An analysis of farming practices shows that grass does indeed make up the largest share of feed for Swiss dairy cows.[12] However, practices vary significantly depending on location, breed of cattle and type of farming system. The intensive nature of livestock production in Switzerland also sometimes encourages a significant use of maize, cereal mixtures and protein crops, both on the plains and in the mountains. Dairy cattle feed practices raise the question of production techniques in a more general context, especially because they contrast with the image of a pasture-based agriculture and because Switzerland is increasingly importing fodder, in particular Brazilian soya. Here, coexistence takes place in a network of strong interdependencies: a supposedly sustainable mountain peasant agriculture (the image of a grazing cow), an intensive and 'profitable' agriculture that supports the existence of processing infrastructure (which mountain farmers also need), and a Brazilian monoculture (soya) that provides the proteins that are lacking at the scale of a Swiss agricultural system. There is a tension between an overproducing dairy sector and the import of foreign fodder that makes it possible to exceed the production limits dictated by the availability of local fodder. This tension is coupled with a contradiction between the ecologisation of agricultural practices within the country's borders and what can be perceived as a form of externalisation of the environmental impact of fodder production through its relocation. In this case, the territorial limits of coexistence become blurred.

2 Beyond the Hybridisation of Agricultural Models: The Multiplicity of Assemblages

This overview of the Swiss dairy sector through the prism of its multiple facets confirms the relevance of the concept of coexistence itself. This sector is indeed marked by a diversity of agricultural models which constitute its reality and which interact with each other. The few discussions developed above of the relevance of the usual categories of differentiation of models—whether analytical-descriptive (industrial milk vs. cheese milk) or programmatic ('quality')—in the specific case of the Swiss dairy sector allows us to make some more general comments on the concept of coexistence and to propose an interpretation inspired by assemblage theory (Deleuze & Guattari, 1980).

---

[12] According to a study, the proportion of grass in cattle feed exceeds 80% on average in more than half of the country's dairy farms (Schmid & Lanz, 2013).

2.1 Some Comments on the Concept of Coexistence

First is the matter of the scale of analysis. What are we referring to when we speak about 'agricultural models'? Individual farm strategies? Regional agri-chains? General orientation at a country scale? Every agricultural model, as an abstraction, is embedded in a larger framework of an agricultural and food landscape. This is illustrated both by the diversity of the fates of models that at first sight seem similar (between cheese PDOs, for example) and by the predominant role of structures common to different models, such as the direct payments system in agricultural policy. Various agents such as the State, supermarket chains, consumers (in all their diversity), international agreements or even topography and climate contribute to shape models in practice, in their lived form, anchored in a territory and a temporality. This has several consequences that may well be known, but which are worth recalling, especially in the context of territorial development projects. Thus, an agricultural model cannot be considered or designed in isolation from its societal and environmental framework. The models are not simple options between which one can choose freely according to one's aspirations and individual visions: the situation of each actor, farmer or non-farmer, is constrained and made up of local specificities, access or lack of access to agri-chains and their infrastructure, access to markets (e.g. obtaining of production rights), and the possible presence of legal frameworks (e.g. synergies with an agricultural policy) or economic frameworks (e.g. a PDO or industrial infrastructure), to mention but a few examples.

Second, it seems important to emphasise the interdependence between coexisting models. As shown for the Swiss case, the fates of systems often categorised as binary opposites (cheese/industry; conventional/quality) are in fact inevitably linked. The differentiation of the alternative depends on the existence of a predominant model (what would organic be without conventional?). Complementarity between agri-chains is also evident in the sharing of processing and marketing infrastructure (e.g. PDOs that mobilise industrial structures for production or for managing surpluses), or in the sharing of tasks between the embodiment of an image based on mountain tradition, which is whole-heartedly used in advertising for the entire sector, and the economic viability of intensive production on the plains, which is essential for the survival of processing structures. Furthermore, the cases presented above tend to show that it is difficult to confine agents to exclusive models: networks and agri-chains overlap and intersect. These observations contradict the idea of an agricultural model as a distinct (and transposable?) unit. In other words, the concept of coexistence of models is not as useful for reflecting on parallel strategies in their specificity as it is for focusing our attention on the relationships and dynamics between 'models' with fuzzy boundaries.

Third, the concept of coexistence of agricultural models as a tool for reflecting on the diversity of today's agriculture at a territorial scale contrasts with the reality of globalised agricultural and food systems. The dependence of a part of Swiss dairy production on soya imports from Brazil is a good example. Indeed, the current Swiss model of intensive milk production is reliant on the existence of a complementary model, several thousand kilometres away, of intensive (non-GMO) soya production. Other examples can also easily be mentioned (e.g. around international seed or animal networks), but we can cut straight to the question that arises: How can we analyse coexistence and territory in such a way that these long-distance interdependencies are not excluded?

Finally, the temporal dimension should not be forgotten just because a territorial approach makes it easy to think about spatial coexistence. The continuous arrangements of the models, according to a constantly evolving context as well as their own dynamics, in turn raise the question about what defines a model's identity and its permanence over time. For example, the prerequisites and factors necessary for producing industrial milk in Switzerland have continued to change over the last few decades (to speak only of the short term), both in terms of agricultural techniques as well as in terms of relations with professional organisations, economic partners and the State. Looking beyond the permanent elements that maintain their identity, how can we integrate this highly dynamic and evolving dimension of the models into their concrete applications? After what degree of transformation and variation, and according to what criteria, will we decide that one model has given way to another?

2.2 Coexistence of a Multiplicity of Assemblages

The concept of coexistence fundamentally calls into question that of the agricultural model, and in particular the spatial and temporal delimitation of the model in question. For a response, I propose to draw on the notion of assemblage proposed by Deleuze and Guattari (1980). More specifically, I will focus on a few central aspects that are especially useful here for rethinking the concept of coexistence, namely the concepts of multiplicity and territorialisation.

2.2.1 Assemblage

An assemblage[13] is a complex of lines and relationships, organised in such a way that any element or point can and should be connected to another point. An assemblage does not really have a centre, nor a hierarchy between the elements that constitute it, and it always remains open to the addition or removal of elements. To understand an agricultural model as an assemblage, according to these few defining characteristics, is to allow oneself to think of it in terms of its openness and its insertion into wider networks of relationships, and to place the emphasis not on the specificities of the elements that characterise it, but on the links that form it. An agricultural model,

[13] This definition of assemblage is especially valid for rhizome-like assemblages, in the vocabulary of Deleuze and Guattari. For the sake of clarity and conciseness, I avoid the detailed terminology of these authors in this chapter as it is not essential to the argument and rely instead on a simplified reading.

understood as a particular organisation (an assemblage) of elements pertaining to the production, processing, distribution and consumption of agricultural products, is thus defined not by what happens at a specific point, but by the set of links that constitute it—including the links that join it to other models.

Similarly, considering models as assemblages resolves the dilemma of the scale at which coexistence should be considered. In an agricultural model, everything can be broken down into a web of relationships, without any imposition a priori of territorial barriers. Thus, the coexistence of agricultural models cannot be defined merely by their co-presence in a geographically delimited territory. It is instead defined more by the links that connect (or separate) the models, uniting them in a broader assemblage, which itself is not on a different scale, but only in a more extensive framework, in a set of links without any real end.

2.2.2 Multiplicity: Beyond Hybridisation

Defining agricultural models in terms of assemblages requires us to stop thinking of them as exclusive categories with well-defined boundaries. Such an approach also means that we can no longer characterise the fundamental ambiguity of models as hybridisation. Indeed, the notion of hybridity relies on the assumption of the existence of fixed and clearly defined boundaries, without which it loses its meaning (Pieterse, 2001): there can be no mongrel without the thoroughbred. If the models are conceived from the outset as imperfect, interdependent and interconnected—which is theoretically and empirically verifiable—then the heuristic utility of the concept of the hybrid disappears, except to defend the relevance of ideal-type categories.

Deleuze and Guattari's (1980) reflections on the notion of multiplicity, which allowed them to free themselves from a binary thinking characterised by the opposition between the multiple and the unique, can lead us to this kind of conclusion. In a way, as an assemblage, a model, in its practical application, is several things at the same time. To confine it to a single category would be to distort it in some way, to distance it from its own reality.

2.2.3 'Territorialisation' and Temporality

The concepts of territoriality and deterritorialisation, central to assemblage theory, are useful here, perhaps counter-intuitively, to provide an answer to the question of the temporality of agricultural models. Indeed, these concepts do not refer to territory in the geographical and spatial sense, but rather to the anchoring, the fixing of an assemblage (territoriality) and the constant tensions that tend to dismantle and recompose it along other lines of relationship (deterritorialisation and reterritorialisation). These tensions make assemblages 'multiplicities of becoming, or transformational multiplicities' (Deleuze & Guatarri, 1980, p. 631). An assemblage is constantly subjected to transformative forces (elements that leave it, changing power relations, etc.) that run up against capacities for resistance and inertia. It reproduces itself over time while transforming itself. Following the example of the Swiss cheese speciality agri-chains, which incorporated the principles of PDO at the turn of the twenty-first century, it can undergo a radical transformation whose effects will be noticeable in the medium term, while still remaining apparently identical in many respects.

Looking beyond the salutary reminder of the inconstancy of all things and the permanence of change, and focusing attention on the tensions between transformative and stabilising forces also makes it possible to integrate more centrally the question of the future of models, their possible evolutions and their potentialities. What makes an assemblage is not only what it is today and was yesterday, but what it is tending towards, with all the uncertainties that this question embodies.

3 Conclusion

In this chapter, I wanted to present a description of the Swiss dairy sector and undertake a more theoretical reflection on the issue of the coexistence of agricultural models, as seen through the prism of assemblage theory. Such an effort encourages us to embrace the complexity of the social world and to be wary of simplifying categorisations, whether derived from common sense or from analytical models proposed by scientific approaches. However, we require categories for functional and analytical reasoning. Furthermore, we need the idealisation and simplification of models as projects in order to look at the world in a way that allows us to determine what is preferable, to make decisions and sometimes to commit ourselves. In my opinion, the assemblage approach offers a compromise between a reification of categories and their total deconstruction. Rather than focusing on the specificities of categories and their delimitation, this approach encourages a search for connections and future potentialities. The category, in our case the agricultural model, becomes a snapshot of a moving object caught in a framework that is itself evolving, and which is constituted by its interactions with other categories. The concept of coexistence offers a prism through which to think about these entanglements. What I propose here is that it will play this role well only if it is focused on the inherent interdependencies between models, if it is turned away from the search for ideal models as an ultimate goal, and if it incorporates the inevitable imperfection and ever-changing character of assemblages as an essential characteristic of social reality.

References

Deleuze, G., & Guattari, F. (1980). *Mille plateaux. Tome 2, Capitalisme et schizophrénie* (645 p). Éditions de Minuit.

Forney, J., Rosin, C., & Campbell, H. (2018). Introduction: agri-environmental governance as assemblage. In J. Forney, C. Rosin, & H. Campbell (Eds.), *Agri-environmental governance as an assemblage. multiplicity, power, and transformation* (pp. 1–16). Earthscan Food and Agriculture.

Gazzarin, C., Kohler, M., & Flaten, O. (2014). Exploitations laitières: Pourquoi la Suisse produit-elle plus cher que la Norvège ? *Recherche Agronomique Suisse, 5*(6), 248–255.

Moser, P. (2003). *Sélectionner, semer, récolter : politique agricole, politique semencière et amélioration génétique en Suisse de 1860 à 2002* (136 p). Hier+Jetzt.

Pieterse, J. N. (2001). Hybridity, so what? *Theory, Culture and Society, 18* (2–3), 219–245.

Ruffieux, R., & Bodmer, W. (1972). *Histoire du gruyère en Gruyère du XVI*e *au XX*e *siècle*, coll. Études et recherches d'histoire contemporaine, Éditions universitaires, Fribourg, Switzerland, 364 p

Schmid, D., & Lanz, S. (2013). Composition de la ration fourragère dans l'élevage de vaches laitières en Suisse. *Recherche Agronomique Suisse, 4*(4), 184–191.

Chapter 16
Neoliberalisation of Japanese Agricultural Policy and Contradictions Between Agricultural Models

Kae Sekine

Japan is an apt case study for analysing the coexistence of different agricultural models, given that the Japanese government has replaced its classical interventionist agricultural policies with neoliberal policies in recent years, making the agricultural space a battlefield for contrasting agricultural models. In this context of neoliberalisation of agricultural policy, with the opening up of markets and international competition leading to lower prices for agricultural products, an economic crisis is confronting the 'family farming model' created after the Second World War as part of agricultural reforms. While the industrial sector, especially the automotive sector, continues to be successful, the agricultural sector is experiencing difficulties in ensuring generational transmission and is suffering from a marked ageing of its actors. Since 2000, due to the deregulation of laws, multinationals have become actively involved in agricultural production, which has led to tensions in the sector and resistance from local actors (Bonanno & Constance, 2008; Sekine & Bonanno, 2016).

Our aim in this chapter is to characterise the process of neoliberalisation of Japan's agricultural policy and to analyse its consequences on its agriculture and territories. To this end, we examine the contradictions between agricultural models induced by these policy changes, on the basis of two case studies: on the one hand, the case of the multinational company Dole, which invested in Japanese agriculture, and on the other hand, that of the process of reconstruction of agriculture in the Fukushima region, Tohoku, after the tsunami of March 2011. The methods we used consisted of the analysis of qualitative data collected through semi-structured interviews conducted between 2004 and 2006, and again between 2011 and 2013, and the analysis of existing statistics and literature. The first section discusses the evolution of Japanese agricultural policy from classical interventionism to neoliberalism, while the second section is devoted to two case studies that show the difficulty of the coexistence

K. Sekine (✉)
Graduate School of Economics, Aichi Gakuin University, Nagoya, Japan
e-mail: kaesekin@dpc.agu.ac.jp

P. Gasselin et al. (eds.), *Coexistence and Confrontation of Agricultural and Food Models*,
https://doi.org/10.1007/978-94-024-2178-1_16

of agricultural models in the rural space. We conclude by discussing the limits of neoliberal policy in meeting societal expectations.

1 Neoliberalisation of Agricultural Policy: A Family Farming Model in Crisis

1.1 Japanese Agriculture and the 'Family Farming Model'

After the Second World War, Japanese agricultural policy remained interventionist for several decades and it was successful in stabilising the rural economy and eliminating hunger (Sekine & Bonanno, 2016). The policy also led to affordable food prices for urban consumers. The dominant agricultural model in Japan is the 'family farming model', characterised by smallholders,[1] which has been following Green Revolution practices since the 1960s, much like other world regions (Mazoyer & Roudart, 2006; Teruoka, 2008). Instead of expanding the size of their farms, most farmers have decided to diversify their economic activities in order to maintain their farms (Jussaume, 1991).

With the growth of the industrial sector and its exports, the Japanese government began to liberalise its domestic market, including the agrifood market by changing the laws pertaining to this sector (Teruoka, 2008; Sekine & Bonanno, 2016). The laws and institutions established after the Second World War in order to create, maintain and protect the family farming model have been gradually revised, if not abolished altogether in some cases, such as high import tariffs, the prohibition of agricultural production by stock companies, and government support for rice production through high procurement prices. Thus, the revamping of agricultural policy that began in the 1980s has led to major transformations of Japanese agriculture.

Statistics from the 2015 agricultural census provide a quick overview of Japanese agriculture to date: agricultural land covers 12% of Japan's land mass, amounting to 4.5 million hectares, while forest areas occupy 68%. Urban spaces cover 20% of the territory. Japan has 1.4 million farms, a figure that has declined by 30% over the last ten years. Of these, 98% are family farms.[2] Group farms, including company farms, account for only 1.3% of the total number of farms. The average farm size remains modest, at 2.5 ha per farm. However, 2.1 million agricultural jobs have been created by these farms, representing 3.3% of the entire Japanese workforce. Furthermore, 84% of farms produce only rice, the staple food of the Japanese, or with a few other

[1] In 2018, the average size of a farm in Japan was 2.98 ha.

[2] The Japanese agricultural census distinguishes between two main categories of farms: 'family' farms and 'group' farms. Farm categorisation depends on the perception of the farmers who answer the questionnaires. Each of these two categories (family and group) can have one of two legal statuses: 'company' or 'non-company'.

products such as vegetables or livestock. Less than 30% are professional farms,[3] while 71% are part-time farms.[4]

Japanese agriculture is in crisis for a number of reasons, including declining farm incomes, ageing farmers, a very low rate of food self-sufficiency (only 39% of food in calories equivalent is produced domestically), and an increasing share of abandoned farmland (9.4% of agricultural land, or 0.42 million hectares).[5] A significant proportion of the agricultural population is aged, with 77.5% of farmers over 60 years old. This observation reveals the younger generation's lack of interest in the sector and the difficulty in ensuring generational transmission of farms. The rural exodus continues, but concomitant with an urban exodus in the opposite direction that has increased in recent years. The sparsely populated rural area is now a haven for wildlife (wild boar, deer, monkeys, bears, birds, etc.), which caused damage to agricultural production to the tune of 17 billion yen in 2015 (MAFF, 2019). Land that is most suitable for agricultural production is being requisitioned for the development of recreational areas, and more recently for the installation of solar power plants, following changes in energy policy after the Fukushima nuclear disaster in 2011.

1.2 Processes of Neoliberalisation of Japanese Agricultural Policy

Since the 1980s, Japanese agricultural and economic policies have been reshaped in line with the principles of neoliberal economics (Sekine & Bonanno, 2016). As far as trade is concerned, market liberalisation under the GATT (General Agreement on Tariffs and Trade) and WTO (World Trade Organisation) regimes has led to an increase in imports, especially of agrifood products, and an increase in exports of Japanese industrial products. The rapid appreciation of the yen after the Plaza Accords[6] in 1985 accelerated this trend. The deregulation of agricultural land law in 2009, the creation of 'special economic zones'[7] and the granting of subsidies have mainly favoured stock companies investing in agricultural production and large farms, to the detriment of small and medium-sized family farms. Agricultural policy is now based on the idea that small family producers are unable to sustain agricultural production and therefore corporate intervention has to be encouraged. This idea is being propagated in political discourse and the media since the 2000s.

---

[3] A professional farm is one in which no one works outside the farm for more than 30 days per year.

[4] A part-time farm is one in which at least one person works outside the farm for more than 30 days per year.

[5] 2015 agricultural census.

[6] The Plaza Accords, signed by the G5 (US, Japan, West Germany, UK, France) at the Plaza Hotel in New York in September 1985, were aimed at stabilising exchange rates.

[7] A special economic zone (SEZ) is an area created by the State with laws to attract investment from national and international companies, with a view to promoting employment and boosting economic activities.

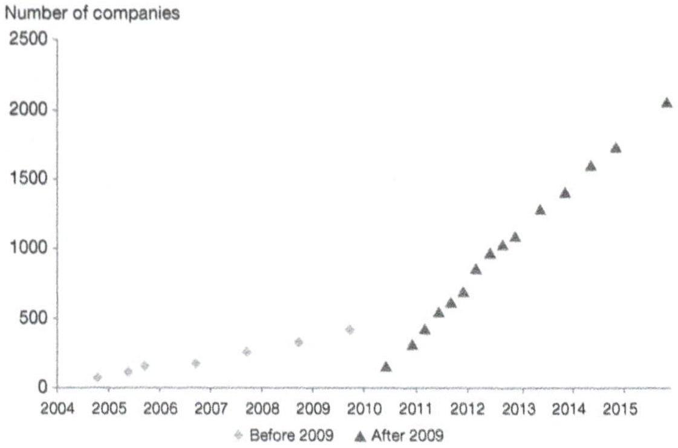

Fig. 1 Increase in the number of companies in the agricultural production sector. *Source* Sekine (2016), based on Muroya (2007) and MAFF (2015)

Figure 1 shows the rise in the number of companies investing in agricultural production. There is a clear increase from 2009 onwards, after the deregulation of agricultural land law.

Figure 2 summarises the characteristics of the companies that have invested in agricultural production: 62% are stock companies, while 23% are food industries and 41% produce vegetables.

Table 1 lists the main companies investing in agricultural production. It includes not only agribusiness companies, but also manufacturing companies such as Toyota, and trading companies such as Mitsubishi, electronics companies such as IBM and Toshiba, communication and finance companies, etc. Some are Japanese or American multinationals. In the case studies presented in the chapter, we analyse companies belonging to Dole, Kagome, Mitsubishi and IBM.

Since the neoliberal turn in the 1980s and its radicalisation in the 2000s, a new agricultural model dominated by multinational companies has emerged in the rural space and has in fact run up against the traditional, family-based agricultural model. Tensions have arisen between these two agricultural models each underpinned by a different rationale, and local actors have often resisted the entry of multinationals.

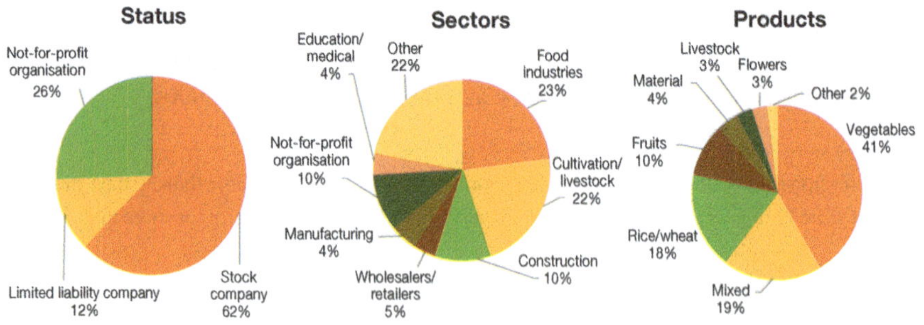

Fig. 2 Characteristics of companies in the agricultural production sector. *Source* Sekine (2016), based on MAFF (2015)

Table 1 Main companies investing in agricultural production

Sector	Company names
Agriculture	Japan Tobacco, **Dole Japan**
Food processing	**Kagome**, Mercian, Kewpie, House Foods, Kyusai
Food service	Mos Food, Monteroza, Watami, Saizeriya, Lawson, Yoshinoya, Skylark
Retailer	Seven & I Holdings, AEON, Coop Hiroshima, Hankyu Department Stroe
Trading	Itochu, Sumitomo, Mitsui, **Mitsubishi**, Marubeni, Nissho-Iwai, Toyota-Tsusho
Beverages	Sapporo Beer, Suntry Holdings, Kirin Brewery
Manufacturing	Toyota Motor, Nittobo, Showa Denko, Omron, Secom
Electronics	**IBM**, Toshiba, NEC, Hitachi
Communication	NTT Communications
Railroads	Japan Railroad
Chemicals	Sumitomo Chemical
Finance	Promise

Sources Sekine (2016), based on Tsutaya (2000), Muroya (2007) and Taniwaki (2011)
Note these companies invest in agricultural production directly or through their subsidiaries

The following section discusses the ways in which these agricultural models coexist in Japanese rural society and their contradictions.

2 Contradictions in Agricultural Models Induced by Neoliberalisation Processes

2.1 A Case Study of Dole Food Company

Dole Food Company is one of the largest multinational fruit and vegetable production and trading companies. It was established in the nineteenth century in the United States (Dole Food Company, 2013). A subsidiary was established in Japan in 1965 and imported and traded tropical fruits, such as bananas and pineapples, from the Philippines (Dole Japan, 2013; Sekine, 2017).

In parallel with the deregulation of the country's agricultural legislation and growing political support from the Japanese government, Dole started to set up franchise farms in 2000. The company has ten farms of over 800 ha in total (Sekine & Hisano, 2009; Sekine, 2016; Sekine & Bonanno, 2016; Farmind, 2018; Fig. 3). Dole's franchise farms sell products under the brand name I LOVE, which stands for 'I live on vegetables'. Through this franchise in Japan, Dole controls the production and sales of produce from its farms by indirectly investing through a Japanese partner

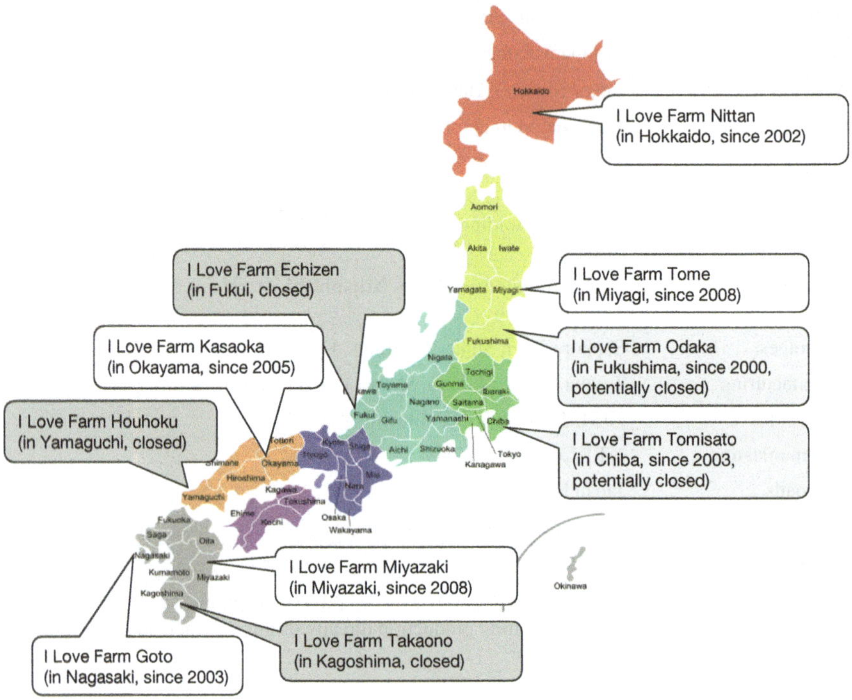

Fig. 3 Dole Japan's franchise farms. *Source* Sekine (2016). Prepared by the author based on interviews with Dole Japan and its website (http://www.dole.co.jp, consulted on 18 June 2012)

company and by guaranteeing market outlets in Japan. Dole was the first company to use a franchise model in the agricultural sector but this model is now being replicated by other agricultural companies.

Although considered a success story in the agricultural sector, following the global economic crisis of 2009, Dole's Asian fresh fruit and vegetable division was bought out by Itochu, a Japanese multinational trading company. Five of the ten franchised farms were closed, sold or halted production until 2018. Following the reorganisation of Dole's international businesses, its subsidiary responsible for the franchise farms, Hokkaido Sanchoku Centre, was sold in 2018 to Farmind, a major banana distributor and one of Dole's Japanese partners (Farmind, 2018). As of 2018, there were five remaining franchise farms producing vegetables such as broccoli, cauliflower, maize, pumpkins, soya beans, etc., on farmland of over 800 ha in total. Their field labour force consists of those who usually find themselves marginalised in the local labour market. Thus, most of the employees are elderly people, housewives and students working for minimum wage.

What does the case of Dole tell us about the issue of coexistence? What are the conditions propitious to the coexistence of the multinational model with the family model? Our analysis is based on a combination of the agro-environmental, socio-economic and cultural points of view.

At first glance, Dole's agricultural model seemed compatible with local agricultural models, for example by helping to solve local agronomic problems. Dole organised rotations that combined the land of its franchise farm with the land of family farmers in the Hokkaido area. In this way, Dole expanded the effective size of its farm by more than 300 ha. However, an incompatibility between the two models arose when Dole used excessive mineral nitrogen on farmland in the Kyushu region, and seemed to use large quantities of pesticides in the Tohoku region. These practices appeared detrimental to local agriculture and posed a water pollution risk. Eventually, the Kyushu farm was closed and relocated just three years after its opening, while the Tohoku farm was sold to a local producer a few years after it was set up. These examples show that environmental/agronomic compatibility is an essential prerequisite for the coexistence of agricultural models.

Furthermore, for the models to coexist, a key factor is the coherence of economic and social expectations of the different actors. In the case of Dole, there was a disconnect between local expectations and its practices. The local actors expected technical transfers, the creation of jobs and the revitalisation of local agriculture. But Dole only offered the possibility of low-skilled jobs. Furthermore, these jobs were insecure with low salaries, very dependent on the farm's activity, and were subject to cut-backs or complete elimination. Indeed, the farms' sustainability and stability remained fully under Dole's control. At the same time, some local actors found themselves competing with Dole for access to markets and local resources, especially agricultural land. At several of Dole's production sites, its presence drove up farmland rents. In Kyushu, a Dole franchise farm entered a dispute with an agricultural cooperative, to which it refused to pay a commission despite agreeing to do so. The co-presence of agricultural models is thus not always peaceful. In this case, the conflict was due to the absence of third-party intermediation and/or public sector intervention.

Finally, another important issue is cultural incompatibility. In Japan, the local culture is characterised by collective action and by the importance accorded to consultation processes and collective agreements. The local actors perceive the multinational's culture to be very profit-oriented and imposed unilaterally, without negotiation with the local actors. Dole has thus remained alien to the local community.

These environmental, socio-economic, and cultural incompatibilities have led to resistance from local actors to Dole's presence in agricultural production. Some producers have expressed their dissatisfaction through their refusal to rent their farmland at the rent proposed by Dole. Local authorities have refused farming permits[8] and local agricultural cooperatives have decided not to collaborate with the company. This shows the importance, whenever possible, of seeking reconciliation between agricultural models and the philosophies that underpin them.

---

[8] In Japan, a municipality's agricultural committee grants farming permits. Some committees refused to issue permits to Dole farms because of lack of trust or the risk of conflict with local producers.

2.2 *Reconstruction of the Tohoku Region After the March 2011 Tsunami*

The second case study pertains to the restoration of agricultural activity near the city of Sendai, the capital of Miyagi Prefecture,[9] after the 2011 tsunami. Sendai is a city with a population of one million inhabitants and the economic, political and cultural centre of the Tohoku region in north-eastern Japan. In the triple disaster that hit the region (earthquake, tsunami and the Fukushima nuclear reactor accident), 80% of Sendai's agricultural land was flooded by the tsunami. In order to accelerate agricultural reconstruction and to encourage business investment in the region, Sendai created a SEZ in 2012. At the same time, it adopted a land use model based on the coexistence of different agricultural models. The land-use plan segmented large, specialised zones favourable to land concentration and accumulation, rather than the coexistence of a diversity of small and large farms in a landscape mosaic.[10]

Given this context, local producers decided to collaborate with three American and Japanese multinationals, IBM, Mitsubishi and Kagome, in a project to restore the region's agriculture. In 2012, they set up a farm, which they named Michisaki (meaning 'indicator'), to produce crops such as tomatoes, lettuce and strawberries in greenhouses. The farm employs local people affected by the disaster, 20 permanent and 25 temporary employees, and receives subsidies for these reconstruction efforts. The subsidies cover 77% of the farm's costs.

What does the Michisaki case tell us about the issue of the coexistence of agricultural models? What are the conditions that are propitious to or hinder the coexistence of the multinational and family farming models? As for the Dole case study, our analysis is based on a combination of the agronomic, socio-economic and cultural points of view.

First, the Michisaki project turned out to be a competitor to the projects of local producers to restore agricultural production. Michisaki's investments created discontent among local actors, since its decision to set up greenhouses on flooded and salted land (following the tsunami) competed with the wishes of the local producers who wanted to develop rice production. This illustrates the clash of interests between the multinational and the family farming models on crop choices for the use of agricultural land.

Second, a similar situation as in the Dole case prevails in terms of socio-economic competition and cultural incompatibility between local actors and multinational companies. Moreover, here too, local expectations were high in terms of technical transfers, employment creation and the revitalisation of local agriculture in order to repair the damage resulting from the Fukushima disaster. And again as in the Dole case, the multinational model has provided low-skill jobs, precarious employment with low wages, and dependence on public subsidies. Rent for farmland became three times higher than before the disaster, led by competition for local resources,

[9] Miyagi Prefecture neighbours Fukushima Prefecture.

[10] According to the city of Sendai, this development plan eventually failed in 2016 due to territorial conditions, the wishes of agricultural producers, etc. (interview on 10 November 2020).

including agricultural land and subsidies for reconstruction. The situation reflects the socio-economic tension between agricultural models in the territory.

The resistance observed was more violent than in the Dole case. Not only did producers refuse to rent their farmland at the rent proposed by Michisaki, and agricultural cooperatives refused to support it, but its employees, tired of routine work, resigned, and local residents set fire to some greenhouses. These were manifestations of a profound discontentment, especially as these communities are usually very peaceful. The Michisaki case illustrates the fundamental contradiction between agricultural models, and between the expectations of local actors regarding agricultural reconstruction and the public policies that were implemented.

3 Conclusions

The tilt of Japanese agricultural policy, starting in the 1980s, towards neoliberalism has encouraged multinationals and industrialists to invest in agricultural production. In other words, agribusiness and non-agribusiness multinationals have benefited from the institutional context to branch out in their activities and develop new ways of creating value for their shareholders. However, their investments have increased the agro-environmental, socio-economic and cultural tension between agricultural models at the territorial level since the early 2000s.

These case studies show the contradictions between the individual or independent family farming model and that of multinationals and stock companies. These contradictions manifest as mechanisms of resistance by local actors. For the most part, this resistance is not organised, but it is widespread and is expressed in a variety of ways, as presented in this chapter. It leads to a crisis of legitimacy of neoliberal agricultural policy.

Therefore, is the coexistence of agricultural models really possible? And if so, under what conditions? Corporate strategies are often driven by short-term profits, which raises issues about the sustainability and stability of agricultural production. The case of Dole shows that multinational companies make decisions to close down, relocate and sell farms not only for reasons that are internal to the company, but also according to their relationship with local stakeholders. This may be one of the rationales of local actors who are resisting the penetration of the multinational model in the agricultural production system. It seems that local actors only accept multinationals' investments when they contribute to territorial agricultural systems and are in line with local expectations. However, there is no guarantee that this will always happen. It is thus essential for local authorities and the State to intervene to monitor the activities of multinationals and other companies. The situation seems to call for a reconsideration of the neoliberal agaricultural policy implemented in Japan.

References

Bonanno, A., & Constance, D. H. (2008). *Stories of globalization: Transnational Corporations, Resistance and the State* (p. 321). The Pennsylvania State University Press.

Dole Food Company. (2013). About us.

Dole Japan. (2013). History of Dole.

Farmind. (2018). *Hokkaido Sanchoku center became a consolidated subsidiary.* News release on 9 May 2018. (in Japanese).

Jussaume, R. J. (1991). *Japanese part-time farming: Evolution and impacts* (p. 212). Iowa State University Press.

MAFF. (2015). *The engagement of companies in the agricultural sector* (3 p). MAFF.

MAFF. (2019). *The evolution of damages in agricultural sector by wildlife.*

Mazoyer, M., & Roudart, L. (2006). *A history of world agriculture from the neolithic age to the current crisis* (p. 524). Routledge.

Muroya, A. (2007). Status quo and challenges in corporate involvement in agriculture. *Journal of Agriculture and Forestry Finance (norinkinyu), 26*(2), 13–26. (in Japanese).

Sekine, K. (2016). Transnational agribusinesses and transformation of Japanese agriculture: Neoliberal reform and resistances. In K. Kitahara, & M. Ando (Eds.), *Transnational agribusinesses and their control over agriculture and food* (pp. 104–127). Akashi Publishing (in Japanese).

Sekine, K. (2017). Resistance to and in the neoliberal agri-food regime: A case of natural bananas trade between the Philippines and Japan. *The Journal of the Research Institute of Business, 55*(3), 15–33.

Sekine, K., & Bonanno, A. (2016). *Contradictions of neoliberal agri-food: Corporations, resistances, and disasters in Japan* (p. 240). West Virginia University Press.

Sekine, K., & Hisano, S. (2009). Agribusiness involvement in local agriculture as a 'White Knight'?: A case study of Dole Japan's fresh vegetable business. *International Journal of Sociology of Agriculture and Food, 16*(2), 70–89.

Taniwaki, O. (2011). Background and status quo of stock companies' involvement in agriculture. In S. Harada (Ed.), *Regeneration of local agriculture and farmland systems* (pp. 201–217). Nobunkyo (in Japanese).

Teruoka, S. (2008). *Agriculture in the modernization of Japan (1850–2000)* (375 p). Manohar.

Tsutaya, E. (2000). Status quo of corporate involvement in agriculture and agricultural corporations: Questioning agricultural cooperatives' existence. *Journal of Agriculture and Forestry Finance (norinkinyu), 53*(5), 32–49. (in Japanese).

Chapter 17
What Future for the Food Systems Development Model that Emerged at the End of the Twentieth Century?

Gilles Allaire

Since the 1990s, the global economy has undergone major changes in terms of the organisation of markets and the institutions that govern them. The quest for quality, with all the vagueness and ambivalence of the term, has become a reason for cooperation as well as competition, a source of conflict and arrangements between producers and between links in supply chains, and the focus of initiatives taken by retail chains and consumers. Quality is the basis of new demands and supply of services, resulting from changes in human lifestyles.

During this period, we have seen a proliferation of so-called 'quality' standards and the organisations that establish and manage them, of guarantee systems, particularly in the form of third-party certification, as well as of participatory systems, and of a range of evaluation mechanisms. This applies to all the sectors, including that of finance. From the point of view of 'regulation theory', these changes affect the capitalist 'development model' and are reflected in a rearrangement of social relations, especially with new modalities of 'forms of competition' (Allaire, 1995; Petit, 1999).

As far as agriculture and food are concerned, a large amount of literature has noted a 'quality turn' with, on the one hand, the emergence of alternative production systems and supply chains, organic shops, short supply chains, and fair trade, and on the other, the proliferation of quality assurance schemes regulated by the State, by collectives or by private entities, whether with regard to B2B (business to business) relationships, i.e. within value chains, where large distributors impose specifications and controls on their suppliers, or in end markets for public consumption, segmented by quality labels. A third aspect, the publicising of quality crises, must be added to characterise the new market regime, which we call 'media regime' (Allaire & Daviron, 2008).

G. Allaire (✉)
INRAE, Toulouse, France
e-mail: allaire.gilles@wanadoo.fr

P. Gasselin et al. (eds.), *Coexistence and Confrontation of Agricultural and Food Models*,
https://doi.org/10.1007/978-94-024-2178-1_17

Since the crises of 2008, the high degree of instability of agricultural commodity markets has become a permanent feature, due largely to the continuing development of liberalisation policies. This explains some of the conversions to organic agriculture and short supply chains. Faced with the instability of wholesale agricultural markets, producers have two solutions: either to opt for insurance (as proposed by the European Union and the United States), which is in line with the financialisation of the economy, or to avoid competition through differentiation.

We situate the issue of coexistence in this context, and in relation to the characterisation of changes in the development model that have been observed from the 1990s onwards, following the crises of the 1970s, the crisis of Fordism and Fordist agriculture (Allaire, 1995, 2002). Does this coexistence manifest as diversification, in the different ways both of producing and eating, and thus in an institutionalisation of the coexistence of production and food models, and in a giving up of the idea of a better way?

In this chapter, I respond to the invitation of this book's editors to present, with regard to the question of coexistence, my work conducted using an institutional economics approach (and more precisely, regulation theory). This work deals with the crisis in Fordist agriculture and the emergence of new regulatory mechanisms, notably in the form of institutions that supervise and regulate markets. I focus here on the sense to assign to the concept of quality by first examining the notions of coexistence and competition. The conclusion returns to the characterisation of the new capitalist development model from a macro-economic and historical perspective.

1 Coexistence

The editors of this book present the idea of coexistence in two ways at the same time: positively (or analytically), and normatively. This is reflected in the dual meaning of the notion of 'agricultural or food model', which can either be, on the one hand, an ideal-type or the 'archetype of an observed reality' (positive vision), or, on the other, a type of ideal (normative representation), a 'desired future' or a standardised procedure. While this notion can have a wide use as an analytical tool, as a normative proposition, it needs to be resituated in a socio-political context.

From a positive point of view, coexistence can involve different objects and scales, such as types of investments (economies of scale vs scope) and types of farm activities (see the debate on the possibility of a partial conversion of a farm to organic farming); types of crops in a terroir; production systems in territories, according to their development logic (specialisation vs diversification) or according to sectors; quality-related market conventions; social movements and political projects; property regimes (private, collective, communal, public, social), etc.

An analysis of the types of coexistence challenges the idea of a convergence of productive models that has to take place in an economic world of perfect competition. It introduces dialectics and invites us to consider social realities as topographies created by forces under tension. Although agrifood systems are diverse at different scales and from different points of view, the trends of segmentation and expansion (by upscaling) concern both 'conventional' and 'alternative' systems, that is if this distinction retains any meaning today. Questions of differentiation based on levels of quality and of coexistence arise both between systems and for each one of them.

The analytical point of view also encompasses the interactions between models, which result in hybridisations, complementarities, synergies and co-evolutions, as well as confrontations, exclusions and marginalisations. This leads to the introduction of normative and political questions.

The identification of coexisting realities already implies a normative point of view. In order to analyse the diversity and coexistence of production systems in a territory, an observer will mobilise his or her representations of local development issues and distinguish, for example, between investments that create employment and those that do not, between local or imported resources, or between specific or generic types of outlets.

The positive approach to coexistence leads to the question of the causes or conditions of coexistence, and the political meaning to be given to phenomena of hybridisation or of marginalisation. Allaire and Wolf (2004) distinguish two cognitive paradigms or rational myths that orient the differentiation of markets: one is the categorisation into distinguishable qualities (material or immaterial), the other is based on the principle of identity associated with 'transcendent resources' (naming a quality refers to an overall, holistic dimension). These two rationales are at work in the institutional hybridisation of food systems and contribute to their qualification.

The normative vision of coexistence is, first of all, the recognition of a diversity of models. In terms of agricultural and rural development in France, this vision dates back to the 1990s. In 1989, the General Assembly of the National Association for Agricultural Development (ANDA[1]) recognised the diversity of farm development models (the 'end of models' was alluded to). Later, the European Commission's Standing Committee on Agricultural Research (SCAR) highlighted a new intellectual and political context in which the productivity narrative is contrasted by a 'sufficiency' narrative (Freibauer et al., 2011). This hesitantly introduced the idea that productivity in itself may not be a reasonable goal.

The normative vision of the coexistence of models has been infusing the so-called 'transition' policies for a decade. One example is the conception of double, then triple performance (economic, environmental and social), which can be understood in a broad or restricted sense, with one or other of the valences dominating. The agroecology promoted by the 2014 Law on the Future of Agriculture[2] in France embeds a broad diversity of systems and options in a same 'transition' process. This type of discourse is also today espoused by the FAO. We are thus moving from the coexistence of normative goals in confrontation to a normative goal of coexistence.

---

[1] French: *Association nationale de développement agricole (ANDA).*

[2] https://www.gouvernement.fr/action/la-loi-d-avenir-pour-l-agriculture-l-alimentation-et-la-foret.

2 Coexistence and Competition Regimes

Together with Marie Dervillé, we developed the concept of competition regime (Dervillé & Allaire, 2014) as the institutional arrangement of mechanisms of coordination delimiting areas of cooperation and competition in a market, or more generally in a system of actors. A competition regime corresponds to rules validated at different government levels. It forms part of national sectoral policies and multilateral agreements (World Trade Organisation (WTO), World Intellectual Property Organisation, (WIPO)), of regional or municipal policies and of the games played by economic actors who themselves are formulators of collective rules, ranging from the organisation of local solidarity and exchanges to global mass markets. These rules are stabilised though institutionalised compromises which exhibit a certain resilience.

According to this definition, a competition regime is also a cooperation regime. Institutions, which 'enable actors in markets to organise themselves, to compete and cooperate, and to exchange' (Fligstein, 1996, p. 658), help distinguish the domains of cooperation and competition. These institutions are specific to each market segment and depend on the economic and political power relationships as well as the institutional and legal context. Changes in power relationships and ideas, in particular conceptions of quality, as well as those that underpin professions and those that are currently prominent in the media (which represent a great diversity of experiences and points of view) cause tensions and changes in competition regimes.

Competition regimes mobilise collective (common) resources of cooperation, which are intangible and created by the organisation of industries or territorial actors, with one or other of these industries or territorial actors being able to take precedence. Sectors that are organised with quality labels, and this concerns all sectors today, have specific competition regimes that organise a diversity of quality regimes. At a territorial level, e.g. in the case of geographical indications, a demand, if it exists, for high-end products is not enough to create a territorial rent; the organisation of the actors must help maintain the collective reputation that supports it, with a stabilised cooperation regime. This must enable the organisation and control of the supply, as well as provide a degree of autonomy for producers and of competition, and consequently legitimate differentiations, by mentions, other quality labels or brands. The notion of competition regime thus offers an original angle of analysis of coexistence, both within a specific qualification system and within a sector.

3 Alternatives and Coexistence of Political Projects

The debate on positioning alternatives has shifted in the last two decades (Allaire, 2021). First, a debate arose around the thesis of 'conventionalisation', which concerns the different 'alternatives': organic farming (Darnhofer et al., 2010; Poméon et al., 2019), short supply chains (Allaire, 2016a) and geographical indications (Allaire, 2011). The question of the coexistence of political projects was then raised, since to refer to the conventionalisation of systems that claim to be alternative is to call into question the possibility of innovations that break with the past and which would retain a radical scope over the long term. Nevertheless, the so-called 'alternative' systems retain their specificity even as they differentiate themselves, while the norms in the so-called 'conventional' models evolve.

Second, a thesis was developed based on van der Ploeg (2014) that aims to explain the coexistence of large mass agrifood markets as well as 'nested markets' which are territorially anchored and based on collective initiatives. This thesis is in contrast to the first in a certain way, as nested markets preserve their identity while being nested in what van der Ploeg calls 'capitalist food empires'; they are nested in that they fill structural holes in this overall system. There is here a structural basis for the coexistence of economic forms that can be considered universal. The business world coexists with that of the commons, as Ostrom (2005) noted.

These theses are complementary, in the sense that if the first is a trend in the evolution of the new agrifood development model, that of its regeneration through the absorption of alternatives, the second explains a counter-trend which is the resilience of alternative social movements, which thrive (and clash) in the failures within the system rather than at its margins. Institutional economics can provide a common theoretical foundation for these theses. No hegemonic system or regime is a closed system without degrees of freedom. On the contrary, such a system tends, not without crises, to incorporate deviances, creations and criticisms (Chiapello, 2009), for if it does not, it is bound to collapse sooner or later.

The differentiation of immaterial qualities first appears in the form of 'alternatives', which consequently assume a market value. This market expands based on the global trend of higher living standards and an expanding middle class (at least until the 2020 pandemic crisis). Conventionalisation occurs with a change in scale of markets and of the stakeholders' functioning.

A paradox of organic farming is that it is a doctrine that defends far-reaching values[3] and a designation reserved for products meeting a market standard that has now become transnational. This has allowed the market to expand, with a significant contribution from public policies, especially in Europe. The latter are mainly justified in the name of the environment, which limits their political scope. Organic farming has become a solution for the future for, among others, consumers and public policies (Allaire, 2016b).

The International Federation of Organic Agriculture Movements (Ifoam) plays a role in the coexistence of organic farming development projects. This association does not produce organic farming standards that have been instituted at the national level (or European level for the EU), but it pursues a double objective: on the one hand, to reaffirm values and major principles (justice, equity, care, etc.) through its 2014 charter and to support participatory guarantee systems based on these values

[3] The four principles stated by Ifoam (International Federation of Organic Agriculture Movements) are: health, ecology, fairness and care (https://www.ifoam.bio/sites/default/files/2020-03/poa_english_web.pdf).

and, on the other hand, to support the expansion of the organic farming market through the standardisation of national systems based on third-party certification.

In Argentina and other countries of the Global South, certified organic farming is a niche export market, through large distribution networks. Values such as 'save the planet, ensure social justice' are upheld by agroecology, which is not subject to certification, and by a movement of family farmers that shuns differentiation through market tools (as they engender exclusion). However, agroecology is tending to become institutionalised, and is gradually finding a place in the markets, e.g. the case in Brazil, with participatory guarantee systems.

Geographical indications (GIs) are another example. Recognised as intellectual property rights as early as the nineteenth century, they are now governed by the WTO and national laws. Starting in the 1980s and 1990s, local development projects in France and elsewhere in Southern Europe, sought an alternative in GIs to mass production agri-chains, aiming to revive local collective heritage. During the same time, to get large supermarket chains to distribute GI products, specifications were standardised in a quality-oriented vision (a sort of conventionalisation) to achieve a representation based on 'superior quality', like the example of 'Label Rouge'. This approach proved unsuitable for certain new projects supported by local authorities concerning small production volumes, and alternatives without GI certification subsequently appeared, often based on participatory systems (Garçon, 2015). The notion of 'products of the terroir' has been used by anthropologists in the context of inventorying local heritage (Bérard & Marchenay, 1995), as also in support of various local initiatives. This term, however, lacks an institutionalised definition.

To what extent do certified organic farming or geographical indications remain 'alternatives' once their markets have expanded, especially after the 1980s and 1990s due to the supermarket chains, and to a greater extent in the last two decades? Let us not fall into the trap of the common but false debate: 'Is it better to have a lot of "partially organic" or a little of "highly organic"?' If we go by Cochoy (2017), the more research we carry out, the more we will conclude that alternatives offered as solutions have flaws, but this, however, is not sufficient reason to reject them! We must maintain a dynamic vision that links conventionalisation and nestification.

In order to inform this debate, it is necessary to distinguish between the institutions that regulate quality in markets, and the definitions of quality in doctrines originating in different social spheres.

4 Quality as an Institution and in the Markets

Proponents of general equilibrium, which is based solely on the assumption of scarcity, find quality to be a strange issue. Thus, markets cannot exist if differences in quality are not known. For the neo-institutionalists (Ménard, 2004), if the object of a transaction is specific, it takes place in a 'hybrid' framework. From an institutionalist (regulationist) point of view, qualities are social evaluations before being market evaluations. They are judgments that take institutional forms; institutions that have repercussions on markets.

4.1 Quality as a Judgment and as an Institution

Quality is not the property of a thing.[4] It is a relationship between resources and an end. It is a judgment on the capacity of a person, a product or a process to be a resource, i.e. to provide a service that satisfies a desirable end. A judgment of quality is contextual and refers to a hierarchy of values that result from customs, jurisprudence and routines. It is not a private (intimate, emotional) judgment, but a popular expression found in narratives, common knowledge, media opinions, scientific works and legal rules. Quality judgments sort and order resources, competences and products or effects, as also the status of companies and people (their occupations, their place in the organisation of work, etc.). Quality institutions are based on doctrines that are shared justifications of quality judgments.

We make a distinction between two dimensions of quality, whose evaluation involves different processes, but which are jointly involved in market evaluation:

- the intrinsic value of quality, which expresses the usefulness of a product, a service or a situation as a *resource*, by evaluating the relationship between means and ends based on efficiency criteria;
- extrinsic or intangible value, which corresponds to the (more or less) general evaluation of the system of resources involved in the activity of production or consumption by considering the different consequences of this activity in relation to desirable ends of collective interest, such as the planet's well-being or the preservation of common heritage.

We see the appearance of 'goods per se' (in French 'bien en soi') via the movement of criticism, which concern particular communities or have a more universal scope, 'It is good for me, but also good for others.' We must reverse the implicit scheme that considers the formulation of the ideal as preceding the conception of the good. On the contrary, the ideal, the 'goods per se', proceeds from successive evaluations and 'the successful empowerment of certain ends' as a result of 'critical operations' (Dodier, 2005, p. 22). Institutionalised qualities (or 'quality regimes'), i.e. prevailing conceptions at a given moment of what is healthy, what is 'sustainable', what is a job well done, etc., are at the crossroads of several 'goods per se'.

The standardisation and the institutionalisation of quality doctrines are not, in themselves, new processes. They have developed over the long term within professional worlds and communities managing common resources, with the circulation and control of quality being ensured by commercial intermediaries and public regulations. Consequently, debates on quality became social issues, occupying all media spheres and reconfiguring the issue of coexistence.

[4] This section draws on ideas presented in Allaire (2012, 2013).

4.2 Quality in the Markets

Quality in markets is the *world of standards*. Standards are tied to many of the institutions that govern markets, property rights, governance structures, and conceptions of control. According to Fligstein (1996), 'conceptions of control' are institutions that structure participants' understanding of a market's functioning and positioning. They are representations of the qualitative hierarchies between firms, regions, products, occupations and statuses, people and knowledge, which structure economic activities.

Commons (1934) distinguishes three kinds of property and property rights to which quality can apply: *corporeal* property, as opposed to *incorporeal* property (debts), and *intangible* property, which pertains to values that will be created in the future. *Corporeal* qualities refer to intrinsic qualities. *Incorporeal* qualities are attributed to an entity by standards. These extrinsic qualities have two facets, distinctive sign and promise. The promise facet does not pertain to services provided at the time of consumption or use, but to indirect and future services, tied to the specificity of processes of production or use and their impacts. Such attributions are not self-evident; they are social constructs, public judgments. The designation of incorporeal qualities is generally protected by intellectual property rights. This raises a question of responsibility in the legal sense (obligation of means) and a question of social responsibility, which refers to the promise.

Indirect services associated with extrinsic qualities have both an effective aspect (subject to a control system), for example, in the case of a commitment to respect labour rights (which is only an obligation of means), and an ideal aspect referring to a goal, for example, improving the living conditions of producers or preserving the health of consumers or users. The expected service is then situated in the future and is therefore called intangible. *Intangible* qualities (such as 'organic farming saves the planet') have a weak link with responsibility and raise issues of accountability.

Quality standards that introduce intangible qualities into markets have two facets: on the one hand, a list of specifications that can be monitored, and, on the other, principles, a doctrine that justifies the standard's objectives. There is a permanent and natural doubt about the links between specifications and doctrine, which constitutes a threat to the credibility of the standard.

A 'tripartite regime of standardisation' has rapidly developed, i.e. a regime of governance that consists of standards-setting, accreditation, and certification (Loconto & Busch, 2010). It affects all areas of economic activity and establishes 'a system of global private government that extends far beyond individual companies', and 'a huge international bureaucracy of standardisation bodies, certification companies, accreditation bodies, developers of measures, and data collectors and analysts has been set up to create global markets' (Busch, 2019).

This new standardisation regime corresponds to what we call the 'media market regime'. It refers to a standardisation of procedures of qualification, but corresponds to the spread of 'conceptions of control' based on the individualisation of social activities and beliefs, which disconnect individuals from domestic, community and professional socialisation frameworks. Rather than eliminating competition between quality doctrines, it organises their coexistence by deploying itself in various fields governed by different conceptions of control.

5 Quality Crises

A 'quality crisis' (Allaire, 2010) occurs when doctrines are challenged. In the case of the mad cow disease crisis, good animal husbandry practices (what people had in mind) were called into question. They could no longer be trusted to distinguish healthy cows from diseased ones. The initial response was to set up national labels, but this was not backed by any real knowledge, since the prion disease does not respect borders. This crisis resulted not only in institutional revisions of the guidelines of market organisation and of the functioning of European health agencies, but also changes in risk perception and lowering of confidence in the techno-structure. This example shows that quality crises do not result from competition between doctrines, but instead fuel it.

Quality crises are a feature of a market regime controlled by forums in which quality judgments are made, ranging from local collectives to the arenas of international negotiations, via the various settings offered by social networks and the media. There is competition between quality doctrines, and each of them can be challenged, no doubt from the outside but by revealing internal ambiguities.

Should there have to be mediatisation, and should the consumer have to be a participant in the debate (as in the case of the mad cow disease) for there to be a quality crisis? This is indeed what characterises a quality crisis. It pertains to representations of quality and the relevance of a doctrine (in this case that of health safety before the manifestion of the prion). We distinguish between local quality crises, which do not challenge the doctrine's fundamentals, and structural crises, which do. The conventionalisation of organic farming is not a quality crisis as long as the market does not collapse. However, the criticism of conventionalisation presents this theoretical possibility as a threat, an existential threat, and therein lies a cause of differentiation within organic farming between long agri-chains, with a generic public standard, and alternative systems with private labels (Poméon et al., 2019).

The case of organic farming is more general. A doctrine of quality, as we understand it, is the expression of a social movement that supports a representation of goods per se. As long as this movement is alive, solutions to quality crises can be reinvented locally.

We should not equate the 'industrial regime crisis' or the 'Fordism crisis' with a quality crisis. Quality crises go hand in hand with the new regime. They are a consequence of the new forms of competition as much as of the lasting variety of productive configurations and collective heritage.

6 Conclusion

The dynamics of the development model that emerged in the 1990s, especially in the agricultural and food sectors, can be analysed using the concept of the competition regime. The general trend is towards the development of transnational standardisation systems, under the pressure of a conflicting plurality of visions of the future. Standardisation is a time-tested process to ensure compatibility between technical objects and economies of scale in the industrial world. We are more concerned here with standards that control conduct, with reference to the assessment of the implications of production and use in the future on our health, our lifestyles and the environment. What is at stake in the economic sphere is the evaluation of investment projects according to criteria that go hand in hand with the financialisation of economies.

The new logic of competition first permeated the different agricultural sectors at different rates, running up against the resilience of the institutionalised compromises of the previous period, that of the integration of agri-chains. Technical standards are now supported by representations that are variously valued and debated. Judgments of quality are formed and circulate beyond the technical spheres, whose borders, in any case, are also being transformed by the digital age.

Information and the new digital technologies play a fundamental role in this new capitalist development model. The information we are referring to includes both so-called 'scientific' knowledge and the opinions and representations that circulate in public spheres. Thus quality, with all the vagueness and ambivalence of the term, has become an issue in the restructuring of activities and markets. The growing complexity of the media world's systems drives the repercussions of quality issues in the market.

The dynamics set in motion over the last three decades have had consequences on the transformation of regional and sectoral production configurations and a reconfiguration of political projects on agriculture and food. In France, the political project of agricultural modernism based on technical intensification is now undergoing a profound renewal. In the professional organisations of the 'conventional' system, we are seeing the emergence, alongside the new role of digital technology, of a rediscovery of professional common resources or collective productive heritage, while the alternatives that appeared at the beginning of this period have split into several models.

References

Allaire, G. (1995). Le modèle de développement agricole des années 1960 confronté aux logiques marchandes. In G. Allaire, & R. Boyer (Eds.) *La grande transformation de l'agriculture* (pp. 345–377). Inra/Economica.

Allaire, G. (2002). L'économie de la qualité, en ses territoires, ses secteurs et ses mythes. *Géographie, Économie, Société, 4*(2), 155–180.

Allaire, G. (2010). Applying economic sociology to understand the meaning of "quality" in food markets. *Agricultural Economics, 41*, 167–180.

Allaire, G. (2011). La rhétorique du terroir. In C. Delfosse (Ed.), *La mode du terroir et les produits alimentaires* (pp. 75–100). Les Indes Savantes.

Allaire, G. (2012). Signes de qualité et marchés. In J. -P. Poulain (Ed.), *Dictionnaire des cultures alimentaires* (pp. 1262–1271). PUF.

Allaire, G. (2013). The multidimensional definition of quality. In L. Augustin-Jean, H. Ilbert, & N. Saavedra Rivano (Eds.), *Agriculture and international trade: The challenge for Asia* (pp. 71–90). Palgrave Macmillan.

Allaire, G. (2016a). De quelles crises les circuits de proximité sont-ils le nom? In P. Mundler, & J. Rouchier (Eds.), *Alimentation et proximités. Jeux d'acteurs et territoires* (pp. 405–418). Educagri.

Allaire, G. (2016b). Que signifie le «développement» de l'Agriculture biologique? *Innovations Agronomiques, 51*, 1–17.

Allaire, G. (2021). Alternative food networks and the socialization of food. In A. Maurer (Ed.), *Handbook of economic sociology in the 21st century—New theoretical approaches, empirical studies and developments* (pp. 221–235). Springer.

Allaire, G., & Daviron, B. (2008). Régimes d'institutionnalisation et d'intégration des marchés: le cas des produits agricoles et alimentaires. In F. Dreyfus, Y. Chiffoleau, & J. -M. Touzard (Eds.), *Nouvelles figures des marchés agroalimentaires: Aapports croisés de la sociologie, de l'économie, et de la gestion* (pp. 113–125). Quæ.

Allaire, G., & Wolf, S. (2004). Cognitive representations and institutional hybridity in agrofood systems of innovation. *Science, Technology and Human Values, 29*(4), 431–458.

Bérard, L., & Marchenay, P. (1995). Lieux, temps, et preuves: La construction sociale des produits de terroir. *Terrain, 24*, 153–164.

Busch, L. (2019). The new autocracy in food and agriculture. In G. Allaire & B. Daviron (Eds.), *Ecology, capitalism and the new agricultural economy* (pp. 95–109). Routledge.

Chiapello, E. (2009). Le capitalisme et ses critiques. In *4e congrès du Riodd: La RSE, une Nouvelle Régulation du Capitalisme, 25 to 27 June, Lille.*

Cochoy, F. (2017). Le paquet de cigarettes neutre; les dangers de l'intelligence en sciences sociales. In G. Allaire, & B. Daviron (Eds.), *Transformations dans l'agriculture et l'agro-alimentaire. Entre écologie et capitalisme* (pp. 325–338). Quæ.

Commons, J. R. (1934). *Institutional economics*. Macmillan.

Darnhofer, I., Lindenthal, T., Bartel-Kratochvil, R., & Zollitsch, W. (2010). Conventionalisation of organic farming practices: From structural criteria towards an assessment based on organic principles: A review. *Agronomy for Sustainable Development, 30*(1), 67–81.

Dervillé, M., & Allaire, G. (2014). Change of competition regime and regional innovative capacities: Evidence from dairy restructuring in France. *Food Policy, 49*, 347–360.

Dodier, N. (2005). L'espace et le mouvement du sens critique. *Annales. Histoire, Sciences Sociales, 1*, 7–31.

Fligstein, N. (1996). Markets as politics: A political-cultural approach to market institutions. *American Sociological Review*, 656–673.

Freibauer, A., Mathijs, E., Brunori, G., Damianova, Z., Faroult, E., Gomis, J. G., O'Brien, L., & Treyer, S. (2011). *Sustainable food consumption and production in a resource-constrained world*. In *The 3rd SCAR (European Commission-Standing Committee on Agricultural Research) Foresight Exercise, Brussels, SCAR.*

Garçon, L. (2015). *Réinventer les pommes et les pommes de terre : une géographie de la qualité à l'épreuve des produits ordinaires*. Thèse de doctorat, spécialité Géographie, Lyon-2.

Loconto, A., & Busch, L. (2010). Standards, technoeconomic networks, and playing fields: Performing the global market economy. *Review of International Political Economy., 17*(3), 507–536.

Ménard, C. (2004). The economics of hybrid organisations. *Journal of Institutional and Theoretical Economics, 160*, 345–376.

Ostrom, E. (2005). *Understanding institutional diversity*. Princeton University Press.
Petit, P. (1999). Les aléas de la croissance dans une économie fondée sur le savoir. *Revue D'économie Industrielle, 88*, 41–66.
Poméon, T., Loconto, A., Fouilleux, E., & Lemeilleur, S. (2019). Organic farming in France: An alternative project or conventionalisation? In G. Allaire, & B. Daviron (Eds.), *Ecology, capitalism and the new agricultural economy* (pp. 207–226). Routledge.
van der Ploeg, J. D. (2014). Newly emerging, nested markets: A theoretical introduction. In P. Hebinck, S. Schneider, & J. D. van der Ploeg (Eds.), *Rural development and the construction of new markets* (pp. 16–40). Routledge.

Chapter 18
Agricultural and Food Models: Not to Believe Too Much in Them, but Believe in Them All the Same!

Ronan Le Velly

'What are, in fact, these models we hear so much about?' ask the coordinators of this book in their general introduction. Their answer is very enlightening. Working on the coexistence of agricultural and agrifood models compels us to look at the three acceptations of the word 'model'. The first refers to the 'archetypes', i.e. the ideal–typical forms that researchers and experts develop to analyse the diverse ways in which agricultural and food systems are organised. The second, which I associate with the notion of 'project' (Le Velly, 2019), refers to the models that groups of actors define themselves in order to orient their action towards a 'desired future'. And the third refers to 'norms for action', rules that aim to frame action, such as standards of organic farming or administrative rules that define who can officially be called a farmer. We need to distinguish between these three acceptations of the term 'model': ideal-types, projects, and norms. As this book's coordinators write, the peasant agriculture of the French sociologist Henry Mendras is not the same as that of the Confédération Paysanne (a major French farmers' union). Nor, we can add, is it the agriculture of the administrative category of family farming defined in Argentina in the early 2000s (Goulet, 2019). The fact that in practice these three kinds of models share relationships with each other and have porous boundaries does not mean that there is no value in distinguishing between them. Even though, as the sociology of science has shown (Latour, 1987), these three types of models form and strengthen mutually, their contents never completely overlap.

This important clarification offers a very useful starting point. The book abundantly expresses it by working from this perspective on fundamental issues of territorial planning, transitions towards more sustainable food systems, maintenance of a diversity of agricultural systems, etc. That said, its contributions also make it possible to debate the use of models for the purposes of analysis. Thus, several of the chapters show that the observed dynamics are diverse, complex and hybrid, to the point of

R. Le Velly (✉)
UMR Innovation, Institut Agro, Montpellier, France
e-mail: ronan.Le-Velly@supagro.fr

P. Gasselin et al. (eds.), *Coexistence and Confrontation of Agricultural and Food Models*,
https://doi.org/10.1007/978-94-024-2178-1_18

never fully corresponding to the characteristics of the models. It is therefore advisable not to put too much faith in models, not to trust them blindly as to their capacity to portray reality or to guide action. Do not believe in them too much ... but believe in them all the same! This book's chapters also aim to demonstrate the extent to which it is still necessary to focus on agricultural and food models, as understood by their three meanings. In this chapter, I will first present the cautionary reasons for not relying too much on models. Then I will visit the valid arguments for indeed identifying these models.

1 Practices that Are Always More Complex than the Models

The idea that agricultural and food practices are always more diverse and complex than the models has been explored in social science research on alternative agrifood networks. I will make a quick detour to this literature before returning to the contributions of this book.

A new field of research appeared in the 1990s on 'alternative agrifood networks', or 'alternative food systems'. These terms encompassed approaches as diverse as fair trade, organic farming, designation-of-origin products and short supply chains. At that time, the aim was not only to highlight these initiatives, in which the research community had until then shown scant interest and which were little recognised by public authorities, but also to emphasise their common capacity to respond to the many injustices of the dominant food system and to establish a new model of agricultural and rural development (Kloppenburg et al., 1996; Renard, 1999; van der Ploeg et al., 2000). The first studies on these various initiatives then naturally put forward a set of oppositions marking the divide between the alternative model and the conventional model: artisanal versus industrial, natural versus artificial, proximity versus distance, diversification versus specialisation, quality versus quantity, 'moral economy' pursuing ethical values versus 'market economy' focused on profits, etc.

This first step was probably necessary to create this field of research. However, from the early 2000s, researchers working on these issues began to view them in a more nuanced way. With an improved knowledge gained through experience on the field, they pointed out that, in reality, the practices associated with the alternative model did not function in a totally different way from those associated with the conventional model. The oppositions previously identified were therefore called into question with regard to the oversimplifications they suggest. Several researchers thus emphasised the urgent need to move away from a 'dualist', dichotomous and oppositional approach to a 'dual' one that acknowledges that alternative initiatives combine alternative and conventional features (Hinrichs, 2003; Kneafsey et al., 2008). This idea has also been expressed through idea of the 'hybrid' nature of alternatives. For example, in a pioneering paper, Ilbery and Maye (2005) showed that alternative meat and dairy chains in the north of England borrowed numerous links from conventional chains, such as abattoirs, wholesalers and supermarkets. This desire to recognise hybridities is not limited to the alternative/conventional opposition. It is part of a wider movement, inspired in particular by actor-network theory, to overcome the all too well established dichotomies between urban and rural, local and global, production and consumption, and nature and society (Goodman, 1999; Le Velly & Dufeu, 2016). As Woods (2009) noted, the challenge for the research community is to 'make connections' and 'blur boundaries'.

This strategy is convincingly pursued in many of this book's chapters. Indeed, Jérémie Forney writes, echoing Woods, that the 'concept of coexistence of models is not as useful for reflecting on parallel strategies in their specificity as it is for focusing our attention on the relationships and dynamics between "models" with fuzzy boundaries.'

Indeed, several authors, notably Rosalia Filippini and Claire Lamine, point out that some farmers are simultaneously involved in alternative and conventional chains. Not only do they not subscribe to a single model, their farms' very sustainability is based on the complementarity of different models. Jérémie Forney also provides the example of the Swiss 'quality' dairy model, exemplified in products like Gruyère, which depends on Brazilian soya imports and the possibility for farmers to sell a portion of their production to industrial chains.

Other contributions allow us to understand that innovative collective actions take place at the crossroads of alternative and conventional rules. A good example is the quality cheese agri-chain developed by the Carrefour supermarket chain in Auvergne, France. Virginie Baritaux and Marie Houdart point out that 'production methods based on traditional know-how (use of grass and hay, processing of raw milk), which were abandoned in the wake of the industrialisation of agriculture and processing, are being implemented by an actor that is emblematic of the agro-industrial system (huge volumes, standardised products, low production costs).' In a similar manner, Vanessa Iceri explains in detail how a community in south-central Paraná, Brazil, has developed an original agricultural model by combining production methods that are inspired by tradition and yet conform to current commercial and health requirements. The resulting form is both traditional and modern; it is impossible to assign it to either category. The same conclusion can be drawn for the integrated pig and poultry production system that was dominant in the 1970s in the Brazilian state of Santa Catarina, with a mix of industrial processing, scientific organisation of labour and small multi-crop livestock farms, such as those studied by Claire Cerdan.

The wide range of practices within each model also makes it impossible to continue to support dualistic reasoning. On this point, the book's chapters confirm the thesis of the unequal alternativity of the different forms of alternative agrifood networks (Guthman, 2004; Raynolds et al., 2007; Kneafsey et al., 2008). This is particularly clear for organic farming and short supply chains in Claire Lamine's chapter on southern Ardèche. The detailed observation of two food buying groups in Montpellier carried out by Emmanuelle Cheyns and Nora Daoud also reveals very heterogeneous methods of functioning.

All this leads to questions about the relevance of working with models, as understood in their three meanings. An 'ideal-type [that] is accorded too much importance', writes Jérémie Forney, can make us lose sight of complexity and diversity. Indeed, the above examples confirm that efforts at categorisation made by researchers should not lead to an hasty fitting of complex realities into simple categories. Models should not be a substitute for an effort to observe and understand practices. On another level, it is also important to remember that projects or standards never automatically determine action. It is not enough to listen to the reasons and aims of fair trade, as stated by its promoters in 'charters' or 'principles', to understand its practices; and reading its certification standards must be seen also only as a point of entry. For fair trade, as for many other initiatives, the same project can lead to very different rules, which in turn structure a space for action in which different strategies can be expressed (Le Velly, 2019; for community supported agriculture, see Mundler, 2007). Let us emphasise: this last statement means that the diversity of practices is not only due to hybridisation between models. Even a single model has to be seen in terms of the diversity it tolerates.

In short, given these conditions, should we even continue to believe in models? Without backtracking on the relevance of the preceding elements, it seems to me that the answer must nevertheless be in the affirmative.

2 How to and Why Study Models?

The arguments in favour of taking models into account, despite the risks set out above, differ slightly depending on the three forms of model considered. I will present them in turn.

Let us begin with models defined by researchers and experts. For this type of model, it is useful to bring the notion of the model closer to that of Weber's ideal-type. In this perspective, the model is an intellectual creation consciously built by researchers while putting emphasis on certain features of the object considered. This stylisation of reality is seen by Weber as a necessary strategy for research: faced with the impossibility of describing the world in all its complexity, it is necessary to accentuate certain features in order to bring differences and relationships to light (see Coenen-Huther, 2003). The definition of the model is then guided by a purpose, and it is normal that two researchers working on different issues propose two different ideal–typical definitions of the same object. In this respect, the four ideal–typical forms of agroecology proposed by Philippe Baret and Clémentine Antier in their chapter must be seen as a frame of reference to understand the mechanisms of transition. The relevance of this typology must be assessed in this light. Besides, the usefulness of such typologies is reinforced whenever their use enables us to identify peculiar forms. For example, Roberto Cittadini and Agnès Coiffard manage, in their chapter, to characterise three original forms of Argentinian agriculture using Fournier and Touzard's (2014) typology.

With Coenen-Huther (2003), we can even go further: 'While the possible approximations of the modelled reality with the real world can undoubtedly be a source of satisfaction for the researcher, it is the differences observed and their analysis that have the decisive heuristic virtues.' In this case too, because they will encourage further analysis, models are vehicles to make sense of the world. Rosin and Campbell (2009) rightly explain that we cannot be satisfied with a binary representation opposing small-scale, localised and authentic organic farming on the one hand, and large-scale, globalised and conventionalised organic farming on the other. However, the virtue of such a model is to encourage the researcher to understand how, in certain cases, actors manage to develop organic farming that deviates from the predictions of this model (Le Velly et al., 2016). In short, while models should not be seen as a faithful representation of agricultural and food realities, they do represent tools to observe and analyse them.

We now turn to the second acceptation of models: models of desired futures. I refer to these models as 'projects' and define them as 'the reasons and ends that a collective gives itself to orient its action towards a desired future' (Le Velly, 2019, p. 7). It is necessary, first of all, to state that a same model/project can give rise to a diversity of practices. For example, in the research I have carried out, I have often observed that the project of proponents of fair trade is vague, ambiguous and open to multiple interpretations. It is therefore not surprising to see its actors defending different strategies, each of which is as legitimate as the others. For example, the increase in a minimum purchase price can, on the one hand, be defended by the desire to cover the costs of sustainable production, and, on the other, be contested because of the risk of reducing the commercial opportunities available to producers, with these two competing justifications applying to the same project. This blurring of the project is confirmed in many other cases: Wald (2015) does so, for example, for the food sovereignty project, and Aurélie Toillier, Saydou Bancé and Guy Faure suggest it in their chapter with what they call the 'paradigm' of ecological intensification.

This ambiguity in projects does not detract from the relevance of examining them in depth. Collective action, especially when it is innovative, involves a project dimension that must not be overlooked (Bréchet, 2021). Reporting on projects is akin to reporting on the actors' capacity to imagine and design broad contours of new global states of the world that they consider more desirable. The simplifications they make, especially through clear-cut discourses such as those that contrast conventional prices and fair prices, reflect this. As Jérémie Forney states in the conclusion of his chapter, 'We need the idealisation and simplification of models as projects in order to look at the world in a way that allows us to determine what is preferable, to make decisions and sometimes to commit ourselves.' The major dichotomies we mentioned earlier (artisanal vs. industrial, etc.) must then be understood in terms of their capacity to give rise to collective action.

Projects serve as a reference or a compass to orient and evaluate action. In the 2000s, I observed how activists of the French Artisans du Monde network assessed the impact of the relationships they had established with their 'partner' producer organisations in the Global South. It was remarkable to see how these activists had a common compass to guide their judgement. Even those who were only remotely familiar with the network's reference documents (e.g. 'Criteria in the North and South') knew that failing to work with highly marginalised producers was a concern and had to be weighed against the desire to 'do business, not charity'. Even in the case of debates and disagreements, everyone referred to the same project. Finally, while the models of desired futures do not determine action, they are essential for assessing and guiding it.

I think it is important to add that the project is a compass that the actors *give themselves*. In this respect, it is a matter of defending a theoretical framework that claims an endogenous normativity: the aims, values and reasons that actors pursue are not laid out in advance but are created in the course of action (Callon, 1986; Reynaud & Richebé, 2009). Even if the project constitutes a form of external reference point on which the actors rely, this reference point is not imposed on them. It is the actors themselves who determine their project, as they go along, in the course of their experience. To illustrate this idea, we can think of the image of the magic rope, proposed by Mische (2009), which the actors throw up in front of them before climbing on it. This preoccupation with the recognition of an endogenous normativity aims to push back a deterministic notion of the project, which would be imposed on the actors because of their position in the social space. Nevertheless, several contributions in the book, like others published elsewhere (Samak, 2013; Joltreau & Smith, 2020), remind us of the extent to which projects are also the expression of instituted social relations. In my view, the strong compartmentalisation of the three agricultural advisory subsystems observed by Aurélie Toillier, Saydou Bancé and Guy Faure in Burkina Faso is able to be interpreted in this sense. In France, the historical opposition between old and new agroecological actors observed by Claire Lamine, as also the tensions between organic and non-organic camps described by Véronique Lucas and Pierre Gasselin in their study on cooperation between farmers, also provide good entry points for explanation. Even though organic farming or short supply chains are developing and are now supported by very diverse networks, and even though organic and conventional farmers have many opportunities to cooperate, the established oppositions in French agriculture continue to weigh on professional identities.

From this perspective, one challenge is to understand the link between the different scales of determination of the desired future models. For example, while it is relevant to understand the project being undertaken by the farmers of Bio Loire Océan as being specific to this group of fruit and vegetable producers in the Loire basin, this project must also be understood in the light of the larger debates structuring the transformation of the French organic sector (Dufeu et al., 2020). Similarly, when the French National Federation of Organic Farming reaffirms its project by reworking its charter and values, it actually aims to offer normative resources that can be taken up by farmers and other actors in the sector (Chance et al., 2018). Viewed in these terms, the case of Argentinian farmers, who are described by Christophe Albaladejo in his chapter as 'silent', is especially noteworthy. Although they are implementing a specific agricultural model, these farmers have so far been unable to articulate the project. The pervasiveness of the opposition between the competing models of agroecology and agrobusiness in this country may explain this. Unless, as Roberto Cittadini and Agnès Coiffard envisage, their project can assert itself in the future by drawing on certain agroecological references.

Let us conclude with the 'norms for action' models, where the purpose of norms, standards and rules is to frame behaviour. In this respect, the reasoning on the relationship between norms and practices is fairly comparable to the one we have used in the case of the model as a project. These relationships are, moreover, relatively well known. In France, research by Crozier, Friedberg and Reynaud has shown that no normative system is capable of completely framing and determining actions. Because they are incomplete, ambiguous or contradictory, the rules leave considerable room for manoeuvre, which the actors use to develop their strategies (Crozier & Friedberg, 1980; Reynaud, 1997). Studies by English-speaking researchers focusing on neo-institutional sociology have also emphasised this point: any norm, even formally codified, is 'ambiguous', subject to interpretation, debate and contestation (Mahoney and Thelen, 2009). The existence of multiple forms of organic farming, despite a single standard and certification, and of strategies that run counter to the project that inspired these rules, is therefore in no way specific to this sector. Such observations can be made in any domain of organised action (Bréchet, 2021).

Having said this, we must not lose sight of the other side of the rules. Even if they do not totally constrain action, they do have a structuring character. They define the perimeter of possible actions and determine what Reynaud (1997) called 'the rules of the game'. Balancing both ends of the analysis is essential. For example, the circumvention of rules that can be observed in certified fair trade must be understood at the intersection of actors' strategies and the room for manoeuvre that the standards allow them (Le Velly, 2017). The engineering of more sustainable agricultural and food systems ultimately involves not only the establishment of standards, but also the taking into account of the way in which actors will react to these standards (appropriate them, reject them, circumvent them, etc.).

Paradoxically, this importance of norms comes to the fore not when actors apply them, but when they seek freedom from them. Innovation processes aimed at creating new norms are constrained by those that already exist. For example, actions aimed at creating short supply chains often run up against the rules that organise agricultural and food chains. When a manager of a central kitchen asks his usual distributor to supply him with products of local origin, he will quickly discover that the distributor will find it very difficult to do so because his organisational structure is not designed for this. The manager will probably be similarly turned down by his territory's agricultural cooperative, which will explain to him that he represents volumes that are too small to carry out the necessary reorganisation for his delivery. And when he will start working with local artisanal farmers, he will also realise that the health standards in force, designed for the industrial model, are particularly restrictive for them (Le Velly et al., 2021). Even if some spaces for innovation exist, the existing organisational models limit the possibilities. And even if the dominant model is probably no more homogeneous than are the alternative models, its main characteristics are strongly asserted when one tries to deviate or free oneself from it. This observation echoes the multi-level perspective approach (Geels, 2004) presented in the chapters of the book's fourth part, and reminds us that we should not underestimate the lock-ins generated by the norms that organise and frame the dominant sociotechnical regime.

3 Conclusion

How can we not believe too much in agricultural and food models, but believe in them all the same? Three statements summarise this chapter's arguments. First, even though models/ideal-types only provide a broad-strokes vision of observable practices, they are a necessary tool to make these practices intelligible. Second, even though models/projects can inspire very heterogeneous practices, they are an essential component of collective action and should accordingly be studied. Third, even if models/standards do not fully frame actions, they structure the space of what is possible in ways that also need to be carefully studied. Ultimately, the subtlety of the relationships between models and practices calls for abundant caution in analyses. That said, the importance of models in understanding and structuring practices invites us to be proactive and ambitious in deepening this analysis.

References

Bréchet, J. -P. (2021). *Collective action*. Presses universitaires de Provence.

Callon, M. (1986). Some elements of a sociology of translation. Domestication of the scallops and the fishermen of St. Brieuc Bay. In J. Law (Ed.), *Power, action and belief. A new sociology of knowledge?* (pp. 196–223). Routledge.

Chance, Q., Le Velly, R., & Goulet, F. (2018). How to influence the trajectory of organic agriculture's development when you are not in the driver's seat? The case of the French National Federation of organic farming. *Open Agriculture, 3*(1), 632–643.

Coenen-Huther, J. (2003). Le type idéal comme instrument de la recherche sociologique. *Revue Française de Sociologie, 44*(3), 531–547.

Crozier, M., & Friedberg, E. (1980). *Actors and systems: The politics of collective action*. Chicago University Press.

Dufeu, I., Le Velly, R., Bréchet, J. -P., & Loconto, A. (2020). Can standards save organic farming from conventionalisation? Dynamics of collective projects and rules in a French organic producers' organisation. *Sociologia Ruralis, 60*(3), 621–638.

Fournier, S., & Touzard, J. -M. (2014). La complexité des systèmes alimentaires : un atout pour la sécurité alimentaire? *VertigO—La Revue Electronique en Sciences de L'environnement, 14*(1). Online.

Geels, F. W. (2004). From sectoral systems of innovation to sociotechnical systems: Insights about dynamics and change from sociology and institutional theory. *Research Policy, 33*(6), 897–920.

Goodman, D. (1999). Agro-food studies in the 'Age of Ecology': Nature, corporeality, bio-politics. *Sociologia Ruralis, 39*(1), 17–38.

Goulet, F. (2019). *Faire science à part*. Presses Universitaires de Liège.

Guthman, J. (2004). *Agrarian dreams: The paradox of organic farming in California*. University of California Press.

Hinrichs, C. C. (2003). The practice and politics of food system localization. *Journal of Rural Studies, 19*(1), 33–45.

Ilbery, B., & Maye, D. (2005). Alternative (shorter) food supply chains and specialist livestock products in the Scottish-English borders. *Environment and Planning A, 37*(5), 823–844.

Joltreau, T., & Smith, A. (2020). Short versus long supply chains in agri-food sectors: Peaceful coexistence or political domination? The case of Foie Gras in South-West France, *Sociologia Ruralis, 60*(3), 680–697.

Kloppenburg, J., Jr., Hendrickson, J., & Stevenson, G. W. (1996). Coming in to the foodshed. *Agriculture and Human Values, 13*(3), 33–42.
Kneafsey, M., Holloway, L., Venn, L., Dowler, E., Cox, R., & Tuomainen, H. (2008). *Reconnecting consumers, producers and food: Exploring alternatives*. Berg Publishers.
Latour, B. (1987). *Science in action: How to follow scientists and engineers through society*. Harvard University Press.
Le Velly, R. (2017). *Sociologie des systèmes alimentaires alternatifs*. Presses des Mines.
Le Velly, R. (2019). Allowing for the projective dimension of agency in analysing alternative food networks. *Sociologia Ruralis, 59*(1), 2–22.
Le Velly, R., & Dufeu, I. (2016). Alternative food networks as "market agencements": Exploring their multiple hybridities. *Journal of Rural Studies, 43*, 173–182.
Le Velly, R., Dufeu, I., & Le Grel, L. (2016). Les systèmes alimentaires alternatifs peuvent-ils se développer commercialement sans perdre leur âme? Analyse de trois agencements marchands. *Économie Rurale, 356*, 31–45.
Le Velly, R., Goulet, F., & Vinck, D. (2021). Allowing for detachment processes in market innovation. The case of short food supply chains. *Consumption Markets and Culture, 24*(4), 313–328.
Mahoney, J., & Thelen, K. (2009). A theory of gradual institutional change. In J. Mahoney & K. Thelen (Eds.), *Explaining institutional change: Ambiguity, agency, and power* (pp. 1–37). Cambridge University Press.
Mische, A. (2009). Projects and possibilities: Researching futures in action. *Sociological Forum, 24*(3), 694–704.
Mundler, P. (2007). Les Associations pour le maintien de l'agriculture paysanne (AMAP) en Rhône-Alpes, entre marché et solidarité. *Ruralia*, (20). Online.
Raynolds, L. T., Murray, D. L., Wilkinson, J. (2007). *Fair trade. The challenges of transforming globalization*. Routledge.
Renard, M. -C. (1999). The interstices of globalization: The example of fair coffee. *Sociologia Ruralis, 39*(4), 484–500.
Reynaud, J. -D. (1997). *Les règles du jeu. L'action collective et la régulation sociale* (3rd ed.). Armand Colin.
Reynaud, J. -D., & Richebé, N. (2009). Rules, conventions and values: A plea in favor of ordinary normativity. *Revue Française de Sociologie, 50*(5), 3–35.
Rosin, C., & Campbell, H. (2009). Beyond bifurcation: Examining the conventions of organic agriculture in New Zealand. *Journal of Rural Studies, 25*(1), 35–47.
Samak, M. (2013). Quand la «bio» rebat les cartes de la représentation des agriculteurs. L'institutionnalisation de l'agriculture biologique dans les Alpes-Maritimes. *Politix, 103*(3), 125–148.
van der Ploeg, J. D., Renting, H., Brunori, G., Knickel, K., Mannion, J., Marsden, T., De Roest, K., Sevilla-Guzmán, E., & Ventura, F. (2000). Rural development: From practices and policies towards theory. *Sociologia Ruralis, 40*(4), 391–408.
Wald, N. (2015). Towards utopias of prefigurative politics and food sovereignty: Experiences of politicised peasant food production. In P. V. Stock, M. Carolan, & C. Rosin (Eds.), *Food utopias. Reimagining citizenship, ethics and community* (pp. 107–125). Routledge.
Woods, M. (2009). Rural geography: Blurring boundaries and making connections. *Progress in Human Geography, 33*(6), 849–858.

Chapter 19
Confrontation Between Models: Coexistence to Navigate Between the Naivety of Consensus and the Violence of Polarisation

Patrick Caron

It is with complete humility that I admit that I cannot answer the main question posed by this book's coordinators: Do the coexistence and confrontation of agricultural and food models open the way to a new paradigm of territorial development?

At a time when the UN Secretary General convened a Food Systems Summit (September 2021) to accelerate the implementation of the 2030 Agenda for Sustainable Development, this approach raises a number of extremely relevant and useful questions. It is to be wholeheartedly welcomed, especially as we are witnessing a growing polarisation of positions on food, with there being no doubt in anyone's mind that the future of the planet and of humanity is at stake. Thus, there is a sometimes violent opposition between the proponents of local or organic food, who proclaim the need for quality, human and environmental health, and social justice, and the defenders of economic interests and the efficient organisation of supply chains, who raise the spectre of shortages. The former often demonise the latter, considering them vile poisoners of the planet and humanity. The latter, in return, denigrate the former, calling them irresponsible 'lefties' and 'champagne socialists'. The divides continue to grow between producers and consumers, between rural dwellers and urban ones, between defenders of ecological causes and advocates of economic pragmatism, between localists and globalists, all accentuated by the hyper-mediatisation of subjects and the functioning of social networks, without any structured spaces for dialogue.

The issue of coexistence therefore immediately raises that of confrontation. These two terms, brought together in this book's title, do not have the same status, and the 'and' that links them raises some questions. This detour is all the more relevant in the context of growing tensions, where divergent visions of the world and of society are pitted against each other. Is one or other of these visions the 'best', inviting each

P. Caron (✉)
Montpellier Advanced Knowledge Institute on Transitions (MAK'IT), University of Montpellier/Cirad, Montpellier, France
e-mail: patrick.caron@cirad.fr

P. Gasselin et al. (eds.), *Coexistence and Confrontation of Agricultural and Food Models*,
https://doi.org/10.1007/978-94-024-2178-1_19

of us to pick up the banner, join the fight and delegitimise the other by means of caricatured arguments? Or do these archetypal visions invite us to find ways and means of rethinking development by articulating or generating a hybrid trajectory through their confrontation?

The book implicitly raises the question of the opposition between the different worldviews and the way in which they are manifested. The use of the term coexistence thus transcends the objective of a renewed look at diversity. From the outset, the notion of coexistence implies a hybrid dimension, both analytical in order to account for diversity, and normative in presupposing, even if it means exposing oneself to refutation, that coexistence would be preferable to uniformity. In any case, it implies being able to exist in the first instance.

After looking at the revitalisation of diversity, I propose to examine what the goal of coexistence entails and how it can be constructed, especially from a political point of view. In the conclusion, I will return to the notion of territorial development and to the way in which the detour through confrontation and coexistence makes it possible to inform a multiscalar engineering of transformation.

1 Diversity's Welcome Return

Let us start by justifying the inclusion of diversity in the agenda, exploring its genesis and laying it out in some detail. The title of the book implicitly affirms the plurality of models. It was high time, too, after decades of advocacy of homogeneity and the promotion of a single model! Based on the need to control nature, increasingly so since de Serres (1603), on the one hand, and adopting a neo-Malthusian stance on the primacy given to the population explosion of the twentieth century and the increase in the availability of food, on the other, the promotion of a standard model was not, until recently, called into question. And this model worked well, allowing the world's population to double between 1960 and 2000, and, during the same period, leading to an increase in life expectancy and an increase in food availability per person (2500–3000 kcal per day per person between 1960 and 2000; Paillard et al., 2010).

The ingredients of the cocktail are well-known in some detail. Increased productivity of land, labour and capital, the absorption of labour into other economic sectors, and the use of fossil energy and chemical and genetic technologies are this model's main pillars. Processing of food products has also been based on the organisation of long supply chains to regulate supplies, ensure their diversity and achieve economies of scale through the concentration of resources in the agrifood sector. This has been accompanied by an organisation of the market based on lower consumer prices and on competitiveness as an engine for growth. In fact, what we have seen is a process of industrialisation, focusing on growth, efficiency and risk reduction. These developments could take place because of the low costs of so-called 'natural' resources. It was assumed that nature could and would provide the resources needed for production indefinitely, thus allowing, through technology and the use of cheap energy, to increase productivity and the volume of production and to fuel growth. Thus, the ease with which capitalism allowed accumulation is largely due to the ecological surplus. Marx (1867) had already stated that the expansion of capitalism could only take place if abundant raw materials remained cheap.

This transformation, which we will call modernisation here, also known as the Green Revolution in the Global South, took place without any consideration of its detrimental side effects. These well-documented effects—the subject of increasingly frequent and strident warnings—have now become unacceptable to some people. Whether it is in response to environmental crises that point to the agricultural sector as the main culprit for climate change and biodiversity erosion, health crises linked to the sector's industrialisation, or social crises that have set the countryside alight, the need to change the model surfaces repeatedly. This is all the more true because increased food availability has not solved the problems of malnutrition. The number of people going to bed hungry every night is not decreasing despite the abundance of food (Caron, 2020), and the number of people suffering from pathologies associated with obesity is increasing dramatically and is fast becoming the number one public health problem (HLPE, 2017). The emergence of environmental conventions, following the Earth Summit in 1992, reflects the need for change at a global scale, sometimes provoking violent reactions. This change has been successfully embodied in innovations claiming to take care of externalities, such as ecological intensification (Griffon, 2013), but the so-called 'dominant' model remains, well, dominant.

Looking beyond the paradox of the growing divide between the calls for change and the impression that nothing is actually changing, we can observe an increasing number of so-called 'alternative' initiatives emerging in reaction to what mainstream agriculture today represents. We are in this way returning to the diversity of development models. Whether it an emerging reality or merely seems like one to us—after all, we do find it difficulty to grasp all that deviates from the norm—, these initiatives are taking shape, becoming visible, federating, and seeking to lead. The examples of urban food policies, the explosion of 'organic' farming, and new behaviours with respect to the consumption of animal products are striking in this regard. Many other examples have also been presented in this book, showing how diversification, innovation, adaptation and transition contribute to the processes of differentiation.

However, such initiatives often run up against a threefold obstacle. First, they struggle to be recognised for the environmental and social benefits they offer and generate. They thus base themselves on criteria and indicators that are very different from those of production or productivity, which are the ones usually mobilised by mainstream agriculture and the only ones considered 'serious'. Second, they find it hard to convince those who are not already convinced. Third, they are unable to influence the development of public policies and more global frameworks of thought and action so that their effects can be translated at scales large enough to make a significant difference in the face of global challenges. These initiatives therefore tend to remain on the fringes and to be described as 'radical' by their detractors, and do not appear to be capable of driving structural transformations of food and agricultural systems at a significant scale.

2 Perceptions of Diversity: Scales, Debates and Instrumenta(lisa)tion

The debate on coexistence therefore leads us first to question the way in which diversity is perceived. It should be noted that the terms used to describe diversity vary, including in this book. They can refer to one or more of its facets and insist sometimes on the state—diversity, stylised model, coexistence—, sometimes on the process that makes it possible to achieve it—diversification, specialisation, hybridisation—, and sometimes on the implementation and articulation of the processes of action through innovation, adaptation or transition.

As the book's coordinators point out, the abstraction process that makes it possible to characterise diversity relies on the identification of ideal types and possibly on the development of typologies that differentiate between several of them. It is indeed a matter of undertaking a process of segregation, in the analytical sense of the term, aiming to distinguish and dissociate two or more objects of the same nature, whether they are spaces, resources, actors, goods, ideas, etc., often with a view to organising interactions or confrontations.

This abstraction process depends on the scale at which the analysis is conducted, and it is therefore necessary to agree on this scale and on the focus adopted. Indeed, what appears heterogeneous at one scale may appear homogeneous at another, and vice versa. The example of the diversity of farms is sufficient to convince us of this. The specialisation inherent in any production basin, whether it be animal products, export crops or non-food crops, for example, projects an appearance of homogeneity. On closer examination, the choice of a single production is most often accompanied by a wide diversity of structures, forms of organisation, practices and even productions, especially at the farm level.

We can also look at the example of the tensions that have accompanied the rise of environmental concerns in agricultural development thinking. Whereas in the nineteenth and early twentieth centuries, especially in America and Africa, isolation in nature reserves was seen as the best way of preserving nature and in particular emblematic species, this practice is now accompanied by an inclination to promote biodiversity in so-called 'ordinary' natural areas. Over the past 20 years, this debate has been reflected in the land sparing/land sharing controversy (Phalan et al., 2011), which links local transformations to global food and environmental issues. In order to combat the erosion of biodiversity attributable to agricultural activity, is it better to differentiate, through zoning, between areas to be isolated and those where agricultural production can be carried out by promoting an increase in production, in order to limit deforestation, curb the expansion of agricultural areas and spare protected areas? Or would it be better, in contrast, to limit or reduce intensification processes, even if it means that farming areas have to have a larger expansion? This debate requires a combining of views at different scales, as illustrated by the analysis of the impact of ecologisation measures implemented in the Amazon. For example, the archipelago of protected areas created locally to the south of the Amazonian agricultural frontier in Brazil, in northern Mato Grosso and southern Pará, constitutes an obstacle to the expansion of development at the scale of the Amazon basin and in turn modifies the trajectories of local transformation in contact with each of these areas (Duheron, 2006; Caron, 2011).

The specialisation/diversification debate has already been expressed through the formulation of two antagonistic visions of agricultural development in reflections on the multifunctionality of agriculture and rural areas (Caron et al., 2008), especially in the Netherlands, a country in which the spatial dimension of processes and the need to segregate have historically been so important. Thus, there has been a clash between the proponents of a model known as 'conventional' in terms of the intensification and specialisation that it embodies, and the promoters of an alternative, so-called 'integrated' agriculture. Whereas the former rely on the capacity, if necessary, to compensate for the externalities generated and to organise the renewal and recycling of resources through circularity, and on the establishment of protected areas, the latter rely on diversification and environmental management of agricultural areas through agroecology. A similar distinction can be observed in the opposition between the advocates of differentiated policies based on the leveraging of local products in areas suffering from so-called 'natural' handicaps, such as mountain areas, and on reliance on the market and competitiveness elsewhere, and those who advocate that such policies should apply also to other agricultural spaces.

While the distinctions are indeed germane for each of these illustrations, the question that arises in these different cases is how—and at what scale—to recognise, organise and manage diversity, taking into account the effects—and externalities—that it generates locally or at a distance. In turn, diversity highlights the importance of the scale at which an analysis is conducted.

As we can see, agreeing to look at diversity means recognising and grasping it, and this exercise in abstraction is closely tied to the intention to act. It lends itself to very many forms of instrumentation. We can take the example of the opposition classically described between industrial agriculture and family farming. A third category, family business farming, identified by Sourisseau (2015) and his colleagues, and Bosc et al. (2018), leads us to think in a new way about the provision of agricultural support and the design of public policies. Defined by the use of permanent wage labour and by a partial disconnect between the farm and the family, it differs from corporate farming by the family control of capital. It is also fully integrated into the agro-industrial system. 'As diverse as the typical family farming forms, [the family business farming forms] also have a role to play in the future of family farming' (Sourisseau, 2018).

This instrumentation is therefore a vector as well as a support for policy design. Segregation, in the sense of marking a difference, opens the door to exclusion on the one hand, and to integration on the other. These two extremes drive a permanent dialectic made up of power relations built on duality and which contribute to it. History is replete with examples, such as South African apartheid (Lhopitallier & Caron, 1999) and the very existence of the Palestinian territories (Caron, 2011). Bouard et al. (2014) show how the integration/segregation dialectic offers a key to understanding the recompositions in New Caledonia.

By comparing the spatial translation of political segregation in latifundian Brazil, in South African rural areas and in the Palestinian territories, I have shown, however, that limiting the reading of social dynamics to the two obvious factors of the partition of spaces, on the one hand, and the exclusion-appropriation pair, on the other, is not sufficient (Caron, 2011). While segregation structures precariousness, the control of flows of people, goods and merchandise through the porosity of borders between segregated spaces makes it possible to organise complementarity while forging and maintaining political control. Migration flows in apartheid-era South Africa or the closure of the Palestinian territories illustrate both the political dominations at work as well as the complementarities that transcend them, driven by the circulation of goods or the labour market. This analysis invites us to think about coexistence and thus to go beyond the Manichean dualism generated by confrontation and segregation.

3 Coexistence: The Challenge of Managing Diversity

Accepting the challenge of coexistence means choosing integration, as opposed to a segregationist vision of development, which is considered negative. Integration, guided by the principle of 'common destiny', should make it possible to better respond to the many development challenges (integration of spaces, populations, cultures, etc.).

Choosing coexistence also marks a commitment to a path of negotiation with the supporters of dominant positions and models, in order that alternatives can survive and flourish. This choice rejects both the status quo as well as the imposition of an alternative option through a revolution marked by confrontation and force. It is therefore a choice of mediation that is made, which, of course, cannot ignore the context in which it is embedded and in which it participates. It is just not possible to envisage such an option when one of the parties involved has no other view than to eliminate the other.

So what are the arguments that underpin and confirm the choice of living together? First, it is what I will call 'heterosis' by analogy with evolutionary biology, namely the increase in capacities and the gain in performance resulting from a confrontation between alleles. Second, this option makes it possible a priori to avoid the loss of control over trajectories inherent in any revolution, or the inertia generated by dominant power relations. By not putting all its eggs in one basket, it also relies on a building up of resilience (Bousquet et al., 2016) and thus on the capacity to adapt and find solutions to the shocks that are bound to occur. Finally, it reflects a rejection of exclusion, including for normative, ideological and even moral reasons. Integration has a positive connotation in current thinking and is perceived as necessarily more favourable than a segregationist vision, which leads to exclusion.

Thus, coexistence appears at first sight to be desirable and beneficial. However, it is necessary to analyse it closely, and in particular the performances and effects it generates. Similarly, the political positions of the actors involved are important, since they may condition the possibility of coexistence. Coexistence is a gamble that cannot be taken for granted, and this examination can thus help to choose between various possibilities. We may find that positions and power relations may be such as to make any form of coexistence unimaginable, leaving only the possibility of revolution to bring about change. On the other hand, coexistence can be chosen as an option in which a progressive trajectory of change is undertaken in a targeted manner, step by step, consisting of transitions during which each of the coexisting components are transformed, thus contributing to the reconstruction of new configurations. Finally, motivated by their common destiny, a group of actors can also target a desired situation, and agree to organise, at the relevant scale, the best way to reach it together.

Thus by choosing coexistence, we are referring to a construction. Such a process presupposes that the terms of the confrontation between the elements present are clearly explained, whether they be actors, forms of organisation, actions, etc., and, in particular, of what may be controversial. It is therefore necessary to see, recognise, name, qualify and affirm the existence of these elements which we want to organise in a coexistence in order to characterise the synergistic and contradictory interactions—and the disagreements—that link them, and to identify the ways of organising and managing their coexistence. These paths are based on the design and implementation of incentivising, arbitration-based, regulatory and investment mechanisms at broader and more legitimate levels. It is therefore a dual process that has to be put in place: of regulation, as we have just seen, and of mediation to trigger a maieutic effect. As coexistence is not self-evident, it is necessary to clarify the positions of each party and to establish or re-establish the conditions for dialogue between them, and to identify the obstacles that need to be overcome. Once the disagreements have been clarified and recognised, the terms of an agreement can be worked out.

What is at stake concerns several registers, all of which have been illustrated by numerous examples in this book, and this in different regions of the world. Sometimes it is a matter of mobilising, facilitating access to and distributing resources—land or water, for example—or products to ensure the cohesion of the project and the community. In other cases, the main issue is the organisation of complementarities, by acting on flows, to renew resources and guarantee the sustainability of living together. The priority sometimes is to regulate competition, often expressed in a violent way in short supply chains at the local level, and other times is to prevent negative externalities. Finally, it may be a question of organising the production of positive amenities and of thus creating a heritage or an asset that can be leveraged collectively.

4 The Territory as a Supporting Framework: Yes, but Not Only

Management of diversity, collective projects, regulation mechanisms, articulations of innovation, adaptations and transitions: we have laid the foundations that make the territory an appropriate framework for organising coexistence, when it is desired and possible. Territories, which are forms of anchoring for living together, are indeed relevant frameworks, at the scale that defines them, for strengthening the capacity of multiple actors to coordinate and define together the orientations to pursue (Caron, 2017). As Valette et al. (2017) state, and as illustrated by numerous examples in this book, 'The territory is more than a mere framework mobilised for innovation. Localised agrifood systems illustrate this capacity of territories to stimulate the emergence of organisational and institutional innovations, to themselves become drivers of change .... Because of the proximities and the forms of social capital that constitute it, the territory is, in fact, a form of organisation that permits the internalisation of certain transaction costs, the minimising of economic risks, the facilitation of learning processes, the leveraging of know-how and traditional knowledge, the guaranteeing of the application of quality criteria to a product or a form of production ..., all the characteristics that make it an asset that can be mobilised in the processes of production....'

The social capital and the 'living together' issue that underpins it make the territory the vector and the active framework for the development of a pact based on diversity and its management to orient the future. I am indeed saying here that intentional management is required in the case of coexistence, whereas in many cases the territory itself is not managed, its transformations resulting from the distributed action of a large number of actors (Lardon et al., 2008).

In other words, even if the exercise is not free of pitfalls, particularly that of identity-based exclusion, nor of deceptions, such as the disguising of opportunistic greenwashing practices, it is at the level of the territory that the global challenges of climate change, renewal of resources, anticipation of migratory processes, the organisation of exchanges, and food security—if not overall security—, can be won. It is at this level that we can remake the world.

But organising coexistence at a given scale, that of the territory in question, is not sufficient. A significant transformation at the scale of global challenges cannot be achieved solely by the infinite reproduction of local initiatives. Several decisions that condition the behaviour of actors have to be taken at other scales or in other spaces: legislation, policies, organisation of markets, etc. These decisions pertain, in particular, to scales at which public policies are designed and implemented to stimulate local innovation, resolve tensions and conflicts, regulate processes of differentiation and competition, guarantee respect for rights and justice, and ensure territorial planning and cohesion. The transformations hoped for in order to meet the challenges of sustainable development are based on a combination of factors and processes, which constitute a regime (Garel & Rosier, 2008), some of which take place at a local scale, others at national, regional or international ones.

This observation invites us to call into question the myth of being able to scale up and out by replicating successful local processes, which are necessarily contextual. In contrast, a pact built locally can be exported to other places and to other scales in order to enable a project, a vision of the world, or a process of transformation. It can contribute, for example, to the design of appropriate national public policies, whether it is a question of supporting local dynamics or making relevant choices and addressing trade-offs. It becomes the basis for a global transformation process to be undertaken by relying on the complementarity of local innovations, territorial dynamics, national policies and international frameworks. Such a pact thus modifies the terms of coexistence and its political management at other scales, infra and supra, including through the traces of clashes and confrontations whose marks it leaves behind. As an iconoclastic proposal, we can even suggest that desirable transformations can be initiated by the implementation of mediation processes at the scale at which alliances and coexistence are possible, before influencing the processes taking place at other scales and coming up against irreducible clashes.

5 Conclusion

As we can see, coexistence leads to a renewed relationship with diversity. Given that it implies a relationship to action, it even transcends the sole objective of a renewed look. While recognising this diversity in all things and at all scales, coexistence suggests the capacity to act on it, to manage it, to make it the basis of 'living together'.

It thus invites us to clarify the categories of analysis and biases, to enrich the dialogue between disciplines, to structure the interfaces between science and policy, and to (re)define the role of the researcher in the transformations underway. By highlighting the polysemy of the term 'model', which is at the same time an analytical archetype, an expression of a desired future, and a standard for action, the coordinators of this book pose in particular the challenge of the interface and the interactions that have to be promoted between these three acceptations. This is indeed a major issue that calls the researcher's posture into question and which the challenge of coexistence also raises.

By affirming diversity and the need to grasp it, coexistence sets the stage for confrontation. However, the goal is not so much to generate coexistence as to organise and manage it, and, for scientists, to specify what science and its disciplines can say about it.

Coexistence's political acceptance repudiates a dual vision of the world and the affirmation of opposing extremes, whether they be stylised representations or concrete realities. Without denying the possibility that this duality may indeed correspond to a fruitful stage of political implementation and organisation of confrontation, thinking about and constructing coexistence is in some ways an alternative to the major revolution that we will have to urgently undertake in the face of planetary challenges. Taking the path of coexistence is to bet that the world can be built by transcending the polarisation promoted by merchants of doubts and certainties, a polarisation exacerbated by the current hyper-mediatisation pervading our societies. Coexistence offers an alternative to this polarisation, the outcome of which will certainly be either a procrastination resulting from power relations or a revolution with unpredictable effects. The challenge of coexistence is the goal of a utopia based on a trajectory that refuses to founder, on the one hand, into the naivety of a consensus incapable of overcoming the status quo and, on the other, into the ease, violence and uncertainty of dual confrontation.

References

Bosc, P. -M., Sourisseau, J. -M., Bonnal, P., Gasselin, P., Valette, E., & Bélières, J. -F. (Eds.). (2018). *Diversity of family farming around the world. Existence, transformations and possible futures of family farms* (341 p). Springer.

Bouard, S., Sourisseau, J.-M., & Zenou, B. (2014). Intégration/ségrégation: Une clé de lecture des recompositions des modèles de développement? Le cas de la Nouvelle-Calédonie. *Natures Sciences Sociétés, 22*(4), 305–316.

Bousquet, F., Botta, A., Alinovi, L., Barreteau, O., Bossio, D., Brown, K., Caron, P., Cury, P., d'Errico, M., DeClerck, F., Dessard, H., EnforsKautsky, E., Fabricius, C., Folke, C., Fortmann, L., Hubert, B., Magda, D., Mathevet, R., Norgaard, R. B., Quinlan, A., & Stave, C. (2016). Resilience and development: mobilizing for transformation. *Ecology and Society, 21*(3).

Caron, P. (2011). *Ré-concilier agricultures et sociétés: dévoiler les territoires et repenser les limites.* Accreditation to supervise research (HDR), Université Paris Ouest Nanterre La Défense, Nanterre, 3 vol., 39 p., 254 p., 242 p.

Caron, P. (2017). Why and how the concept of 'territory' can help in thinking rural development. In P. Caron, E. Valette, T. Wassenaar, C. G. d'Eeckembrugge, V. Papazian (Eds.), *Living territories to transform the world* (pp. 15–22). Quæ.

Caron, P. (2020). Nourrir 10 milliards d'êtres humains et assurer leur sécurité alimentaire: une question dépassée? *Raison présente, 2020/1*(213), 11–20.

Caron, P., Reig, E., Roep, D., Hediger, W., Cotty, T., Barthelemy, D., Hadynska, A., Hadynski, J., Oostindie, H., & Sabourin, E. (2008). Multifunctionality: Epistemic diversity and concept oriented research clusters. *International Journal of Agricultural Resources, Governance and Ecology, 7*(4–5), 319–338.

Duheron, E. (2006). L'aménagement du territoire par l'environnement en Amazonie. Étude de la région du complexe Cristalino (Mato Grosso, Brésil). Mémoire de master, EMTS, 58 p.

Garel, G., & Rosier, R. (2008). Régimes d'innovation et exploration. *Revue Française de Gestion, 2008/7*(187), 127–144.

Griffon, M. (2013). *Qu'est-ce que l'agriculture écologiquement intensive?* (224 p). Quæ.

HLPE. (2017). *Nutrition and food systems.* A report by the High Level Panel of Experts on Food Security and Nutrition. Committee on World Food Security, Rome, 152 p.

Lardon, S., Tonneau, J. -P., Raymond, R., Chia, E., & Caron, P. (2008). Dispositifs de gouvernance territoriale durable en agriculture. Analyse de trois situations en France et au Brésil. *Norois, 2008/4*(209), 17–36.

Lhopitallier, L., & Caron, P. (1999). Diversité et recomposition de l'espace rural dans le district d'Amatola, province du Cap de l'Est. *L'espace Géographique, 28*(2), 170–183.

Marx, K. (1867, 1985). *Le capital. Critique de l'économie politique.* Flammarion.

Paillard, S., Treyer, S., & Dorin, B. (Eds.). (2010). *Agrimonde. Scénarios et défis pour nourrir le monde en 2050* (296 p). Quæ.

Phalan, B., Onial, M., Balmford, A., & Green, R. E. (2011). Reconciling food production and biodiversity conservation: Land sharing and land sparing compared. *Science, 333*(6047), 1289–1291.

Serres, O. (de). (1603). *Le theatre d'agriculture et mesnage des champs* (2nd ed., 907 p). Abraham Saugrain.

Sourisseau, J. -M. (Ed.). (2015). *Family farming and the worlds to come* (361 p). Springer.

Sourisseau, J. -M. (2018). On the boundaries of family farming: Examples of family business farming. Introduction. In P. -M. Bosc, J. -M. Sourisseau, P. Bonnal, P. Gasselin, E. Valette, & J. -F. Bélières (Eds.), *Diversity of family farming around the world. Existence, transformations and possible futures of family farms* (pp. 132–135). Springer.

Valette, É., Caron, P., d'Eeckenbrugge, G. C., & Wassenaar, T. (2017). General conclusion and outlook. In P. Caron, E. Valette, T. Wassenaar, G. C. d'Eeckenbrugge, & V. Papazian (Eds.), *Living territories to transform the world* (pp. 255–265). Quæ.

Chapter 20
Governing the Coexistence and Confrontation of Agricultural and Food Models in a Territory: Paradigm, Postures, Methods

Pierre Gasselin, Sylvie Lardon, Claire Cerdan, Salma Loudiyi, and Denis Sautier

This book is based on the premise that an improved understanding of the coexistence and confrontation of agricultural and food models, and thus of their interactions at different spatial and organisational scales, facilitates recognition and support for combinations of these models that can potentially be useful for sustainable territorial development. Indeed, territories are both the substratum and the result of new forms of agriculture and food production, some of which are instituted as models, whether they are analytical archetypes, desired futures or standards for action. These agricultural and food alternatives are being invented and asserted as responses to the environmental, health-related, nutritional, economic and social criticisms of a long legacy of productivist growth and heavy urbanisation. But it is not enough to categorise, compare or even support these technical, organisational and institutional innovations as independent and juxtaposed elements. Given the goal of sustainable

P. Gasselin (✉)
UMR Innovation, INRAE, Montpellier, France
e-mail: pierre.gasselin@inrae.fr

S. Lardon
UMR Territoires, INRAE and AgroParisTech, Aubière, France
e-mail: sylvie.lardon@agroparistech.fr

C. Cerdan
UMR Innovation, CIRAD, Saint-Pierre, Réunion, France
e-mail: claire.cerdan@cirad.fr

S. Loudiyi
UMR Territoires, VetAgro Sup, Clermont-Ferrand, France
e-mail: salma.loudiyi@vetagro-sup.fr

D. Sautier
UMR Innovation, CIRAD, Montpellier, France
e-mail: denis.sautier@cirad.fr

P. Gasselin et al. (eds.), *Coexistence and Confrontation of Agricultural and Food Models*,
https://doi.org/10.1007/978-94-024-2178-1_20

territorial development, it is important to analyse and govern the conditions of coexistence between these agricultural and food models, where coexistence is conceived as configurations not only of competition, confrontation and power relations, but also of co-presence, co-evolution, complementarities, synergies and sometimes even hybridisation.

Jan Douwe van der Ploeg, Ronan Le Velly and Patrick Caron graciously agreed to read the full book and have honoured us by writing the foreword and two chapters of critical analysis. In this final chapter of the book, we do not intend to conclude or even respond to these transversal analyses. Nor does this chapter aim to circumscribe a research area that has opened up new questions for the scientific community and outlines new strategies for territorial development. We only wish to put into perspective the fundamental elements around which this research effort has been organised. First, we return to the title question of the book. Indeed, are we not being presumptuous in speaking of a new paradigm of territorial development? We then show that the authors of this book assert three different epistemological postures. Then we offer a general overview of our approach, before concluding.

1 A New Paradigm?

As Jan Douwe van der Ploeg points out in the foreword, the coexistence and confrontation of territorial agricultural and food models are no longer the same as in the past. Compared to the simple duality and stability of the configurations observed in the last century, the situations today of coexistence and confrontation are multifaceted, unstable and crisscrossed by hybrid forms. The new diversity of forms of agriculture and food systems is shaping multi-hued mosaics that compel us to undertake a close analysis of local situations. As a result, the imperative transitions that we must think about and accompany are and will be made up of partial processes that combine themselves and move in directions that cannot be predetermined. With this as a point of departure, this book invites us to take a fresh look at two central aspects of thinking about and governance of territorial development—even though it may not rise to the level of a paradigmatic revolution.

First, the hybridisation and articulation of innovative forms of territorial organisation, actors and scales lead to the emergence of new dynamics of territorial development. Organising and combining agriculture systems to address the new food challenges means calling development models into question and thinking about the coexistence and confrontation of these models. For sustainable territorial development, it is not so much the differentiation and juxtaposition of forms of organisation that are important to observe and analyse, but rather their coexistence, confrontation and hybridisation. This makes it possible, on the one hand, to make initiatives visible that are not yet known to development or support organisations, and, on the other, to offer territorial actors new tools for analysing the dynamics at work and for building collective actions.

Second, an improved understanding of the coexistence and confrontation of agricultural and food models encourages the recognition of and support for potentially useful combinations of these models for territorial development. To this end, we have proposed a framework for analysing the coexistence of agricultural and food models structured around four dimensions: specialisation and diversification (as processes and effects), innovation (as a process, a system and an impact), adaptation (as a process, a property and a result) and transition (as a transformation and a project). We show that the sustainability and multifunctionality of agriculture and food systems cannot be examined solely on the basis of the differentiation and heterogeneity of sociotechnical and socio-ecological forms. Of course, this reading of social, technical and environmental heterogeneity is essential not only for thinking about the environmental, social and economic pillars of development, but also for criticising its values as well as its perverse effects (socio-economic inequalities, environmental and health damage, etc.) and for rethinking the governance of our territories. However, too little attention is paid to the interfaces and interactions between the diverse and dynamic forms of agricultural and food systems. It is in this setting of frictions, complementarities and co-evolutions of agricultural and food models, at the scale of rural, peri-urban and urban territories, and in articulation with higher levels of organisation, that a reshaping of tomorrow's challenges and of the 'theories of action' to address them is taking place.

2 Three Epistemological Postures

We have already identified (Gasselin & Hostiou, 2020; Gasselin et al., 2020) the three different epistemological postures that researchers take when considering the coexistence and confrontation of agricultural and food models. These postures reflect the authors' contrasting positions on knowledge, actors and action. We thus distinguish between functionalist coexistence, coexistence based on power relations, and coexistence based on transition.

The first family of studies examine the functional complementarities between systems and the properties that result from these interactions. These studies investigate, for example, how interactions between agricultural models optimise heterogeneous resources, in particular in territories in which the environmental, planning and social organisation conditions vary. Other studies explore how hybridisations between agricultural models contribute to an increase in the number of innovation hubs and how they are—or are not—favourable to sustainable development. Interactions between agricultural models can also build up agricultural systems' adaptation abilities and make territories more resilient, for example in their food supply capacities. In this family of studies, the researcher pursues a functional and systemic analysis of the situations of coexistence of agricultural and food models.

The second family of studies examine the power relationships between actors and the conditions propitious to good governance of a diversity of agricultural and food models. These studies thus assess the effects of domination or the ways to rehabilitate silenced identities and fight against marginalisation. They aim to denounce and resolve situations of exclusion through criticisms of power relations (economic, political and social).

Finally, the third family of studies view coexistence as a situation of transition or transformation of agricultural or food models. The analysis pertains to trajectories of change that have to be planned and managed. The challenges are then to describe and support changes that are more or less radical (as opposed to adaptative), more or less selective (as opposed to inclusive) and more or less specialised (as opposed to diversified). This type of study investigates the management of transitions.

This categorisation of epistemological postures has the merit of providing indications of the level of the researcher's commitment to change and, in so doing, drawing a gradient of greater or lesser politicisation of issues that interest him or her and of the analytical frameworks mobilised. However, these three epistemological postures are not mutually exclusive. For example, Claire Cerdan (Chap. 1) analyses the evolution of power relations between, on the one hand, the actors of the agro-industrial model and, on the other hand, those of the on-farm and artisanal production model. But she also emphasises that the artisanal projects benefit from the know-how acquired during the years of specialisation on farms and in industries. In this way, she highlights the advantages of this functionalist coexistence. Kae Sekine (Chap. 16) follows the same line, showing how the multinational company Dole unilaterally decides to close and relocate its farms in the face of resistance from local people in Japan. She also examines the functional interactions when the multinational uses excessive amounts of mineral nitrogen or large quantities of pesticides that are detrimental to local family farming and pose a risk of water pollution. For their part, Philippe Baret and Clémentine Antier (Chap. 14) propose the use of an analysis grid to assess the intensity of agroecological transitions, according to how radical are the innovations concerned and how inclusive/exclusive is the group of actors promoting them. Thus, these authors, while dealing mainly with transition processes, also recommend an analysis of the power relations between actors in innovation niches and those in the dominant model. These illustrations show that the researchers often adopt hybrid epistemological postures in the studies in this book, even if each of the studies leans primarily towards one of the three postures. Moreover, the epistemological posture of the researchers varies not only according to their studies, but also over the course of their scientific trajectory (Petit et al., 2018).

3 A Renewed Approach

In this way, the analysis of situations of coexistence of agricultural and food models invites us to renew our fundamental thinking on territorial development, and indeed its governance. The principles set out in the previous two sections pose veritable analytical and methodological challenges. We start by reminding ourselves of the importance of use of the concept of agricultural and food model and the conditions under which this is possible. We then show that the analysis of 'situations of coexistence' is organised within a systemic framework and makes it possible to investigate the controversies that run through them. Finally, we present a summary diagram of our approach.

3.1 From the Model to the Analysis of the Concrete System

The case studies presented in this book underscore the importance of identifying the diversity of conflicting agricultural and food models in territories, identifying the actors who promote or criticise them, and characterising the arguments deployed in favour or against these models. The agricultural and/or food model is a frame of reference (political, technical, economic, etc.) constructed by actors or researchers and guides their thinking and actions. Muller (1990) suggests that we consider it as a cognitive and normative framework shared by actors that provides keys for interpreting reality and, as a result, guides action in the field of public action (which refers to the concept of project as mobilised by Ronan Le Velly, Chap. 18). Thus, the concept of the frame of reference proposed by Muller (*ibid.*) is very close to that of the model as defined in the book's introduction in its three acceptations (archetype of an observed reality, desired or criticised future, set of standards for action). Gisclard and Allaire (2012) show us that the frame of reference and the underlying model are embodied in a process of institutionalisation that relies as much on the substantiation of ideas and norms as on the transformation of public policies: 'The institutionalisation of family farming, as a legitimate social form and productive model, is the product of a transformation of the representations associated with small producers, which owes as much to the dissemination of new ideas, frames of reference of rural development programmes, at the international level, or of national political contingencies, as well as to a progressive organisation of the professional interests of Argentinian family producers' (ibid., p. 214). This is also the general sense of Christophe Albaladejo's proposition (Chap. 10), which hypothesises that the model results from the convergence of four changes: in the 'social agenda', in the markets, in the 'public agenda', and finally in science and technology.

This is why the cognitive and/or normative frameworks that constitute models at a given moment provide different types of actors, located at different territorial levels, with elements for interpreting and decoding the complexity of reality. These same frameworks also influence the objectives and measures of public action. Several recent collective studies have shed light on the impact of models[1] on agricultural and environmental policies: international agronomic models are shaping land use (Loconto & Rajão, 2019); modelling is becoming a field of competition between scientific actors seeking to influence policies (Aykut et al., 2019); and interest groups are resorting to modelling to influence public environmental policies (Demortain, 2019).

[1] Understood in these studies as instruments of quantification and prediction.

We therefore agree with Ronan Le Velly's warning (Chap. 18): 'It is therefore advisable not to put too much faith in models, not to trust them blindly as to their capacity to portray reality or to guide action. Do not believe in them too much … but believe in them all the same!' We are firmly convinced of the importance of characterising and interpreting these models, but also of focusing on analysing concrete reality, especially 'systems of concrete action', by paying close attention to practices, strategies and powers (Crozier & Friedberg, 1977). A detailed understanding of these practices and strategies, in their diversity, can shed light on situations of coexistence and their potential for sustainable territorial development. Jérémie Forney puts it well (Chap. 15): 'When an ideal-type is accorded too much importance, it masks specificity and originality. […] once put into practice, a model becomes anchored in space and time, and its limits dissolve in the richness of reality.'

3.2 A Framework for the Systemic Analysis of Situations of Coexistence

In this book, we have proposed a framework for analysing the coexistence of agricultural and food models that is structured according to four dimensions (see the General Introduction and Fig. 1): specialisation and diversification (Part I of the book), innovation (Part II), adaptation (Part III), and transition (Part IV). Each of these dimensions is shown to be relevant and problematised by a state of the art, and then illustrated by case studies, which are summarised and subjected to a transversal analysis in the introductory chapter of each part. Let us recall here the fundamental elements that justify each of these dimensions:

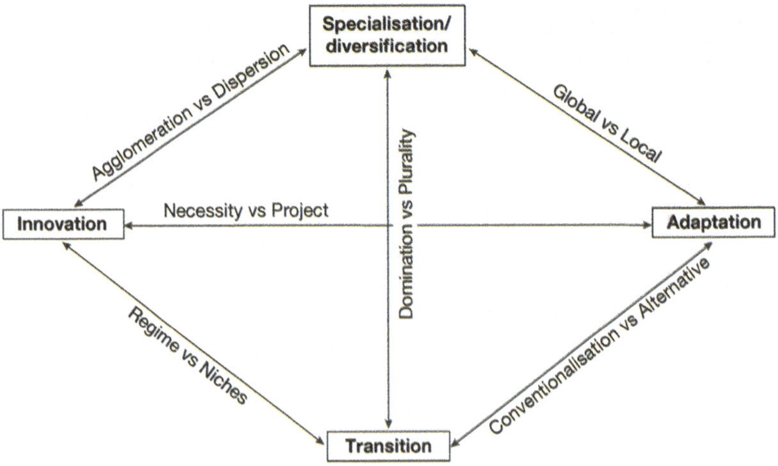

Fig. 1 Framework for analysing the coexistence of agricultural and food models in territories. Based on Gasselin et al. (2020)

- Analysing the specialisation/diversification processes requires us to articulate the scales (time, space and actor organisations) of agricultural and food models, as well as to study the relations between the territory concerned and wider scales (region, nation, world). It is also a matter of exploring the relationships of domination, even hegemony, and marginality of agricultural and food models;
- Paying attention to the innovation processes at work in the interactions between agricultural and food models offers an original view of territorial and social innovations, and reveals useful hybridisations or, on the contrary, the roadblocks to innovation;
- Examining the capacity of agricultural and food systems to adapt is to look for interactions, complementarities or competition between forms of organisation and the way in which they can be combined, or even hybridised, at a territorial scale;
- Finally, considering transitions in terms of the coexistence of agricultural and food models that are institutionalised in science, the political arena, the market and society makes it possible to envisage plural configurations in which various models coexist in a territory, without one eliminating the others.

Each of the four dimensions provides a unique perspective on the conditions under which agricultural and food models coexist, but each of these four dimensions also interacts with the other three. Thus, several authors in this book highlight the transversal aspects between the four dimensions, which we illustrate below on the basis of the findings of certain chapters (see the double-arrowed lines in Fig. 1):

- 'Agglomeration versus Dispersion': Frederic Wallet (Chap. 3) highlights the interfaces between processes of specialisation/diversification and those of innovation in European policies. He shows that specialisation with a coherent diversity of sectors ('smart specialisation') allows the leveraging of knowledge production and diffusion processes and thus the stimulation of innovation between various value chains;
- 'Necessity versus Project': in her study of the Faxinal Emboque community in Paraná state, Brazil), Vanessa Iceri (Chap. 9) shows that innovation processes can promote an increase in the adaptation capacities of actors, productive systems and territories, either through a voluntary project or through an approach that has been imposed to deal with unexpected hazards;
- 'Regime versus Niches': Philippe Baret and Clémentine Antier (Chap. 14) propose a cartography of the dynamics of the agroecological transition, making it possible to classify innovations according to how radical and/or inclusive they are. In so doing, they invite us to move away from a binary reading of 'innovation versus dominant regime' in order to plan collective trajectories capable of orienting the agroecological transition;
- 'Global versus Local': In her study of pig and poultry farming in Santa Catarina state in southern Brazil, Claire Cerdan (Chap. 1) shows that 'diversification and specialisation [of activities and actors] are part of the same process of adaptation of productive spaces to the global system';
- 'Conventionalisation versus Alternative': Claire Lamine (Chap. 11) reports on the recompositions of the territorial agrifood system in southern Ardèche (southern

France) and the ecologisation of practices. These recompositions manifest through a combination and hybridisation of conventional and alternative forms, both of individual farmers' trajectories and in collective action. Thus, 'producers who might have been considered opposites in their production and marketing approach [...] are now converging somewhat in their strategies, practices and visions';

- 'Domination versus Plurality': in their study of the Vietnamese dairy sector, Guillaume Duteurtre and his colleagues (Chap. 13) show that the coexistence of peasant farms and industrial firms in a territory results from pragmatic adjustments in land management and appropriate local partnerships. This balance between domination and plurality is a consequence of 'power relations, while at the same time being part of market dynamics driven by demand, techniques, investments and cognitive models'. This shows how 'transition leads to parallel trajectories, i.e. the superposition of several regimes'.

These transversal aspects underpin the systemic nature of the proposed analytical framework, essential for refining an integrated approach to territorial development. It should be emphasised that the hybridisations generated at the interface of agricultural and food models are sometimes the manifestation of a 'conventionalisation' of innovation niches due to the dilution of the actors' initial principles and the primary aims of the innovation. Hybridisations then take place to the benefit of powerful actors who capture the innovation rent generated by the pioneers, who are often in situations of social, economic, territorial and political marginality. It is therefore necessary to keep a critical eye on the ways in which these hybridisations emerge and function. The forms of political, socio-professional and citizen regulation are essential points of reference for investigating controversies, managing conflicts and pursuing the goals of sustainability, ethics and equity.

3.3 Considering Situations of Coexistence and Investigating Controversies

Any analysis of a 'situation of coexistence' of agricultural and/or food models is predicated on identifying the actors and/or systems, the nature of interactions, the objects and the 'setting' under consideration. This exercise is necessary not only to define the scales envisaged, but also to determine the disciplines that will be best equipped to answer the questions raised. The ambition to formulate a framework for analysing situations of coexistence of territorial agricultural and food models led us to formulate generic hypotheses (see the Introductions to Parts I to IV). Therefore, they have to be fine-tuned and adapted to the contexts and issues of the proposed fields of study. It is then imperative to examine dispassionately each of the agricultural and food models present, something that many researches find hard to do since they are focused on a single model. Finally, the coexistence of agricultural and food models inevitability brings with it controversies in which different actors ally or oppose each other to legitimise their own choices and often discredit those of others. The characterisation and analysis of these controversies at the local level are therefore essential to recognise not only what makes a model coherent, but also its divergences and the conditions of interaction with others (Feuer et al., 2020).

3.4 A Comprehensive but Demanding Approach

Our central premise is that a better understanding of the situations of coexistence and confrontation of territorial agricultural and food models is necessary to govern the ecological, food, social and health transitions that are urgently required. The new diversity of these models requires us to be more lucid about what is happening in their interactions. Figure 2 shows a summary diagram of the methodological principles put to the test in this book.

We can summarise our methodological approach in three main parts: analysing situations of coexistence and confrontation of agricultural and food models according to a four-dimensional analytical framework; combining three postures of thought and action; and encouraging the recognition of and support for combinations of models that are potentially relevant for sustainable territorial development.

4 Conclusion

The successful coexistence of agricultural and food models in territories depends on the satisfaction of demanding conditions. First of all, it is necessary to increase the capacity of actors to control the processes and activities that concern them in their territory (Deffontaines et al., 2001). This applies in particular to those who are marginalised by inequalities in access to resources, the inequitable sharing of wealth and asymmetries in economic, political, media and symbolic power. The suitable resolution of controversies between the proponents of various agricultural and food models depends on this newfound capacity (Sen, 1987; Dubois & Mahieu, 2009).

In this book, we show that even though the coexistence of agricultural and food models in territories is addressed in the scientific literature, it has never been theorised as such. We propose and implement a framework for analysing situations of coexistence and confrontation based on four dimensions (specialisation/diversification, innovation, adaptation, transition) with the goal of taking a fresh look at agricultural and food development in rural and urban territories.

Analysing and supporting territorial development by taking the coexistence and confrontation of agricultural and food models into account reveals new levers for action: promoting complementarity between specialisation and diversification at various spatial and organisational scales; combining innovation and the tangible and intangible heritage specific to the various agricultural and food models; building up the capacity to adapt in the complementarity of agricultural and food models; and undertaking a transition to new territorial development configurations.

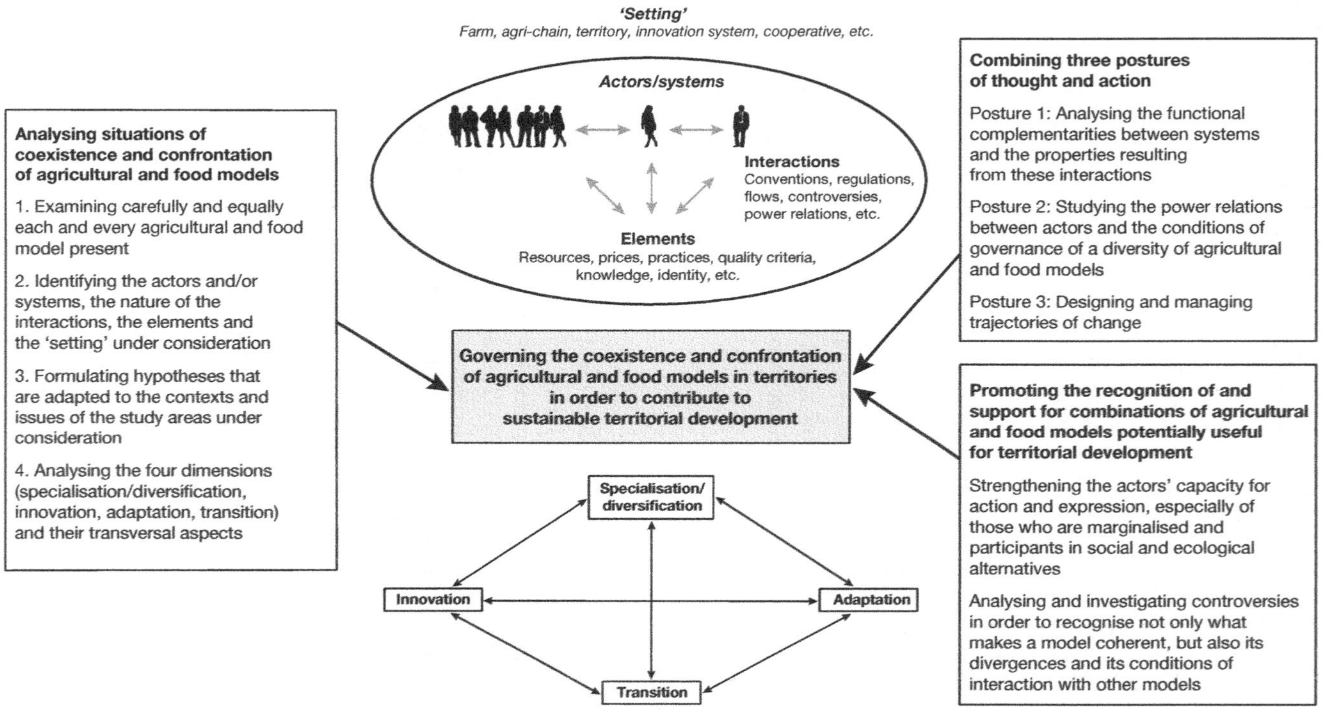

Fig. 2 Governing the coexistence and confrontation of territorial agricultural and food models: analytical and methodological principles

But the coexistence of agricultural and food models also poses the challenge of inventing new territorial development governance systems and building new skills. Indeed, taking the coexistence of agricultural and food models into account amounts to thinking of the place of every individual and of modalities of living together in the territory concerned. It is a matter therefore of thinking about development priorities defined by values (ethics, in particular with regard to future generations, and equity, in particular in terms of social, economic and spatial justice) and sustainable development objectives (peace, food sovereignty, climate change, employment, etc.). The governance of the coexistence of agricultural and food models requires mediation as well as innovations and learning to promote functional complementarities between systems, come up with innovations propitious to sustainable development, rein in the effects of domination and fight against marginalisation, and finally transcend the disparities of the actors' projects in order to facilitate living together.

References

Aykut, S. C., Demortain, D., & Benbouzid, B. (2019). The politics of anticipatory expertise: plurality and contestation of futures knowledge in governance. Introduction to the special issue. *Science and Technology Studies, 32*(4), 2–12.

Crozier, M., & Friedberg, E. (1977). *L'acteur et le système* (p. 500). Éditions du Seuil.

Deffontaines, J. -P., Marcepoil, E., & Moquay, P. (2001). Le développement territorial: une diversité d'interprétations. In S. Lardon, P. Maurel, V. Piveteau (Eds.), *Représentations spatiales et développement territorial. Bilan d'expériences et perspectives méthodologiques* (pp. 39–56). Hermès Science Publications.

Demortain, D. (2019). Les jeux politiques du calcul. Sociologie de la quantification dans l'action publique. *Revue D'anthropologie des Connaissances, 13*(4).

Dubois, J. -L., & Mahieu, F. -R. (2009). Sen, liberté et pratiques du développement. *Revue Tiers Monde, 2*, 245–261.

Feuer, H. N., Van Assche, K., Hernik, J., Czesak, B., & Różycka-Czas, R. (2020). Evolution of place-based governance in the management of development dilemmas: Long-term learning from Małopolska, Poland. *Journal of Environmental Planning and Management*, 1–19.

Gasselin, P., & Hostiou, N. (2020). What do our research friends say about the coexistence and confrontation of agricultural and food models? Introduction to the special issue. *Review of Agricultural, Food and Environmental Studies, 101*(2–3), 173–190.

Gasselin, P., Lardon, S., Cerdan, C., Loudiyi, S., & Sautier, D. (2020). The coexistence of agricultural and food models at the territorial scale: An analytical framework for a research agenda. *Review of Agricultural, Food and Environmental Studies, 101*(2–3), 339–361.

Gisclard, M., & Allaire, G. (2012). L'institutionnalisation de l'agriculture familiale en Argentine: Vers la reformulation d'un référentiel de développement rural. *Autrepart, 3*, 201–216.

Loconto, A., & Rajão, R. (2019). Governing by models: Exploring the technopolitics of the (in)visibilities of land. *Land Use Policy*, 104241.

Muller, P. (1990). *Les politiques publiques*. coll. Que sais-je? (127 p). PUF.

Petit, S., Hostiou, N., Tallon, H., & Gasselin, P. (2018). Faire recherche sur la coexistence de modèles: diversité des regards de chercheurs. In *Séminaire permanent «Élevage et développement durable des territoires»: Coexistence et confrontation de modèles d'élevage dans les territoires, Montpellier, 27 June 2018, Inra-Cirad.*

Sen, A. (1987). *On ethics and economics*. Basil Blackwell.

The manufacturer's authorised representative in the EU is Springer Nature Customer Service Centre GmbH, Europaplatz 3, 69115 Heidelberg, Germany. If you have any concerns regarding our products, please contact ProductSafety@springernature.com

Printed and bound by CPI Group (UK) Ltd, Croydon, CR0 4YY
30/09/2026
02225244-0008